Dynamics of Rockslides and Rockfalls

Springer

Berlin
Heidelberg
New York
Barcelona
Hong Kong
London
Milan
Paris
Singapore
Tokyo

Theodor H. Erismann · Gerhard Abele

Dynamics of Rockslides and Rockfalls

With 120 Figures and 10 Tables

Springer

Authors

Prof. Dr. Theodor H. Erismann
Tannenstrasse 27
8212 Neuhausen
Switzerland

Prof. Dr. Gerhard Abele †
Germany

ISBN 3-540-67198-6 Springer-Verlag Berlin Heidelberg New York

Library of Congress Cataloging-in-Publication Data

Erismann, Theodor H.:
Dynamics of rockslides and rockfalls / Theodor H. Erismann, Gerhard Abele.
p.cm.
Includes bibliographical references and index.
ISBN 3540671986
1. Rockslides. I. Abele, Gerhard. II. Title.

QE599.A2 E75 2001
551.307–dc21

2001018349

Springer-Verlag Berlin Heidelberg New York
a member of the BertelsmannSpringer Science+Business Media GmbH
http://www.springer.de

Printed in Germany

Cover Design: Erich Kirchner, Heidelberg
Dataconversion: Büro Stasch, Bayreuth (*www.stasch.com*)

SPIN 10881848 32/3111 – 5 4 3 2 1 – Printed on acid-free paper

Foreword

Just as this book is going to press, a town in the Eastern Alps is threatened by rockfalls. Several times during the last weeks, rock masses have rushed down from a rock face about 600 m above a residential and industrial area. Although there have been neither casualties nor major damage to date, an area comprising some 50 residential houses and several industrial sites had to be evacuated. In addition to the local and government authorities who are responsible for managing the situation and making the appropriate decisions, politicians and public media are also present. Experts from different geoscientific disciplines have become involved, and they are expected to give competent answers to questions like the following: Will the rock mass break down in several stages, or might a catastrophic event occur by collapse of the whole rock face at once? What is the total volume involved? What velocity could the falling rock mass attain, and how far might individual blocks be projected? Could one issue a warning before major events? Beyond these immediate concerns, the decision makers are also confronted with the following issues: Is there a long-term geological process going on, comprising a rock mass even larger than the worst case considered? Are we experiencing a purely natural geological process, or are we caught up in a hazard which is triggered, enhanced, or even caused by human activities like mining? Experts have to give appropriate advice.

We may hope that the most critical phase of this rockfall process is past, that the rock face will stabilise, and that the dams constructed will be strong enough to guarantee the safety of the people living or working within the area of risk. It is also a fact that hazard and risk assessments by the many experts involved can differ considerably, and we may further hope that the decision makers will follow the best estimate.

Everyone interested and involved in the dynamics of rockfalls and rockslides knows that the work of experts in this field may be an issue. According to Erismann and Abele, the catastrophic failure of the Vaiont hydroelectric project in 1963 was not only a tragedy for the victims, it was also a tragedy for the experts and engineers involved. Comprehensive monitoring and investigations were carried out after the first movements started. The worst possible case could have been realistically estimated on the basis of Heim's findings of 1932. However, under the pressure of the consequences for the operation of the power station, this worst case was not taken into consideration.

Erismann and Abele address this book to experts who are sometimes working under extreme conditions regarding time horizon, spatial accessibility, and responsibility for the consequences of their results. Therefore they emphasise simple, easy-to-apply methods, which are robust under the extreme working conditions of a threatening catastrophe. However, it is not a "cookery book", to be swiftly consulted during the elaboration of exercise. We must read, or even better, study this book thoroughly during quiet times. We should follow with concentration the difficult and complex path from physical analysis to prediction prescribed by the authors. I am sure that, having done this, our work on rockfall and rockslide prediction and mitigation will be of increased quality. Furthermore, by referring to Erismann's and Abele's book, experts will find a more general scientific and methodological basis on which they may come to less divergent conclusions than hitherto.

Abele's contribution to this book was his treasure of morphological observations and his ideas about processes and mechanisms; however, he did not live to see its publication. Erismann has laid the physical basis, and by doing the writing, has put the finishing touches to 25 years of his involvement in this theme. So the reader will be right to expect a mature treatment and some conclusive character within this book. Nevertheless, the observations and physical models presented have the ability to stimulate new research and development of quantitative methods. In this respect, this book has also a youthful character, pointing the way to future work to be carried out by the next generation of scientists and engineers. My personal good wishes for Erismann's and Abele's book are that it may become a standard textbook, to be carefully studied both by scientists as well as by practitioners of the dynamics of rockfalls and rockslides.

Vienna, January 2001
E. Brückl

In memoriam Albert Heim
1849–1937

Preface

To a certain extent, this book is a torso. Originally it was planned by Gerhard Abele and myself to deal, under the title "Rockslides, Rockfalls, and Related Events", with the entire set of problems connected with rockslides and rockfalls. The geomorphological and geological questions – in other words: the lion's share of this work – were intended to be treated by Abele while I only was expected to present a physically and mathematically based idea of the most important mechanisms.

Fate thwarted such far-reaching intentions. Detail work was underway for a couple of months when Abele informed me that he suffered from a serious disease. He was, at the time, optimistic with respect to the continuation and completion of our task. However, things developed at a dramatically fast pace, and, in spite of heroic efforts to overcome the obstacles, only a few of the sections allotted to him could be committed to paper. And then, on the 8th of October, 1994, almost exactly one year after having mentioned his illness, he died.

The future of our mutual work was in suspense.

After the first shock had gone by, I realised that I was not alone to deal with this critical situation. Some friends, familiar with the scientific situation in the particular field, encouraged me not to give up and to envisage a modified concept: the mechanisms, formerly one of several aspects (though, of course, an important one), should become the very essence of the book. Thus it would be possible to treat this side of the ensemble in more detail and thereby close, at least to a certain extent, a gap existing in the literature. In fact, in the course of the last decades various hypotheses have been proposed about the mechanisms governing particular phenomena. Attempts for a comparative assessment of such hypotheses are made sometimes, especially in connection with the presentation of a new hypothesis. Yet an overall view, dealing with different kinds of mechanisms and based, as far as possible, on physically and mathematically solid ground, is lacking.

In this context I have to mention Prof. H. and Dr. A. Heuberger (Salzburg, Austria) who, as long as 20 years ago, had foreseen the necessity of looking at catastrophic mass movements not only from a geomorphologist's standpoint but also from that of an engineer. Thus they had awakened my interest in the matter. And in the crisis caused by Abele's death they were first to present the above-mentioned arguments in favour of a concentration upon the physical (and in particular the dynamic) aspects. Prof. D. Barsch (Heidelberg, Germany), acting as a scientific consultant to the editors, allowed me a most encouraging discussion that issued similar conclusions. And, last but not least, Herta Abele, Gerhard's widow, not only gave her agreement for the free use of what her husband already had written but also handed me his comprehensive collection of literature on the subject. Without these encouragements and this help I should not have been able to do the job.

Thus the narrative turns back to its beginning: there is nobody who contributed more to make a new start possible than my late co-author. It was in the extended discussions with him that the solutions of so many problems faintly began to appear at the horizon; it was in walking or climbing with him over the debris of Köfels, Val Pola, and Tschirgant that opinions were confirmed or rejected; and it was on the basis of his immense field experience (embracing hundreds of events) that the engineer learned to understand how a geoscientist looks at things.

So, in spite of having changed its title and lost part of its volume as the consequence of a tragic destiny, this book remains in more than one respect what it initially was intended to be: the result of the close co-operation of two friends.

Neuhausen (Switzerland), January 2001
T. H. Erismann

Acknowledgements

The present book could not have been written without the support of so many persons who contributed to the final result in various ways.

At the very origin stood nothing but the formation of an interdisciplinary team by the late Professor E. Preuss (Munich University) who saw that his "frictional heat hypothesis" could not be successfully investigated by a single person. As mentioned under Heading 2.4, this team consisted of himself, Professor H. Heuberger (Salzburg University), and the undersigned author.

Heuberger, in addition, was kind enough to follow up the work on the present book, reading and criticising it section by section from the viewpoint of a geomorphologist. Luckily two further distinguished experts agreed to take part in following up: the mineralogist Professor L. Masch (Munich University) and, somewhat later, the geophysicist Professor E. Brückl (Vienna Technical University).

The precious encouragement by Herta Abele (Mainz) and Professor D. Barsch (Heidelberg University) has already been acknowledged in the foreword.

Magnificent first-visit guides in unknown sites were Professor A. Cancelli and Doctor G. Crosta (Milan University) for Val Pola, and Professor G. Patzelt (Innsbruck University) for Tschirgant and some minor Tyrolese events.

Certain publications might have passed uncommented without the recommendations of particular persons. Professor G. H. Eisbacher (Karlsruhe University) drew the author's attention to the Canadian studies which formed the basis for the description of Pandemonium Creek. Doctor A. von Poschinger (Geological Survey of Bavaria, Munich) regularly forwarded new results obtained by himself and others in Bavaria and in Switzerland, especially in connection with the age of certain events. Doctor F. Würsten (Neue Zürcher Zeitung, Zürich), with an article, gave rise to an interest in the work of James Hutton. And the President of EMPA (Swiss Federal Laboratories for Materials Testing and Research), Professor F. Eggimann (Zürich/Dübendorf), made possible the use of his lab's library, thus opening the door to world-wide search of literature. The respective work, sometimes rather tricky, was impeccably and rapidly accomplished by the librarian J. W. Glas and one of his co-workers, Christa Surber.

The acquisition of photographs sometimes was not an easy task, and not all persons who were helpful in this context could be mentioned in the respective legends. In particular, Professor R. L. Shreve (University of California, Los Angeles) provided the addresses of D. R. Hirst (US Geological Survey, Tacoma, WA) and Professor J. S. Shelton (La Jolla, CA) who contributed photographs of Sherman and Blackhawk, respectively. The above-mentioned description of Pandemonium Creek would have been impossible without the excellent photos made by members of the Geological Survey of Canada. Doctor S. G. Evans (Ottawa) took the care of selecting and mailing them.

In analogy thereto, Professor W. Welsch (University of the Armed Forces, Munich) who had visited the site (and climbed Huascarán) a few weeks after the event, generously made available his precious original slides for reproduction. Doctor W. Glanzmann, director of the former Swiss Federal Office for Military Airfields (Dübendorf/Zürich) and his co-worker R. Ryter, after a search for aerial photos taken shortly after the catastrophe of Val Pola, effectively enhanced the outfit of the respective section. Professor H. K. Hilsdorf (Karlsruhe University) not only contributed the illustrations showing fracture mechanical testing of rock; he also made possible Doctor B. Hillemeier's study on the subject. Various photographs taken by Abele and incorporated into the collection of the Department of Geography at Innsbruck University, were made accessible for selection and reproduction by Professor A. Borsdorf. Contributions by members of EMPA (Dübendorf/Zürich) were made by Doctor Jarmila Woodtli using scanning electron microscopy and M. Bosshard with conventional photography. Maps, like photos sometimes difficult to get at, in a particular case were procured in no time by Professor H. Kienholz (Bern University) who, in addition, had initiated the contacts with the University of Milan.

It would go too far to enumerate the persons who were kind enough to give oral information. Still a few exceptions may be permitted. Unforgotten in this context are enlightening fundamental discussions with Professor H. J. Körner (Geological Survey of Bavaria, Munich) almost 20 years ago. Thanks are owed to Professor H. Mostler (Innsbruck University) who, via Professor Heuberger, gave the information about the sulphurous springs in the deposits of Köfels. Doctor H. R. Fuhrer (Zürich University and ETH) communicated precious historical details about the military application of artificial rockfalls. Doctor D. Mayer-Rosa and Doctor M. Baer (Swiss Federal Seismographic Station, ETH Zürich) provided seismograms from Val Pola and useful information about possible (and impossible) conclusions. And an old friend, Professor D. Vischer (ETH Zürich), contributed some precious comments on the current methods of treating waves excited by rockfalls.

Each of these supporting actions was sincerely welcome as a valuable stone in a mosaic, and each, no matter whether short or long, oral or written, photographic or printed, has helped to improve the quality of this book.

T. H. Erismann

Notation

As a principle, the symbols used to express quantitatively treated values are explained immediately before or after the presentation of the respective equations. Still the following letters usually stand for particular items. Double meaning of a letter does not occur in a given context.

Symbol	Dimension	Item (parameter etc.)
a	$m\ s^{-2}$	Acceleration
a	–	Coefficient
A	m^2	Area
B	m	Breadth, width
c	m	Half-crack length
c, C	–	Coefficient
D	m	Diameter
E	Pa	Modulus of elasticity
E	$kg\ m^{-1}\ s^{-1}$	Viscosity
f	s^{-1}	Frequency
f	–	Fahrböschung (overall slope)
F	N	Force
g	$m\ s^{-2}$	Gravitational acceleration (= 9.81 on Earth)
G	$N\ m^{-1}$	Energy of separation
h	m	Partial thickness, partial height
h	–	Relative thickness = $H\ L^{-1}$
H	m	Total thickness, total high
i	–	Ordinal number (usually index)
J	–	Coefficient of restitution
k	–	Ordinal number (usually index)
K	$Pa\ \sqrt{m}$	Fracture toughness

Symbol	Dimension	Item (parameter etc.)
L	m	Length
m	kg	Mass
n	–	Number of items
N,	–	Total number of cycles
p	Pa	Pressure
P	–	Point
q	–	Quotient of two densities
R	m	Radius
s	Pa	Stress
t	s	Time
T	s	Duration of period
u	$m\ s^{-1}$	Velocity
v	$m\ s^{-1}$	Velocity
V	km^3	Volume
w	$m\ s^{-1}$	Velocity
W	J	Energy
x	m, km	Horizontal distance along midstream path
y	m	Horizontal distance at right angles to x
z	m	Vertical distance
β	rad, °	Slope angle
∂	$kg\ m^{-3}$	Density
μ	–	Coefficient of friction
Ω	s^{-1}	Angular velocity

In addition, $A, B, C, \ldots$ and, for particular reasons, N, P, Q, R are used as ordinal symbols. Current mathematical signs are used (e.g. d for differentials, $e = 2.7183$ for the basis of natural logarithms, f preceding brackets for functions, Δ for differences, $\pi = 3.1416$ for Ludolph's number, and Σ for sums).

Contents

Introduction

Couldn't we, to acquire better understanding of big rockslide motions, dig deeper by using a bit more physical reasoning and possibly also calculation?

Albert Heim (original in German)

1.1 Quantification – a Stringent Necessity

In the wide field of scientific problems exists a marked difference between mainly academic questions and those bearing *practical importance.* It may be worthwhile, perhaps, to spend considerable work and money to find out that Mount Everest is two metres lower than assumed previously (and, of course, the value of precise measurements for the observation of tectonic processes is incontestable). But nobody except the involved scientists and their sponsors would consider it a catastrophe if a more accurate measurement would add another couple of centimetres to the actually accepted figure. On the other hand, the question of whether an expected rockslide will reach a certain village or not, involves, in the best case, emergency measures (like evacuation) and the loss of substantial material value, in the worst case the risk of inhabitants being killed.

Given this background, the following chapters will treat the mechanisms of rockfalls and rockslides with a permanent *side-glance to practical application*, and special attention will be paid to the basic parameters of displacement, namely

1. the position attained by the distal elements of the mass,
2. the velocity of the mass at critical points of its travel.

Furthermore, the results obtained from the analysis of known events will be looked at in view of being used for *improved prediction* in the future; and prediction will, as far as possible, be treated from the (rather uncomfortable) standpoint of an expert assisting the local authorities in decision making.

In this context there might be some doubts with respect to the importance of *velocity*. Can it not be assumed that, in determining the expected extreme position of the moving rocks, one automatically defines the area threatened? And is there any difference whether a building is demolished at a speed of 10 or of 100 m s^{-1}? The answers are trivial for damages due to the elements of the mass proper, but they do not cover possible secondary effects among which the most dangerous is that of water (s. Sect. 2.5, 2.6, 7.1, 7.2).

*

It is at once depressing and encouraging that the *analysis of catastrophes* yields the best arguments to stress the importance of well-founded prediction. And, of course, it is easy to find impressive examples. However, the demonstration of the destructive potential residing in rock masses speeding downhill should be complemented by statements about possibilities (availed or missed) to save human lives. The event of Huascarán/Yungay in Rio Santa Valley, Peru (Welsch and Kinzl 1970; Plafker and Ericksen 1978; Patzelt 1983), with a death toll of about 18 000, holder of a macabre record, may serve as an example. Being released unexpectedly by the heavy earthquake of May 31, 1970, it was far beyond the reach of any competent expert. And the heavy, though by far less catastrophic, event of 1962 could, in the local opinion, be interpreted as a discharge promising safe decades rather than a bad omen. Anyhow, it would not make sense to mention the Huascarán event here if there were not one detail of particular interest for the future. As the city of Yungay was completely destroyed, a new one, Nueva Yungay, has been built in the meantime. The location chosen for this purpose lies within the relics of a larger prehistoric event. To the knowledge of the authors, no expertise concerning the risks thus taken has preceded the initiation of construction work. For further information refer to Sect. 2.7, 5.3, and 7.2.

To exemplify the *consequences of wrong prediction* (and simultaneously the importance of velocity), another tragedy is most enlightening. In 1963 it was known for several years that a part of Monte Toc in Vaiont Valley, a tributary of Piave Valley, Italy, would slide into the basin of a hydroelectric power station (Müller 1964, 1968; Broili 1967). On the basis of careful observations of the unstable rock mass, the experts predicted a slow, creeping motion, and it was not considered necessary to evacuate the basin completely. When the mass came down on October 9, however, it must have reached a velocity of about 25 m s^{-1}. The rockslide itself did comparatively little damage, but a secondary effect, a water wave over 200 m high, inundated parts of Piave Valley and destroyed almost completely several villages, the largest of which was Longarone. The author does not share L. Müller's opinion (1964) that the speed of the moving rocks could not have been expected (in the summary of his study Müller even speaks of "*... the unprecedented velocity which exceeded all expectations ...*"). Obviously, Heim's fundamental work (1932) had not been consulted with the required care in spite – or because – of having been on the market for more than 30 years (and being mentioned in the list of references). As will be elucidated in Sect. 2.6 and 6.2, Heim's book contained the necessary information for a useful quantitative forecast of velocity, not very precise, but sufficient to give an idea of the excessive imminent danger. So there are strong arguments in favour of the hypothesis that the death of almost 2 000 persons could have been avoided. In any case, the question remains, how an expected event of this calibre (about 0.3 km^3) could be treated by well-known experts without taking into account all details available in the most prominent publication on the subject existing at the time.

Wrong prediction does not always mean fundamental errors as in the last-mentioned example. And even if a correct view of the envisaged mechanisms is applied in quantifying complex phenomena, it must be borne in mind that the knowledge of the parameters involved never will be an exhaustive one, especially if prediction is aimed at. In other words, no pretensions can be made for a high *accuracy* of calculated figures, and a well-considered estimate of a realistic "safety coefficient" (one of the author's teachers called this "coefficient of ignorance") may be as difficult as the respective detailed calculation. Nevertheless, quantification is stringent, be it to acquire better understanding of a phenomenon, be it to trace the influence or to estimate the order of magnitude of a parameter, be it, in certain cases, to reduce to absurdity a preposterous (though seemingly plausible) hypothesis.

*

The above-mentioned endeavour to help experts in assessing the risks of catastrophic events must find its reflection in an appropriate *presentation of mathemati-*

cal analyses. An expert confronted with a mortal danger, perhaps for hundreds of persons, hardly will be interested in anything but easy-to-understand and fast-to-apply results. On the other hand, the same expert, if interested in quantitative aspects and working far away from the dramatic side of his job, perhaps would like to follow the line of thought leading from geomorphological facts to their mathematical expression. In view of such ambiguous demands certain rules have been established for the present book. In principle, mathematical deductions will be reported by describing the basic ideas and the fundamentals of the mathematical approach; the subsequent intermediate operations will be mentioned only as far as necessary to get at a result without losing the track (the degree of shortcutting being a function of the degree to which the respective operations are generally known or can be found elsewhere, be it in this book or in the literature). These rules are valid mainly for the chapters dealing with particular mechanisms or models (Chap. 3, 4, 5, 6, and 7). On the other hand, mathematical considerations will, as much as possible, be banished from the other chapters. The results obtained by mathematical means, however, will extensively be used in all parts of the book, with compulsory mention of the sections in which more mathematical details can be found.

The highlights of Chap. 8 will in first instance be a briefly commented reference *list of results.*

1.2 Remarks on Nomenclature

From time immemorial mankind has had to deal with the destructive power of rockslides and rockfalls. As a consequence, many words used in this context are old. And when the respective phenomena began to be approached in a scientific manner, it was inevitable that certain *words turned out to be imprecise or even misleading* in the expression of their meaning. To quote the most obvious example: around 1970 the catastrophic displacement of a cubic kilometre of dry rock usually was denominated as a "landslide", and the same term was used for a far slower gravity-driven movement of a thousand cubic metres of wet clay. So it was, of course, an improvement that, more and more, the word "rockslide" was applied for the first-mentioned kind of event, thus offering the advantage of better distinction. In German and French, by the way, a similar, though not identical, situation is given by the broad use of the words "Bergsturz" (literally translated: "mountain fall" or "mountain crash") and "éboulement" ("crash" or "collapse").

So it was necessary to set up *definitions correlated with the respective words* in an unmistakable manner. Varnes (1978) presented a systematically arrayed and well-illustrated typology of rockslides (or "landslides"). A most useful general reference book in five languages was edited by Visser (1980). These and other publications (e.g. Hutchinson 1988) have been very helpful to the authors in their endeavour to use a clear and unequivocal nomenclature.

*

It has, however, to be remembered that this is a book primarily dealing with dynamic mechanisms and not with geomorphological problems, and that, as a consequence, it is necessary to use *physically adequate terms* even if they are in contradiction with accepted traditions.

In such instances particular attention has to be paid to the *title of this book*. The respective definitions in reference lists, although certainly useful for other purposes, are not satisfactory in the present context. Visser (1980), to pick out an example, stipulates for a rockslide (p. 148, Item 2 048) a displacement on bedding, joint or fault surfaces with incidental lubricating function; for a rockfall (p. 148, Item 2 049), undercutting in case of mountain sides is declared as compulsory, but neither guidance by a slip plane nor lubrication are tolerated. Such definitions have little to do with the problems of displacement so that the terms "rockslide" and "rockfall" must be defined anew in an adequate form. And, as adequate definitions are a matter of the geomorphological background, they will be dealt with in the introductory section (2.1) of the geomorphological chapter, and special attention will be paid to the distinctive criteria of the two phenomena.

Another showpiece in this context are the half-synonymous words "*flow*" and "*stream*". No lesser researcher than Albert Heim (1932) used both terms in parallel, of course in their German form ("fließen" and "strömen"), and applied them in an undifferentiated manner both to the motion of water and of dry debris. Of course, around 1900 scientific writing was far more "belles lettres" than it is nowadays, and Heim excelled in a rich vocabulary, frequently using various terms for one and the same thing or process.

Now "flow mechanics" is a well-defined branch of physics treating the motion of fluids (liquid and gaseous media), a motion that can be laminar (if viscosity prevails), turbulent (if dynamic effects prevail) or supersonic (if compressibility plays an important role). As will be elucidated in Sect. 5.1 and 5.6, a fast-moving disintegrated *rockslide mass differs from a liquid* (not to speak of a gas) so profoundly that the use of one and the same word for both kinds of locomotion is at least questionable. It is not intended to enter into a detailed linguistic discussion in the present book. Yet misunderstandings will be avoided by using the word "flow" only in the case of liquids, gases, and genuinely fluidised media, i.e. media essentially moving like a liquid. To quote a complementary pair of examples: on the one hand, terms like "granular flow" (s. Sect. 6.4 for references), "debris flow", "debris stream", "sturzstrom" (as proposed by Hsü 1975, 129–130, and Nicoletti 1989, 420), and even "debris avalanche", though rather unambiguous in its actual use, will be avoided in this book; on the other hand, "mud flow" will readily be used in spite of standing for very different parameter configurations (water-to-solid ratios, velocities, etc.), but in any case expressing materials subjected to the laws of laminar or turbulent flow. In accordance with the above-mentioned considerations the state of disintegration will be expressed simply by "coherent motion" and "disintegrated motion", the intermediate states as well as essential circumstances (e.g. presence of water, state of surfaces, etc.) being signalled by additional attributes. So information will be presented in unambiguous form without the necessity of introducing new terms.

It is well understood that Bagnold's fundamental work, in spite of having made plausible phenomena analogous to *laminar and turbulent flow in dispersions* of solids in liquids (1954), by no means is in contradiction with the philosophy suggested above. In fact, Bagnold considered gravitationless conditions in which the particles of solid neither floated on top nor sank to the ground. So the dominating role of gravity, one of the essential circumstances in the displacement of a disintegrated mass, was ruled out. And in a later study (1956), where gravity no longer was neglected, the central question was the transport of relatively fine granular particles by a moving fluid rather than the influence of a fluid upon moving rock particles.

Furthermore it will be observed that in the optics presented above the *criteria required to describe different types of motion* are other than purely geomorphological. Minor importance, for example, is attributed to the question of whether a slide belongs to the rotary or to the translational type. What counts, are the conditions given by the displacement of the centre of gravity, the friction between mass and ground, energy dissipated within the mass, and other physically relevant parameters. Of course, also the geometry of the ground enters into such considerations, so that differences between rotary and translational slides are not ignored. But these differences appear as normal steps of the calculation as soon as the said geometry is modelled. So it becomes plausible that various terms, though important in geomorphological and geological descriptions of rockslides and rockfalls, will be mentioned only incidentally (if at all) in the following chapters.

Chapter 2

Case Histories, Geomorphological Facts

We possess nothing certainly except the past.

Evelyn Waugh

2.1 Introductory Remarks

It has already been pointed out (Sect. 1.2) that the purpose of this book differs from those of most studies in the field of gravitational mass movements. Therefore the most important chapters (3, 4, 5, and 6) deal in a physical, partly also a mathematical manner with various mechanisms, the lion's share being devoted to the mechanisms of displacement (Chap. 5). This distribution of weights is clearly reflected by reduction of geomorphological information to the amount dictated by necessity. Nevertheless the fundamental *importance of geomorphology* remains untouched: every bit of physical reasoning must be derived from some geomorphological fact, and every mathematical equation must find its correlative in a geomorphologically expressed process.

Probably the best way to back up theoretical considerations with geomorphological reality lays in the form of *coherent case histories* offering, by virtue of their immanent logical ties, far more cogent arguments than the enumeration of unconnected facts picked out from a multitude of events. Still there are limits to such a strategy. It is, for instance, worthwhile to use a case history in order to demonstrate one of the basic mechanisms on which generations of scientists rely in elaborating their hypotheses; but it would mean to break a butterfly on the wheel if the same case history were used to illustrate nothing but a single process of moderate importance. Thus, to fill gaps left in the description of complete events, several of the case histories will be complemented by additional geomorphological facts taken from events not described in extenso. To spot the geomorphological basis thus defined, besides the six main events also six of the last-mentioned additional group are listed hereafter in a table, and their locations are presented in Fig. 2.1. Eventually, it goes without saying that further sporadic bits of geomorphological information will be presented right in the sections dealing with the mechanisms proper.

*

The *choice of the "key events"* to be described in more detail had to be made with due care. In order to be selected, each of them had to fulfil several indispensable conditions. First of all, it had to be what might be called a "key experience", i.e. a case demonstrating in a particularly striking manner one or several mechanisms of high importance. This does neither mean that such mechanisms necessarily must work as they were understood when first proposed, nor does it mean that the described event was the one that gave rise to the first description of the said mechanism. The criterion for the choice rather was the best possible quality expected for making evident the fundamentals of mechanisms. Secondly, it was required that the events be sufficiently well documented (in first instance by appropriate maps) to present them in a more or

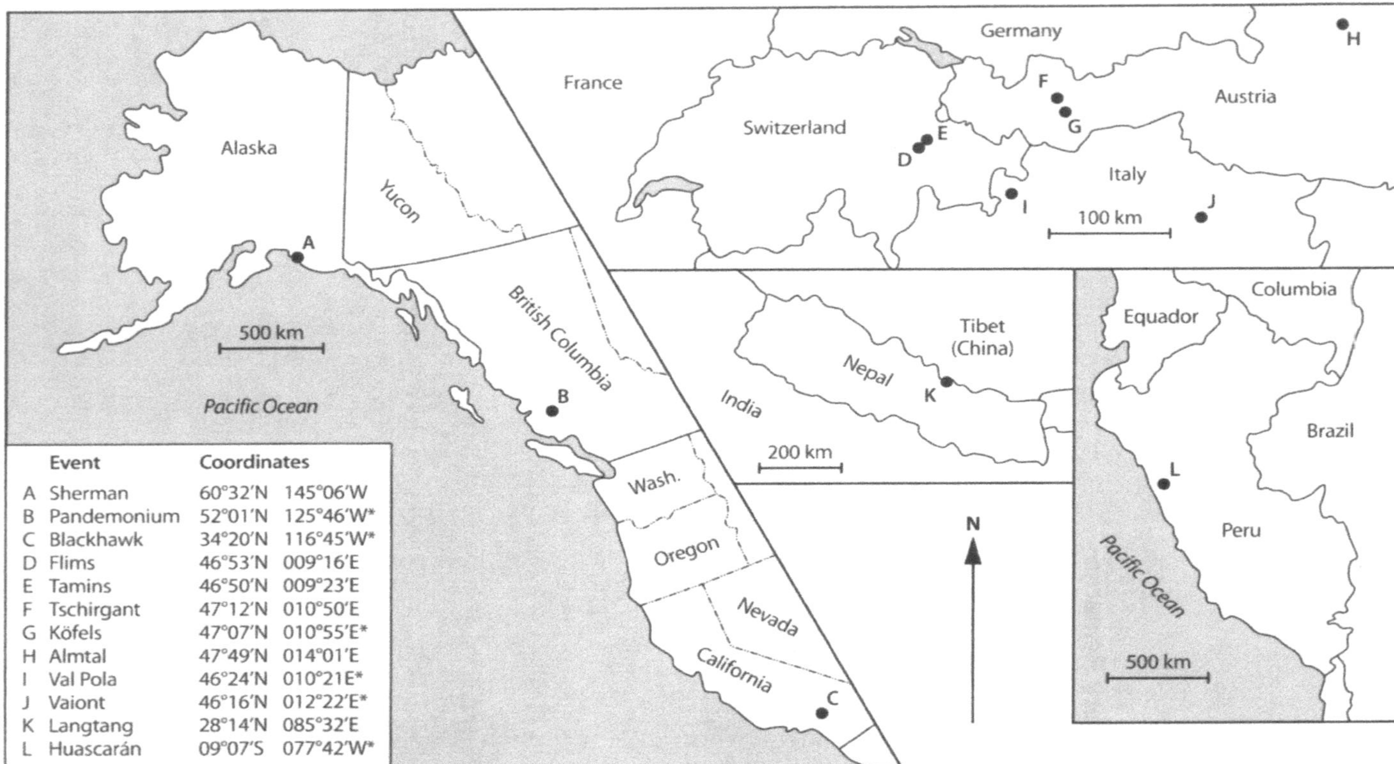

Fig. 2.1. Locations of "key" and some additional events in North American, European, Asian, and South American mountains. "Key" events (names are titles of respective sections) marked by asterisks (*) (sketch by Erismann)

less uniform, easy-to-understand form. Finally, no event, even if it was considered perfectly suitable, could be used without being thoroughly known. The best approach to an intimate knowledge, namely extensive visits to the respective rockslides, was not possible in all envisaged cases. Thus, for events only known from reports a further limitation was given by the necessity not only of good documentation but also of exemplary descriptions giving the details needed to formulate case histories according to the requirements of the present book.

Contrarily to the more theoretical Chap. 5 dealing with displacement, the case histories are correlated with the demonstrated mechanisms in various manners, sometimes one history referring to more than one important mechanism, sometimes also one and the same mechanism appearing in more than one history. So no particular methodological order recommended itself, and the case histories were arrayed in *geographic order* as evident from Fig. 2.1 and Table 2.1 (from north to south and from west to east).

*

Before closing this section, the announced discussion (Sect. 1.2) of the terms used in the book's title has to be brought to an end. From the dominating point of view of possible motional mechanisms, no fundamental *distinction between rockslides and*

Table 2.1. Basic information about "key" and some additional events. The table is compiled from the references mentioned in the text in connection with the respective events (Sect. 2.2–2.7). In view of a compact presentation, remarks are confined to those directly connected with the aims of the mentioned sections

Event	Volume (km^3)	Material	Fahrböschung[a] (s. Sect. 6.2)	Year A.D.	Remarks
Sherman	0.03	Sandstone	0.172	1965	Run-out on glacier
Pandemonium[b]	0.005	Gneissig rock	0.235	1959	"Speedometry included"
Blackhawk[b]	0.28	Limestone	0.098	[c]	Air lubrication?
Flims	9.0	Limestone	0.07 (0.12)	[c]	Slope?[d]
Säsagit/Tamins	1.7	Limestone	<0.23 (0.09)	[c]	Slope?[d]
Tschirgant	0.2	Dolomite	0.16	[c]	Water lubrication?
Köfels[b]	2.2	Augengneiss	0.18	[c]	Frictionite observed
Almtal	0.3	Limestone	0.11	[c]	Water lubrication?
Val Pola[b]	0.035	Gabbro	0.46	1987	Flood wave
Vaiont[b]	0.3	Limestone	0.23	1963	Flood wave
Langtang	2.0<	Migmatite/gneiss	?	[c]	Frictionite observed
Huascarán[b]	0.053	Granodiorite+ice	0.25	1970	Giant bounces

[a] According to Heim's definition of overall slope.
[b] "Key" events (names used in titles of respective sections).
[c] Prehistoric events.
[d] Alternative assumptions: Flims or (for figures in parentheses) Säsagit/Tamins penetrated into Domleschg (s. text and Fig. 2.5; Sect. 5.5; Abele 1997b). In the first-mentioned case determination of slope is difficult.

rockfalls can be presumed. In fact, both are gravity-driven, potentially catastrophic, fast displacements of more or less considerable rock masses. In other words, no mechanism can be active in the one which – at least potentially – cannot be expected to be active in the other (Chap. 5). To quote an extreme example: although it is not very probable that self-lubrication (e.g. by melting crystalline rock, Sect. 5.5) might occur in an average-sized rockfall, its occurrence in case of a singularly large event is by no means excluded. So the comparison of rockslides and rockfalls is mainly confined to an analysis of gradual differences (and similarities) in the importance of mechanisms and parameters and to the deduction of the resulting consequences.

The most characteristic feature of a rockfall consists in the dominance of a rapid descent over a very steep slope, in many cases more or less free from obstacles. Such displacement, commonly known as "falling", although "pseudo-falling" perhaps would be a more adequate term, implies a particular balance between accelerating and decelerating forces: in descending, a high percentage of the weight acts as a motor and only a limited one is able, by generating friction, to produce a braking effect. The result is not only a high rate of acceleration (in extreme cases practically equalling the gravitational acceleration) but also – at least as long as atmospheric drag is not a substantial limiting factor – an almost equal velocity of all moving elements. As one of the essential sources of collisions within the moving mass consists in differences of velocity, the result is a low degree of *interaction between the elements*. In terms of kinetics, this fact is stressed by the low energy exchange resulting from the small differences in velocity between the colliding elements (Sect. 4.2, 5.7).

These effects sometimes are enhanced by another fact. As a rule, in spite of impressing exceptions (example: two consecutive rockfalls near Randa, Valais, Switzerland, 1991, total volume almost 0.03 km^3; Noverraz and Bonnard 1991, 165–170; GEOTEST/BWW 1992), rockfalls are of substantially smaller *size* than rockslides. So, if the width of its track is limited, the mass of a rockslide may be kept together therein while the elements of a rockfall are more or less free to spread laterally, thus further reducing the number of collisions.

In such instances it is obvious that the *cohesion* of rockslides, though neither absolute nor homogeneous or isotropic, normally is distinctly higher than that of rockfalls which in certain cases may be considered as nothing but a sum of individually moving elements. Both in analysis and prediction this is an important point, especially if the probabilistic aspect is taken into account.

It makes sense, for example, to determine the reach of not too large a rockfall by calculating the *paths of single elements* and systematically varying certain key parameters so as to obtain a reasonably realistic scatter band (Sect. 5.3). This means nothing less than the neglect of interaction as a mechanism of minor importance. Needless to say, such an approach would yield completely wrong results in the case of a rockslide where, as the interaction within an immense number of elements results in an efficient averaging process, the conditions of a probabilistically correct calculation are best respected by using well-balanced mean values for the parameters (Sect. 6.3, 6.4). This statement should not be misunderstood as a suggestion never to take into account the motion of single blocks in rockslides (Sect. 5.3, 5.7). The situation as described above might be summarised by the following statement: certainly both rockslides and rockfalls are chaotic to a high degree, but chaos and chaos sometimes are quite different things.

As will be discussed in detail under Heading 5.1, no general statement can be given with respect to the causal relation between *cohesion and reach* of an event. Nevertheless it is with good reason that rockslides, as a rule, are considered as reaching farther than rockfalls. But this is not so much a matter of cohesion as one of size and geometry: rockfalls, after a dramatic descent over a steep slope, very often are completely or partially arrested by hitting a massive obstacle more or less frontally.

As far as small events are considered, the theoretical study of rockfalls can effectively be complemented (and theoretical results verified) by *experiments* (Broili 1976; Bozzolo 1987, Bozzolo et al. 1988). Small rockfalls, released by an explosion and continuously recorded by several cameras, have repeatedly been used for the purpose. One of the author's candidates for a doctoral degree described in his thesis several particularly well instrumented and carefully evaluated experiments with such "in vitro" rockfalls. The term "in vitro" is probably an understatement as in certain cases thousands of cubic metres of rock were released. Thereby most valuable detailed information, in first instance about the motion of single blocks, was obtained (Sect. 5.3).

*

In view of the considerations made above, especially in the context of case histories, it is obvious that the sections following hereafter could not be written without a perceptible *personal aspect*. Other authors, having other key experiences, probably would have chosen other events to illustrate the same mechanisms. This fact is not considered as a disadvantage: would a book not bear the stigma of a computer-made product if it had no personal touch?

The importance of geomorphological aspects is once more stressed in the last section of this chapter (2.8): it contains a reference *list of essential results* having emerged from the case histories (Sect. 2.2–2.7) and considered worth to be taken into account in the subsequent chapters.

2.2 Pandemonium Creek

One of the two particularly important problems discussed in the present section has been raised in first instance by the investigation of so called *long run out* rockslides in which masses are moving far beyond the distance that could have been expected when considering the size effect alone. This effect, according to which large rockslides move more economically and thus reach farther than smaller ones, was first expressed by Heim (1932, 121–130) and quantified in an easy to apply form by Scheidegger (1973). In view of its value for modeling, the size effect is paid due attention in a special section (6.3). The existence of long run out events is an important obstacle for a reliable mathematical modeling of rockslides with the intention of predicting the threatened area. The span of uncertainty can, of course, be reduced by taking into account various geomorphological circumstances; however, very substantial differences still remain, and without the introduction of further mechanisms, effective mathematical modeling remains problematic. One of the possible candidates for an answer to this question is treated hereafter.

The second problem has to do with the remark made under Heading 1.1 that besides the reach of a rockslide, also the *velocity* at any point of its travel may be of outstanding importance. In this context the event of Vaiont/Longarone, one of the heaviest catastrophes of the twentieth century, was mentioned. It will be discussed more in detail in Sect. 2.6, and it will be demonstrated at various occasions (Sect. 2.5, 2.7, 6.2) that a reliable estimate of velocity is a more tricky task than might be assumed at first sight. This is true for post eventum analysis, not to speak of prediction.

With respect to both, explanation of long run out and determination of velocity, the rockslide of Pandemonium Creek is unique. On the one hand, its overall slope (for definition s. Sect. 6.2), being in the range of 0.23, is far below a value of around 0.4 as would be expected for a mass of 5 million m^3 from one of the equations based on the size effect. More than that: there is well founded evidence in favour of an exceptionally low effective coefficient of friction for a well defined portion of the travel. Such differentiated information would be impossible if the slide were not "equipped", over a distance of more than three kilometres, with something like a set of "*built in speedometers*". To understand this fact, it is necessary to look at the event in some more detail. This is possible owing to a most conclusive description by Evans et al. (1989) with excellent illustrations.

*

As far as its *history* can be reconstructed, the rockslide, unobserved by any known eyewitness, took place in the summer of 1959 (Evans et al. 1989, 438) in the uninhabited southern part of Tweedsmuir Provincial Park, British Columbia, Canada, at the divide between the rivers Atnarko and Klinaklini. No discrete fact can be spotted as having triggered the event: neither a particularly heavy earthquake nor abnormal precipitations were observed in the critical period. So Evans and his co authors concluded that very probably the slide was entirely due to long term degradation of the rock material's strength.

The detailed *description* given by Evans et al. (1989, pp. 430–438) may in short be summarised as follows (Fig. 2.1, 2.2a,b, 2.3, 2.4): A spur of diorite, irregular in shape, originally extending between 2 625 and 2 275 m elevation and having a volume of about 0.005 km^3, broke off the north flank of an unnamed peak and slid over a steep slope (total horizontal extension 3 000 m, total vertical extension 1 425 m) to the narrow valley of Pandemonium Creek. For about half of this accelerated travel the slide moved on a glacier (km 0.4–1.6 in Fig. 2.4), then on more or less naked rock. The material, most probably disintegrated already in this initial phase of motion, was, however, efficiently kept together by the channelling effect of lateral moraines (s. contour lines) so that the lateral extension in crossing the creek (km 3.05) was little more than twice that of the spur in its original position. As the mass hit the (equally steep) opposite slope almost at right angles, a considerable run up took place, the topmost clasts reaching an elevation 335 m above the bottom of the valley. Then the debris, guided by the slightly asymmetrical geometry of the ground, turned to the right, slid back to the bottom of the valley, and followed its course, thus descending by about 400 m over a horizontal travel of 3.1 kilometres (km 4.2–7.3). As the slopes on both sides of the creek are steep and practically uninterrupted, the mass was laterally kept together even better than in the preceding descent.

This phase, probably the most interesting of the event, allowed the determination of the velocity of motion at several successive points. The topographic basis for a series of "*recording speedometers*" is given by the simple fact that the valley's course, directed to the east as a whole, contains a succession of approximately sinusoidal curves. At the time of the exploration by Evans and his colleagues the traces of the passing rockslide were fresh enough to allow a rather accurate geometrical evaluation yielding the "superelevation" assumed by the debris in passing through the curves of this natural "bobsleigh run". The details of the mechanism behind the phenomenon and in particular its inherent sources of errors will be discussed later (Sect. 2.7, 5.7). In the present section it suffices to mention its strongest point, namely the repeated possibility to measure the velocity in a series of locations under approximately equal conditions. This means that, even if the absolute figures obtained are systematically erroneous to a certain extent, their mutual (relative) comparison, owing to similarity of the errors, will deliver a rather accurate idea of the manner in which velocity developed during this phase of motion.

More than that: a lucky coincidence arranged the topographical parameters so that, probably by exception of a slightly higher velocity in the first curve (Item I in Fig. 2.4), certainly a consequence of the preceding run up, the discussed phase was distinguished by, roughly speaking, *constant velocity* (from Evans et al., p. 440, an average of 26.17 m s^{-1} and a statistical scatter of ±4.08 m s^{-1} are obtained; for more quantitative details s. Sect. 5.7). This circumstance is due to the fact that most of the kinetic energy acquired in the very rapid initial descent was dissipated in the run up phase. So, as a consequence of an almost "ideal" initial velocity, no substantial change of kinetic energy took place during the considered three kilometres. In other words: the average value of the apparent coefficient of friction (definition s. Sect. 6.2) must have been in the same range as the average slope. And this slope, as easily can be seen from the figures at the end of the second to last paragraph, was as low as 0.129 (slope angle 7.35°). According to Scheidegger's formula, almost three times this slope (0.38) would have been expected for an event of the considered size!

At the mouth of the valley the mass, no longer channelled and moving at a decreasing pace on the comparatively flat fan of Pandemonium Creek, was free to spread laterally. Nevertheless a portion of it turned to the right by another 90 degrees and, after a southbound travel of several hundred metres, finally came to a rest with its *distal end buried in one of the Knot Lakes*. Owing to this fact the total extension of the deposit is not exactly known. The figures as presented here (like those published by Evans et al., though slightly differing therefrom for obvious reasons) are based on the assumption that only an energetically insignificant portion of the debris is excluded by cutting the evaluation on the edge of the lake.

Fig. 2.2 a. Pandemonium Creek rockslide, steep northbound descent, view from north. *Arrow* shows zone of release. Headwall (on top) from which the mass originated and glacier (below) where it accelerated (by courtesy of the Geological Survey of Canada, Ottawa, Ontario, represented by S. G. Evans)

As concerning the resistance during the travel through the "bobsleigh run", the exorbitant discrepancy between theoretical expectation and reality hardly can be explained without the introduction of at least one *particularly effective mechanism* active under circumstances as encountered in Pandemonium Creek Valley. It is, of

Fig. 2.2 b. Pandemonium Creek rockslide, steep northbound descent, view from north. *Arrow* shows zone of release. Below glacier, well-channelled track to (invisible) Pandemonium Creek (by courtesy of the Geological Survey of Canada, Ottawa, Ontario, represented by S. G. Evans)

course, an accepted practice to attribute a certain reduction of frictional resistance to the presence of water in the critical zone of a moving rock mass. However, the conditions usually considered as favourable for a lubricating effect of water (fluidisation of fine grained material, hydrostatic pressure of pore water, etc., s. Sect. 5.5 and 5.6) are

Fig. 2.3. Pandemonium Creek rockslide, "bobsleigh run" through Pandemonium Creek Valley, view from east. *Dotted lines*, background: traces of motion in descending (left) and running up (right). *Dotted lines* in valley: marks of superelevation used as "speedometers". Foreground: first half-kilometre of run-out area (by courtesy of the Geological Survey of Canada, Ottawa, Ontario, represented by S. G. Evans)

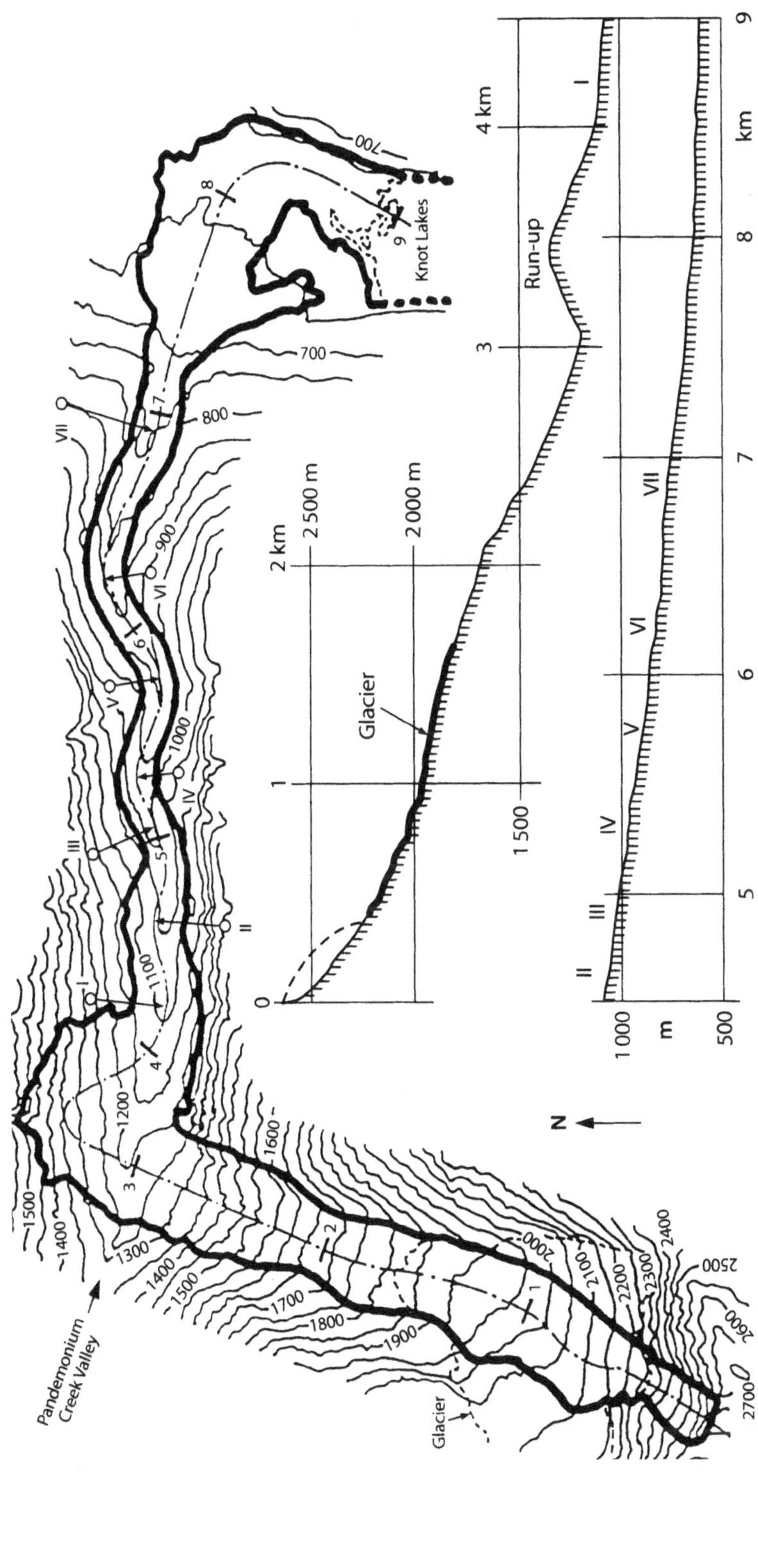

Fig. 2.4. Pandemonium Creek rockslide. Map and longitudinal section along *dots-and-dashes line* with kilometre marks (corresponding to assumed path of centre of gravity). Scales of map and section are equal. *Circles* with *arrows* labelled by Roman numerals: approximate centres and radii of "speedometer" curves. *Dotted line* in section: approximate initial position of mass (from Evans et al. 1989, 432, adapted to the requirements of the book by Erismann)

not (or, at least, not to sufficient extent) evident in the case of Pandemonium Creek. So it becomes essential to find a hitherto unknown mechanism fitting into the described configuration. The importance of this task in two respects reaches far beyond a simple reduction, in a single case, of a parameter depending on a multitude of circumstances. On the one hand, the Pandemonium Creek rockslide by no means was unique in moving most economically under the above cited conditions. On the other hand, the Pandemonium Creek rockslide is, perhaps, unique in offering an extended portion of track with a particularly unchanging character, allowing to be picked out and analysed on a sound quantitative basis almost independently from the rest of the event. And it is the scientific strength of this extraordinary rockslide that the said portion of track was rushed through most economically and at a practically constant pace.

In his extensive book on Alpine mass movements (1974) Abele had established various correlations between geomorphological parameters. This work is outstanding by its completeness: no less than 285 events were listed and evaluated as far as their parameters could be made available, and 190 thereof were definitely recognised as rockslides. It must be pointed out that, owing to the multitude of involved parameters and the width of their scatter bands, quantitative evaluations based on a moderate number of cases may be seriously falsified by excessive configurations in few events. In other words: a large number of cases is essential for reliable results of quantitative analysis. One of the most interesting diagrams in this context shows the *overall slope in function of the volume* for as many as 77 rockslides (p. 45). The main part thereof – after exclusion of the few uncertain cases, reduction of the volume range to events of at least 0.005 km^3, and castling between Flims and Tamins/Säsagit (s. later) – are reproduced in Fig. 2.5 (s. also Table 6.1).

It will be observed in this figure that several Alpine slides (Obernberg, Almtal, Tschirgant, Fernpass, and Flims) are conspicuous by a low overall slope. They are simi-

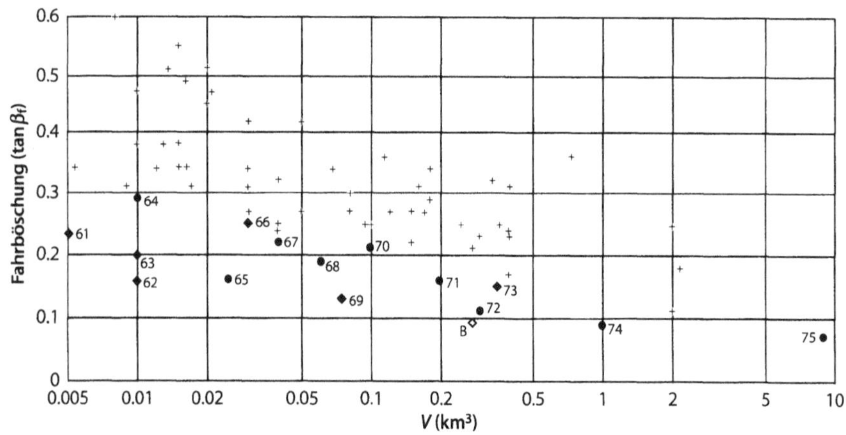

Fig. 2.5. Long run-out rockslides in the Alps (*circles*: Abele 1974, 45) and in North America (*squares*: Eisbacher 1977, 1979b; Evans et al. 1989) compared with other Alpine slides (*crosses*). *61:* Pandemonium Creek; 62: Damocles; *63:* Twin E; *64:* Elm; *65:* San Giovanni; *66:* Frank; *67:* Goldau; *68:* Obernberg; *69:* Nozzle; *70:* Lavini di Marco; *71:* Tschirgant; *72:* Almtal; *73:* Rockslide Pass; *74:* Fernpass; *75:* Flims (including Domleschg tongue in Fig. 2.6); *B:* Blackhawk (not used for further evaluation); *V:* volume (km^3); tanβ_f: Fahrböschung (s. Sect. 6.2) (sketch by Erismann)

lar to Pandemonium Creek in having made a substantial portion of their travel on a ground of loose material which, at the time in question, most probably consisted of *water saturated fluvial sediments* (Abele 1991b, 33, 1997b, 2).

In addition, at least three of the mentioned slides (Almtal, Fernpass, Flims) were to a high degree *laterally confined* in the respective phase of motion. This last mentioned criterion was not fulfilled in the case of the Tschirgant slide which, after a descent from the northern slope of the wide Inn Valley, ran out crossing this valley and partially penetrating into the tributary valley Ötztal. Its low overall slope probably shows that the existence of ample water saturated valley fill can suffice as a long run out condition, while lateral confinement, though effective as a booster, is not a conditio sine qua non. The overall slopes of three further Alpine events (the famous catastrophes of Elm and Goldau, as well as Lavini di Marco and Obernberg), having travelled on water saturated loose material only to a limited extent and/or lacking strict lateral confinement, lay somewhat less clearly below the main cluster of points in Fig. 2.5.

In order to illustrate the main characteristics observed in the mentioned events, Fig. 2.6 shows, in a form reduced to the information strictly essential in the present context, the general shapes of Almtal, Tschirgant, Fernpass, and Flims. Short *comments about these Alpine slides* may suffice (for details s. the references). Almtal and Fernpass probably are almost as obvious as Pandemonium Creek if abstraction is made from the impossibility to perceive the traces of superelevation in the curves of a prehistoric event. The case of Flims was the object of various discussions over the course of several decades (Pavoni 1968; Scheller 1970; Abele 1991b, 1997b). In Fig. 2.6 two possible scenarios are shown for the origin of deposits found more than 10 km upstream in Domleschg Valley (river Hinterrhein). According to the conventional opinion this material came from the Tamins/Säsagit slide, second largest in the region (volume about 1.7 km^3) which is assumed to have barred the confluence of Vorderrhein and Hinterrhein effectively enough to enforce a full stop on the Flims event, the largest known rockslide in the Alps (volume at least 9 km^3). Contrarily, according to Abele's last publication (1997b), the Tamins/Säsagit slide did not penetrate far into the Domleschg, and its dam only was able to deflect the enormous masses of the Flims slide southward and so to force them into the Domleschg valley. For the regression curves established in Sect. 6.2 this last assumption was used. It must, however, be pointed out that the results would not have changed much if the data of the Tamins/Säsagit alternative had been used: in either case the run out of a large slide took place under conditions as stipulated here above. Finally, the Tschirgant rockslide, already mentioned in the context of lacking lateral confinement, will be discussed as containing valuable evidence about pull apart structures in long run out deposits (hereafter and Sect. 5.5). In Fig. 2.6 the tongue penetrating into the Ötztal is visible as a protruding end pointing to the altitude figure "720".

To cross check the results obtained for Alpine rockslides, data of six *North American events* of similar character are added in Fig. 2.5, and five silhouettes are shown in Fig. 2.6. By no means unexpectedly it becomes evident that equal combinations of circumstances are valid for long run out events in America as well as in Europe. The parallel is stressed by the existence of an intermediate group of events with an obviously less effective mechanism as in the cases of Elm, Goldau, Obernberg, and Tschirgant. Most probably, this group is represented by the famous Frank and the Rockslide Pass events. Only the Blackhawk (Fig. 2.13) has to be considered separately as the physical background of its long run out character, in spite of perfectly fitting

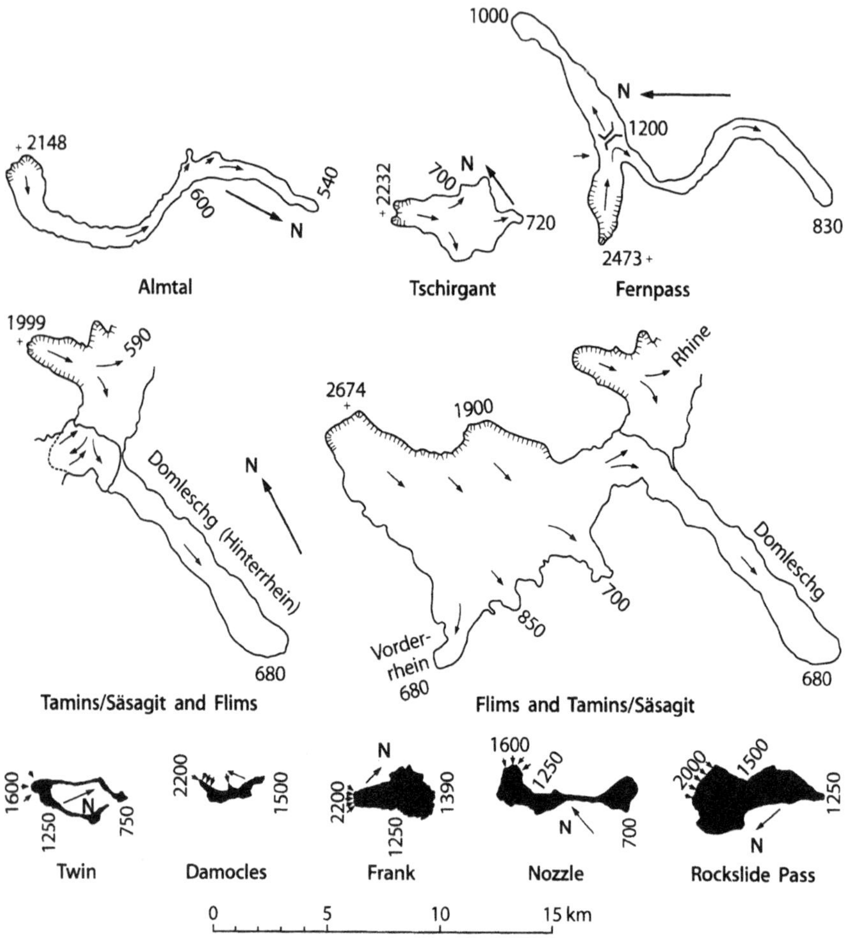

Fig. 2.6. Approximate shapes of four Alpine (Fig. 2.1) and five North American rockslides. For better comparison all events are represented as moving, roughly speaking, from the left to the right. Head scars are symbolised by hatching or (for silhouettes) by short *arrows*. Numbers represent elevations in metres. The tongue moving up Domleschg Valley is shown twice to account for the possibilities proposed (refer to text) (sketch by Erismann)

into the frame of Fig. 2.5, is not necessarily identical to that of the other examples. For this reason the Blackhawk is excluded from the calculation of regression curves (Sect. 6.3) and will be treated in the next following section (2.3).

In considering Fig. 2.5 and 2.6 it becomes plausible that Nicoletti and Sorriso Valvo (1991) attribute a particular importance to the *geometry* of a moving disintegrated mass. This problem will be given due attention in the context of reach prediction (Sect. 6.3).

*

On the basis of the mentioned evidence, Abele presented the idea of a regular correlation between the discussed geomorphological configuration and long run out. He did it first in a rather tentative form (1991b). Later he cast it into the more substantial frame of a theoretically *coherent hypothesis*, backed up by further details of field evidence (1997b). The mechanism, as illustrated in Fig. 2.7, is almost self explanatory. When a rockslide, having acquired a disintegrated state, reaches a valley floor with a water saturated fill of gravel and/or finer material, two mechanisms are started: on the one hand, parts of the fill, mixed with water and distal elements of the slide, are pushed forth ("bulldozed") like a frontal bow wave (Fig. 2.7c); on the other hand, the remaining fill is compressed under the weight of the slide. The first mechanism obviously has a rather turbulent character so that disintegrated rockslide material may be mixed with valley fill, thus generating an unstratified deposit that is more or less graded according to the local circumstances (Fig. 2.8). The second mechanism can be compared with the compression of a water saturated sponge: water is pressed out, and its pressure contributes to support the mass of the slide, thus acting as a pressurised lubricant in the true sense of the word (definition in Sect. 5.5).

As a *precursor* of this hypothesis may be considered the phenomenon designated as "flowslide" by Rouse (1984). Based on ideas expressed some years earlier by Bishop (1973), this mechanism is described as a "*... temporary transfer of part of the normal stress onto the fluids in the void space ...*" (p. 491). This (as well as the critical comments to air cushion and dust lubrication) corresponds quite well to the line of thought presented here. What lacks, is, perhaps, a clear distinction from other fast mass displacements as given in this section and complemented under Heading 5.5.

It is most essential to take into account the trivial fact that the pressurised water tends to *escape* by any practicable passage. Owing to the normally reduced permeability in the sole of the mass and the back pressure of the bow wave mentioned above, the most important (though not unique) possibility of escape is lateral. If the valley is wide in comparison with the moving mass, the lateral displacement of water will induce a secondary lateral motion of the slide riding on its top. The result is lateral spreading with a turbulent zone analogous to the bow wave (Fig. 2.7), and the formation of longitudinal cracks is probable. If, by contrast, the valley is narrow, its channelling effect reduces the possibility of lateral escape so that an ideal configuration for retaining the water in the lubricating zone is created (Fig. 2.7). But even in such

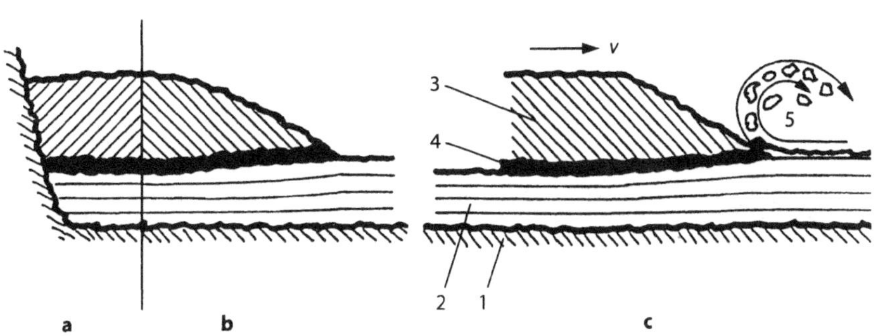

Fig. 2.7. Lubrication by water extruded from saturated valley fill, compressed under the weight of the rockslide mass. For details s. Sect. 5.5. **a, b** Section across asymmetric valley with channelling effect and improved lubricant sealing on the one side (**a**) and increased escape capacity on the other (**b**). Full effects in channelling or escape obviously require symmetrical sections. **c** Longitudinal section. *v:* velocity of slide; *1:* bedrock; *2:* fill; *3:* rockslide mass; *4:* comparatively impermeable bottom layer; *5:* turbulent "bow wave" consisting of water, fill, and rockslide material (s. Fig. 2.8) (sketch by Erismann)

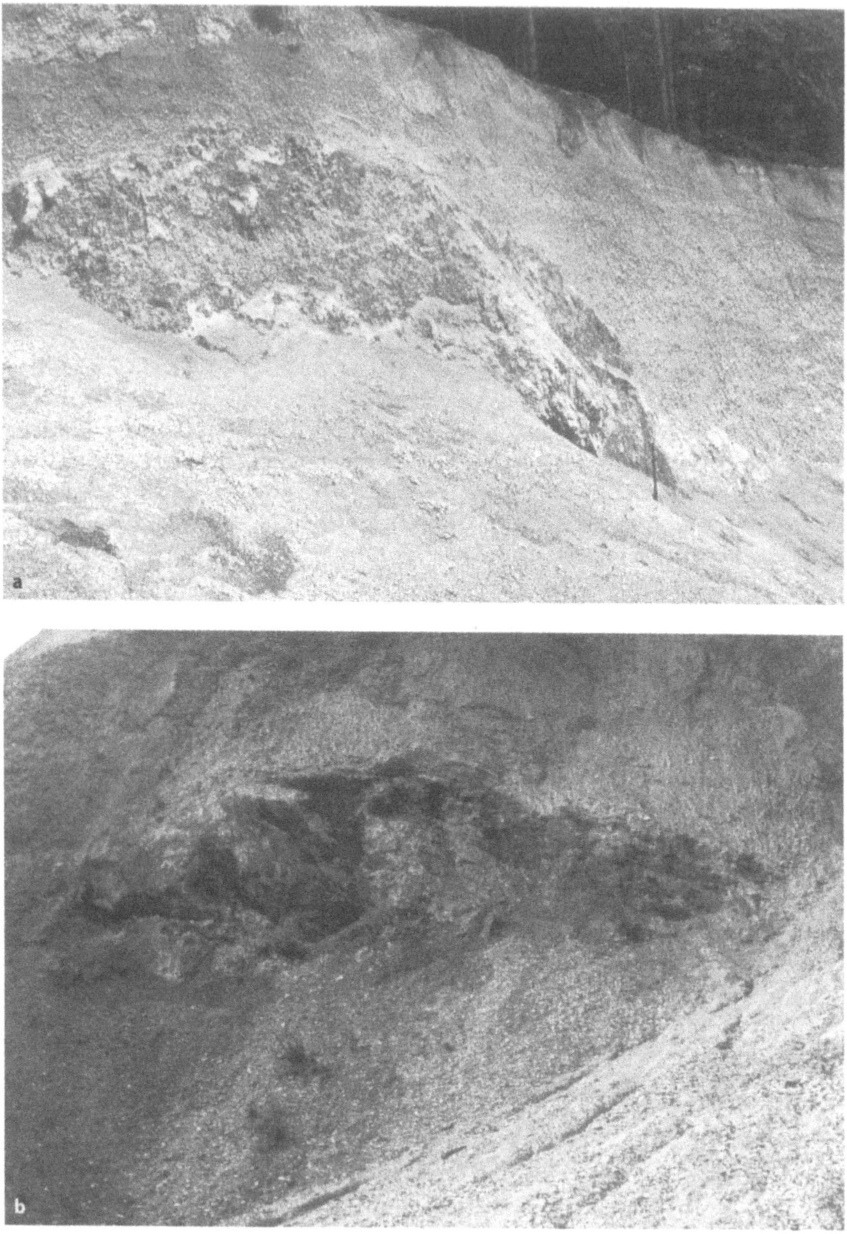

Fig. 2.8. Lenses of rockslide material suspended in valley fill, probably owing to a mechanism similar to that represented in Fig. 2.7c, Item 5 (bow wave). **a** Almtal rockslide (Oberösterreich, Austria, Fig. 2.1), gravel pit of Heckenau (at right end of sketch in Fig. 2.6). **b** Tamins/Säsagit or Flims rockslide (Grisons, Switzerland, Fig. 2.1), hill of Rhäzüns Castle, Domleschg, about 5 km from mouth (s. sketches in Fig. 2.6). Note indistinct gradation of fill both in **a** and **b** (photos by Abele)

instances the continuous escaping of water cannot be sealed completely so that the leading portion of the rockslide mass, owing to its overriding fresh portions of valley fill, is privileged in profiting from lubrication. A consequence thereof is a tendency of the mass to be elongated. Finally, transversal cracks may be formed before the unavoidable collapse of pressure lubrication.

It is obvious that in a relatively small mass like that of the Pandemonium Creek event such *pull apart structures*, after having been formed in motion, tend to disappear with fading thickness. In larger rockslides, on the contrary, impressive examples are known. A particularly well developed piece of evidence is the young Tschirgant rockslide (Fig. 2.1, 2.6, 2.9, 2.10; radiocarbon age about 3 000 years B.P.; approximate volume 0.2 km^3; Patzelt and Poscher 1993) located in Tyrol (Austria) several kilometres north of the famous Köfels area (Sect. 2.4). The evidence represented in Fig. 2.10 is situated in Ötztal Valley, somewhat upstream from its mouth. As Patzelt and Poscher have shown, the slide partially had penetrated as far as that. In fact, the gravel at this site definitely came from the Inn. Later the river Ötztaler Ache cut a gorge through the mass, thus generating an excellent longitudinal section of the deposits. From the photo it appears evident that a transverse crack first had been opened so that a rapid water jet, carrying along gravel, was pressed out in a vertical direction, a process that, by its safety valve function, necessarily weakened (or even cut short) the lubrication. So the jet was overrun and barred by a further portion of rockslide debris. It is nothing but natural that the material in the arrested jet is graded according to the degree of being dragged upward by the water (i.e. coarse on bottom). For details s. Sect. 5.7. Another impressing example of pulling apart might be a phenomenon reported for the Canadian Nozzle slide (Eisbacher 1979b, 325): between the distal portion of the debris and the main deposits was found a gap about one km long; and this zone of "*rarifaction*" was on one side flanked by – perhaps extruded in a similar way as proposed for Fig. 2.10 – relatively fine grained material!

It is an important quality of this concept that, in advancing, the mass continuously finds fresh lubricant, thus allowing the mechanism the ability to work over a theoretically unlimited distance under favourable conditions. On the other hand, a clear limit to its effectiveness is given by the necessity to get through before the lubricant has been able to escape. So sufficient *velocity as a fourth condition* (besides water saturation, low permeability at the bottom, and lateral confinement) turns out to play an important part. As far as conceivable with the information available at the present time, this thrilling competition between acquisition and escape of lubricant is treated in Sect. 5.5. In the particular case of Pandemonium Creek a further source of water may be suspected in melted glacier ice swept along by the descending mass. How far this "single shot infusion" of lubricant carried along by the mass actually did influence the run out, remains an open question.

*

A useful by product of the circumstances observed in the Pandemonium Creek event refers to the *energy dissipating mechanisms*. On the one hand, a run up with a high angle of lateral deflection, wherein energy is transformed from kinetic to potential and back to kinetic form, is recognised as an effective energy dissipating mechanism. Even if the maximum velocity estimated by Evans and his colleagues is somewhat exaggerated (the situation is similar to that of the Val Pola event which will be analysed in Sect. 2.5), a reduction of velocity from something between 81 and 100 m s^{-1} (Evans et al., 440) to

Fig. 2.9. Tschirgant rockslide (Tyrol, Austria, Fig. 2.1), frontal view of twin scar and descent zone (bottom of photo is practically at level of Inn Valley). Photograph taken from run-up debris barring Oetztal Valley at a distance of about 6 km from the scar (above figure "720" in sketch of Fig. 2.6). The floor of Inn Valley, hidden behind run-up crest, is as good as perfectly horizontal and more than 2 km wide (photo by Erismann)

38 m s^{-1} and the respective energy loss between 78 and 86% (not to speak of at least 50 m elevation lost between km 3 and km 3.9) cannot be far from reality. On the other hand, as the energy lost owing to a partial source of dissipation necessarily is smaller than the total loss of energy (expressed by the apparent coefficient of friction), even repeated bobsleigh run like curves in a valley with efficient channelling qualities cannot be considered as an important energy dissipating mechanism (for details s. Sect. 5.7).

In closing the section at this point, various *unanswered questions* are, of course, left aside. More detailed discussions of such questions will be found in those sections promising the best approach for explanation, containing the most important implications of a problem, and/or offering a theoretical basis for understanding. So a short mention of each question as well as of the section(s) allowing to dig deeper will suffice under the present heading. There are the influences of sliding down on glacier ice and being funnelled by lateral moraines in the steep initial descent (Sect. 2.7, 5.1). There is, at the end of this descent, the uncertainty in determining the maximum velocity attained when crossing the creek (Sect. 2.5, 6.2). There is, of course, the amount to which fluidisation of fine grained material or other trivial lubricating mechanisms may have been active (Sect. 5.6). There are the dynamic problems of superelevation based speedometry (Sect. 2.7, 5.7). And there is, last but not least, the dots and dashes line in the map of Fig. 2.4, used as a basis for the respective longitudinal section and following – as insinuated by Evans et al. (1989, 442) – in the best possible manner the path of the centre of gravity of the mass (Sect. 2.5, 6.3).

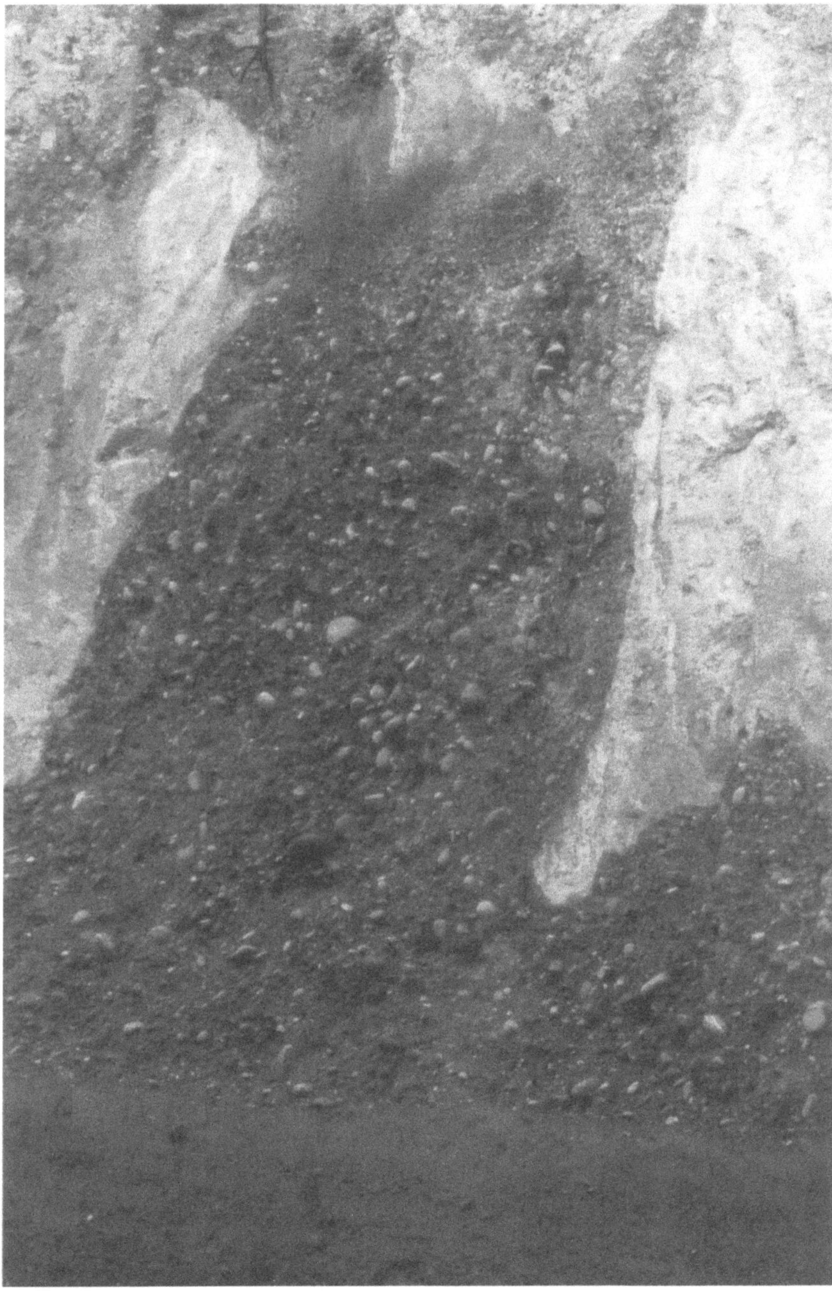

Fig. 2.10. Tschirgant rockslide (Tyrol, Austria, Fig. 2.1), valley fill (*dark*) lifted by water jet through pull-apart crack in the debris. South-western wall of gorge cut by the Ötztaler Ache before flowing into the Inn (protruding tongue at right end of respective sketch in Fig. 2.6). The gradation of the fill material (coarse on bottom) is obvious (photo by Erismann)

2.3 Blackhawk

In a similar way as in the case of Pandemonium Creek (Sect. 2.2) the main hypothesis discussed hereafter is due to the existence of long run-out rockslides. The idea for the presented solution was, however, born in the mid-sixties, more or less simultaneously with the first attempts to discuss heat-generated lubricants (Habib 1967; Goguel 1969) and at least 20 years earlier than Abele's concept of compressed water lubrication. At this time water generally was considered as a possible lubricant, but no consistent hypothesis was at hand to explain how water effectively could contribute to reduce the resistance of a heavy mass moving at high velocity. In addition, for the key event of the present section the possibility of water lubrication was considered as excluded for reasons described hereafter. So another candidate for the explanation of the long run-out phenomenon was envisaged, namely the daring and fascinating idea of air lubrication as proposed by Shreve (1966, 1968a, 1968b). It was conceived under the impression of various geomorphological facts observed in the study of two rockslides and giving a seemingly coherent image of a most spectacular mechanism that was declared as having been active in a large class of slides summed up under the collective term of "Blackhawk Type".

Physical and mathematical *fundamentals about lubrication* are found in Sect. 5.5.

Fig. 2.11. Sherman rockslide of 1964, aerial view from west. The mass originated from the highest mountain visible behind the debris at right (as it lost its top in the event, it is now called Shattered Peak). Note the large extension of the deposits on the glacier (area of ice covered by debris is about 5 km^2) and the slight run-up at the extreme left (by courtesy of US Geological Survey, University of Puget Sound, represented by S. M. Hodge)

In a short description of the *Sherman rockslide* in Alaska (Fig. 2.1, 2.11) which, in 1964, after a steep descent of about 600 m, had run out on a glacier, Shreve (1966) reported several interesting details. One of the most important thereof was the "... *absence of large scale mixing in the motion* ..." (p. 1641), a phenomenon pointing to a remarkably *moderate relative displacement between the particles* in the course of their travel. Like the size effect, Heim, in his admirable book (1932, 104–105), had discussed the fact that rockslides, as a general rule, preserve the sequential order of their components from the start right to the end of displacement. Anyhow, Shreve had a good reason to assume that velocity gradients were to a high degree concentrated near the bottom of the disintegrated mass in motion. So it appeared plausible that somehow the mass was separated from the ground (i.e. from the glacier) by a lubricating medium. It was in this publication (1966) that the possibility of air as a lubricant was first mentioned.

It is, perhaps, somewhat strange that, with the categorical statement "*Water and mud as lubricants ... are ruled out by the subfreezing temperature at the time of the landslide ...*" (Shreve 1966, 1642), an important alternative was excluded. Probably no consideration was given to the possibility of frictional heat having generated a *lubricant by melting* snow or ice just at the points where friction occurred. Otherwise a simple comparison of the kinetic energy dissipated in arresting the mass with the melting energy of ice would have shown that the slide was able to generate about 250 000 metric tons of water.

Two further geomorphological facts reported in the description of the Sherman rockslide are worth mention in the present context. On the one hand the debris was gradated coarse on top: boulder-sized on the surface, finer grained, tightly packed, and relatively nonporous below (Shreve 1966, 1640). This statement, pointing to *low permeability* and low escape rate of a lubricant through the mass, obviously favours any possible lubricating mechanism without special preference for a particular lubricant. On the other hand, the fact that the base of the debris was, at least to a certain extent, in contact with foliated glacier ice, might be interpreted as evidence supporting lubrication by melting ice.

*

In various points the mentioned publication overlaps with two other studies in which air lubrication was presented more in detail. One (Shreve 1968a) deals mainly with the physical background of the hypothesis; the other (Shreve 1968b) demonstrates its geological and geomorphological aspects in the frame of an excellent description of the *Blackhawk rockslide* in Southern California (Fig. 2.1, 2.12, 2.13). Of course, the last-mentioned article repeated the geomorphological facts pointing to an extreme concentration of the velocity gradient at (or even below) the bottom of the moving disintegrated mass and opposed it to a fluid-like displacement where "... *average velocity ... increases from zero at the bed to a maximum at the ... upper surface ...*" with the unavoidable consequence of sequence inversion (1968b, 30; for details s. Sect. 5.6). Accordingly, in the description of the debris special attention was paid to the impression of its not having been massively disturbed in motion. For instance, it was stated that "... *color bands ... continue from clast to clast without ... offset ...*" (1968b, 29) so that the catchword "jigsaw puzzle", already used in describing the Sherman slide, was repeated with emphasis. As a whole the mass looked as if it had been "... *moved as a nearly undeforming sheet ...*" (p. 30).

Fig. 2.12. Blackhawk rockslide, aerial view from north. The mass originated from the noticeably eroded mountains in the near background at right (it is difficult to judge how far the topography of this region has been changed by tectonic activity and/or erosion since the event). Note the large extension of the deposits in either direction (for dimensions s. Fig. 2.13). Blackhawk Canyon is visible both in the debris (ending in their protrusion at the extreme right) and in the hills. At left behind the debris of Blackhawk those of the (older and smaller) Silver Reef rockslide (partly buried) (professional aerial photo supplied by J. S. Shelton)

Shreve was aware of the two crucial problems inherent in the considered mechanism: the *necessity to "trap" compressed air* below the mass and to counteract its permanent *tendency to escape* wherever possible.

The *trapping* was proposed as being the consequence of a large bounce: "*... a huge rockfall ... at a projecting shelf ... or sudden steepening of slope leaves the ground, overriding and trapping a cushion of compressed air ...*" (1968b, 37). The formulation of this "jumping hill" hypothesis betrays a gap in the line of thought. In fact, the description is adequate only for the process of overriding, not for that of compressing. From Fig. 2.14 it is evident that the overridden air volume *V* remains practically constant as long as the debris moves at more or less constant velocity. In other words: compression at the required scale does not take place. In such instances the causal connection between velocity and trapped air volume (p. 38) becomes questionable, including the subsequent speculations about possible velocities. Nevertheless it is worthwhile to reconstruct a plausible energy line using the method of Müller-Bernet adopted by Heim (1932, 143–147; refer to Sect. 6.2): even with conservative assumptions (a coefficient of friction $\mu = 0.2$ during acceleration; an elevation of 25 m of the centre of gravity with respect to the ground), correct application of the method yields a line aiming at a point far beyond the summit of Mount Blackhawk. Taking into account that the cen-

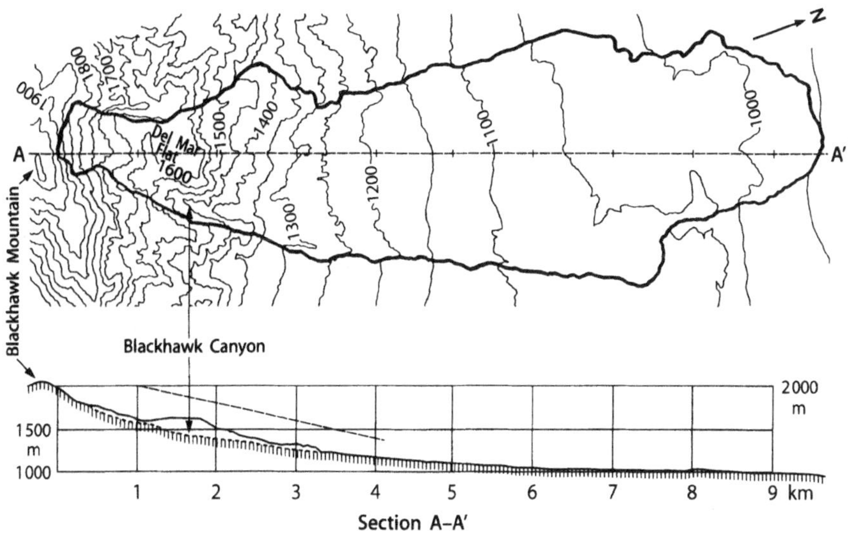

Fig. 2.13. Blackhawk rockslide. Map and longitudinal section. Scales for map and section are equal. Debris begins at the mouth of Blackhawk Canyon (around km 3.3). Assumed take off from low gneiss ridge near km 2.8 (not perceptible in section). Lines in section: *plain:* section *A-A'*; *dotted with hatching:* along Blackhawk Canyon; *dots and dashes:* energy line yielding $v = 76.2\ \mathrm{m\,s^{-1}}$ at take-off, centre of gravity being located 25 m above the ground and $\mu = 0.2$ during acceleration (sketch by Erismann, based on Shreve 1968)

Fig. 2.14. Compression problem in air lubrication. Rockslide mass crossing a "jumping hill". **a** Leading end landing; **b** bulk of mass airborne; **c** trailing end taking off. As long as velocity is not dramatically decreasing, volume *V* below the mass remains approximately constant and no substantial compression of the air is possible (sketch by Erismann)

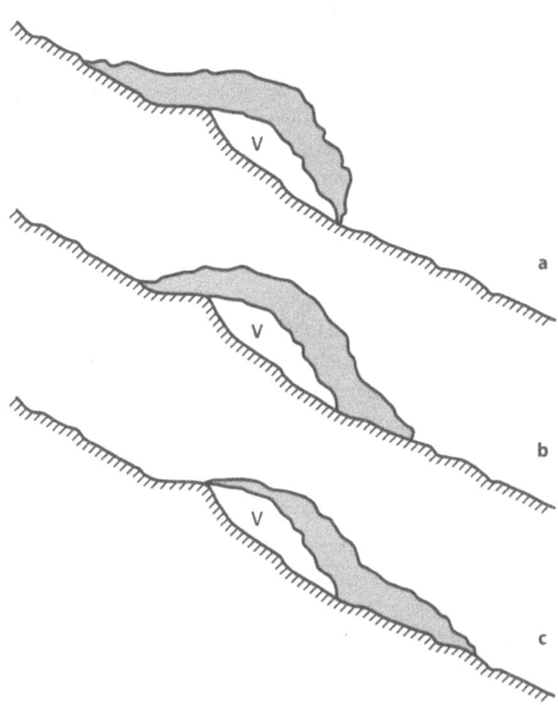

tre of gravity of the ready-to-move mass must have been a point of the energy line, it becomes evident that a velocity of 76.2 m s^{-1} (250 ft s^{-1}) as proposed by Shreve is a substantial overestimate.

In spite of such incontrovertible evidence for the impossibility of trapping and compressing air in large quantities, the considerations following hereafter are made as if, by some unrecognised mechanism, a sufficient volume of compressed air were available below the debris.

Thus, concerning the second problem, *air escape in a vertical direction* was calculated by Shreve on the basis of a laminar flow through the pores of the mass, and the permeability was considered small enough in view of the poorly sorted material (1968a, 653–655). Obviously it was not taken into account that conventional permeability measurements are valid for stationary conditions, the material being kept together by a standard case. Contrarily, the debris, if it were riding on a cushion of compressed air at all, necessarily would be subjected to a two-dimensional tensile stress like an air-supported dome and, having no tensile strength (Sect. 5.1), the interstitial voids would rapidly be enlarged and the flow (as being proportional to the third power of the width of a passage) would be dramatically multiplied. So the "sheet" of debris can be compared with an air-supported dome consisting of a material without substantial tensile strength (e.g. woven rubber band). It is needless to say that such a dome could not be blown up properly.

With respect to *lateral escape,* the formation of ridges (as observed in situ), considered as having been arrested before the rest of the mass, was proposed as a sealing mechanism (Shreve 1968b, 39). It was omitted, however, that owing to the fast relative shearing displacement between arrested and moving material, a complete separation was unavoidable, thus creating a new passage for escaping air. In addition, in order to obtain correct results, it would have been necessary to consider an escape far beyond the laminar-to-turbulent transition. In fact, the air would, under the pressure dictated by the thickness of the material, reach the sonic barrier. A calculation on this basis (Erismann 1979, 30–33) has demonstrated the poor ability of retaining the air under the mass even if perfect impermeability in the vertical direction were assumed.

This last criticism came relatively late. It was preceded by a study in which the question of *how compression could have taken place* was asked and backed up with sensible arguments (Scheller 1970, 62–65). It speaks for the suggestive power of a giant mass imagined as "hovercrafting" on nothing but air – and perhaps also for the moderate interest devoted to Scheller's critical remarks as being published in a doctoral thesis mainly dealing with other problems – that the air lubrication hypothesis subsisted for many years as an accepted school of thought.

This is true although early doubts had come up when immense slides, considered as belonging to the long run-out type, were found on *the Moon and the Mars* where the atmosphere, if any, is obviously unable to act as a lubricant (Hsü 1975; Lucchitta 1978). Certainly the fact was not conclusive for two reasons: air lubrication might be just one of several mechanisms giving rise to long run-out; and it was by no means sure that those giant rockslides, occurring under extraterrestrial conditions, could really be interpreted as genuine long run-out events. In fact, it should be taken into account that many relations between parameters depend on the gravity of the celestial body in question. To state two trivial examples: for the release of a rockslide the strength of equal materials is independent from gravitational acceleration while the weight (and thus the driving force) is proportional thereto; and, of course, the relation between kinetic and potential energy shows a similar difference. So, in order to compare comparable items, it would be necessary to know the details of a lunar slide as well as those of a terrestrial one. Assuming, for instance, congruent geometry on the Earth and on the Moon, a given obstacle might be cut away by the terrestrial slide while it would stop the Lunar event owing to lack of gravitational energy transformed into kinetic energy. Thus the comparison of rockslides on different celestial bodies turns out to be far more intricate then it might appear at first sight.

On the basis of the above analysis it is, perhaps, superfluous to recall the trivial fact that the air lubrication hypothesis is diametrically *opposed to the size effect*: a thin mass will (if at all) be a better hang glider than a thicker one. So, after an exhaustive study of various aspects, the authors, now standing on an even more solid ground, have nothing to add to the conclusion drawn by one of them (Erismann 1979, 32–33) on the basis of a sketch corresponding to Fig. 2.14 and a tentative calculation of lateral air escape: "*... it is perfectly clear that the air cushion hypothesis can no longer be considered as an explanation of the observed low coefficients of friction ...*".

*

It must, however, be admitted that, if the map and the section presented in Fig. 2.13 are correct, the *Blackhawk slide belongs to the long run-out type* and that, for the time being, we only know it was not lubricated by compressed air, yet we do not know how it managed to get as far as it actually did. And it sounds like irony that strict application of Heim's "Fahrböschung" method (1932, 121–130; s. also Sect. 6.2) yields an overall slope of 0.098, substantially less than 0.13 as suggested by Shreve (1968b, 3). Anyhow, it may be useful to recall several facts, hypothetical in part, which might have played a role.

In the first instance, it is somewhat enigmatic that in a description coming up to remarkably high scientific standards practically no details are given with respect to the geomorphology of release. In fact, the most relevant statement in this context is confined to the observation that "*... rapid headward growth of Blackhawk Canyon ... undermined the overthrust block of Furnace Limestone at the highest part of the summit ridge ...*" (Shreve 1968b, 3). Now this block must have had no less than the volume of released rock, i.e. something in the range of 0.28 km^3. And the space eastward of Blackhawk Canyon was restricted so that the material must have been *piled up to a substantial height*. So far things are clear and conclusions compelling: the centre of gravity of the ready-to-start mass must have been higher than might be assumed at first sight. The following consideration, however, although plausible, is not more than speculation.

In view of the arid climate of the region, Shreve categorically excluded any influence of water upon the long run-out character of the event (1968b, 37). This was, perhaps, a somewhat rash dictum. With good reasons it might be asked how Blackhawk Canyon could have grown rapidly without the *influence of water*. In arid areas, the rare precipitation is sometimes particularly intense and in certain cases catastrophic. So water lubrication, probably one of the most frequent causes of long run-out (Sect. 2.2, 5.5), cannot be ruled out a priori.

Anyhow, the "Enigma of Blackhawk" remains an enticing challenge for future generations of geologists and geophysicists.

2.4
Köfels

The *deposits found near the hamlet Köfels* between Längenfeld and Umhausen in Ötztal Valley (Tyrol, Austria) have been the object of repeated discussions in the course of more than a century. The south-to-north course of the valley is barred by an enormous mass with a volume of at least 2.1 km^3 (Abele 1974, 189; Fig. 2.1, 2.15, 2.16, 2.17, 2.18). The river Ötztaler Ache has cut the gorge Maurach (in German "Maurach-Schlucht", about 500 m deep) into the material which, in the respective portion, is heavily disintegrated, the prevailing grain size in certain locations not exceeding 2 mm. This fact is of particular interest as most of the mass displays a very high consistency, giant boulders being frequent, lateral slopes as steep as 40° having subsisted up to now, and the general impression suggesting bedrock rather than debris. A counterpart to this mass is a deep, cirque-like scar of approximately equal volume located in the slope west of Köfels. The geomorphological appearance of a very large rockslide (by far the largest in the crystalline Alps) is completed by a large hill-shaped tongue (Tauferberg) which obviously was arrested at the level of Niederthai (about 1 500 m), thus barring the creek Horlach/Stuibenbach, an eastern tributary of the Ötztaler Ache, and forcing it into Tyrol's most spectacular falls, the Stuibenfall.

In such instances it was nothing but natural that in the first scientific description of the site (Escher von der Linth 1845, 539) the opinion was expressed that the mass was the result "*...of the collapse of near-by mountains...*". This opinion, as far as it was at all known to the scientific world, could no longer maintain its apparent triviality when Pichler (1863, 591–594) introduced the fact that in the debris "*pumice-stones*" were present, an essentially vesicular glass, later ascertained as chemically more or less identical to the main component of the mentioned mass, an augengneiss (Fig. 2.19; Hammer 1924). This material, mostly porous and easy to work, was well-known to the local craftsmen who used it to polish wood and for certain decorations. Pichler's scientific description "Die vulcanischen Reste von Köfels" ("The volcanic remnants of Köfels") was the first step of a long, troublesome, and sometimes dramatic journey which yielded consistent results more than a century later. And even at the present time, 150 years after Escher's report, the discussion about Köfels is not entirely free from controversy. In such instances it is, perhaps, worthwhile to give an abridged outline of the varying hypotheses proposed, so to say the milestones of the mentioned journey.

As can be concluded from the title of his study and in accordance with the knowledge available at his time, Pichler assumed that the fused rock was the *result of volcanic activity*. Trientl (1895), though presumably not familiar with Escher's work, revitalised the rockslide hypothesis, postulating an enormous earthquake as a cause of this and various other mass movements in the Ötztal which, owing to its steep slopes, is rich in such events. Like Pichler, Trientl dated the catastrophe as postglacial and considered a volcanic origin of the fused rock as plausible, suggesting, however, an immediate link between the earthquake and the emerging melt. So it is natural that he expected immense new findings of pumice. But no such findings occurred, and no further evidence pointing to a volcanic eruption was brought forth. So the unsatisfactory situation remained: there was fused material, there was no stringent geomorphological evidence in favour of its volcanic origin, but no other source of energy was known that could have melted the rock.

Fig. 2.15. Köfels rockslide. General view from the north. *Dotted lines* show periphery of deposits with the plateau of Köfels at right, the Maurach gorge in the centre, the Tauferberg at left, and, in front of the gorge's mouth, the separate debris of "Lärchbühl" (connection with Köfels slide uncertain). *Arrow* indicates direction of displacement (by courtesy of E. Preuss)

Fig. 2.16. Köfels rockslide. View from the west (near Schartle Pass). The Tauferberg takes up the centre, showing the steep slope to the Maurach gorge with two bright outcrops. The floor of the gorge is hidden by the plateau of Köfels, the hamlet being visible in the left foreground. Niederthai is in the valley behind the Tauferberg (photo by Erismann)

Nevertheless, several renowned scientists adopted, like Trientl, though independently from him, the hypothesis of a giant rockslide having taken place (Penck and Brückner 1909; Ampferer 1939; Stiny 1939), be it in combination with volcanic activity or not. No noteworthy attempt was made by these investigators to introduce another mechanism of melting.

Such an attempt was made in 1937 when Suess and Stutzer independently from each other proposed an extraterrestrial source of energy: could the material not have been *fused by the impact of a meteorite*? At first sight, this hypothesis seemed to replace the volcanic origin by a mechanism providing ample energy and known for having generated amorphous "impactites" in other cases. Nevertheless, the idea initially was received with a certain reserve as not taking into account the complexity of geomorphological evidence. On the other hand, no definite clue in favour of a volcano came forth. So the impactite hypothesis acquired more and more acceptance and finally was considered by many scientists as the best solution (Milton 1965; Kurat and Richter 1968, 1972; Storzer et al. 1971). Roughly speaking, it was assumed that the melt was generated directly by the impact (the scar west of Köfels played the role of a half-crater) while the mass displacement, if any, was considered as a consequence thereof. Thus the problem of melting energy was ruled out and the general geomorphological situation seemed plausible.

*

One of the investigators who came to Köfels, the mineralogist Preuss, instead of finding further evidence connected with the meteorite hypothesis, ran into serious *minera-*

Fig. 2.17. Köfels rockslide. View from the east (western slope of Tauferberg, elevation about 1500 m). In front of the dominating mountain (Fundusfeiler, 3080 m), the crest of Schartle (2088 m) makes the impression of being somewhat "decapitated". Note the extended sliding surfaces, visible particularly well between Schartle and Köfels (partly hidden by trees in the foreground at right) (photo by Erismann)

logical and geomorphological inconsistencies. In a fundamental publication (1974), Preuss not only expressed his doubts but also presented a new idea of how the material might have been melted. Besides volcanic heat and the kinetic energy of a meteorite there was, if any rockslide had occurred, a third source of energy, namely that of vertical mass displacement in the gravitational field. The existence of large sliding planes west of Köfels (Fig. 2.17, 2.18; Preuss 1974, 3) suggested that *frictional heat* could have done the work.

Aware of the impossibility to deal alone with other than mineralogical problems, Preuss associated himself with a geographer well acquainted with the Ötztal (Heuberger 1966, 1975). There were, indeed, various geomorphological details not fitting into the seemingly coherent puzzle, and critical questions could be asked with good reasons. How could, for instance, the entire fused rock known at the time be deposited near sliding planes that the mass had left behind (Fig. 2.18)? It might be answered that this material was catapulted more or less vertically and fell back just after the mass had passed, i.e. after at least one minute (Erismann et al. 1977, 76). But how, then, could the bouncing material have maintained its low viscosity (and high temperature) so that it was able to penetrate into cracks as narrow as 2 mm (Preuss 1974, 6–9)? And how, if it really came down in a liquid state of low viscosity, could this material avoid being splashed around in minute drops? And where, eventually, were the remnants of the substantial ejecta blanket necessarily accompanying the impact of a meteorite and not likely to disappear completely in less than 9 000 years (^{14}C data, s. Heuberger 1966, 37; Ivy-Ochs et al. 1998)?

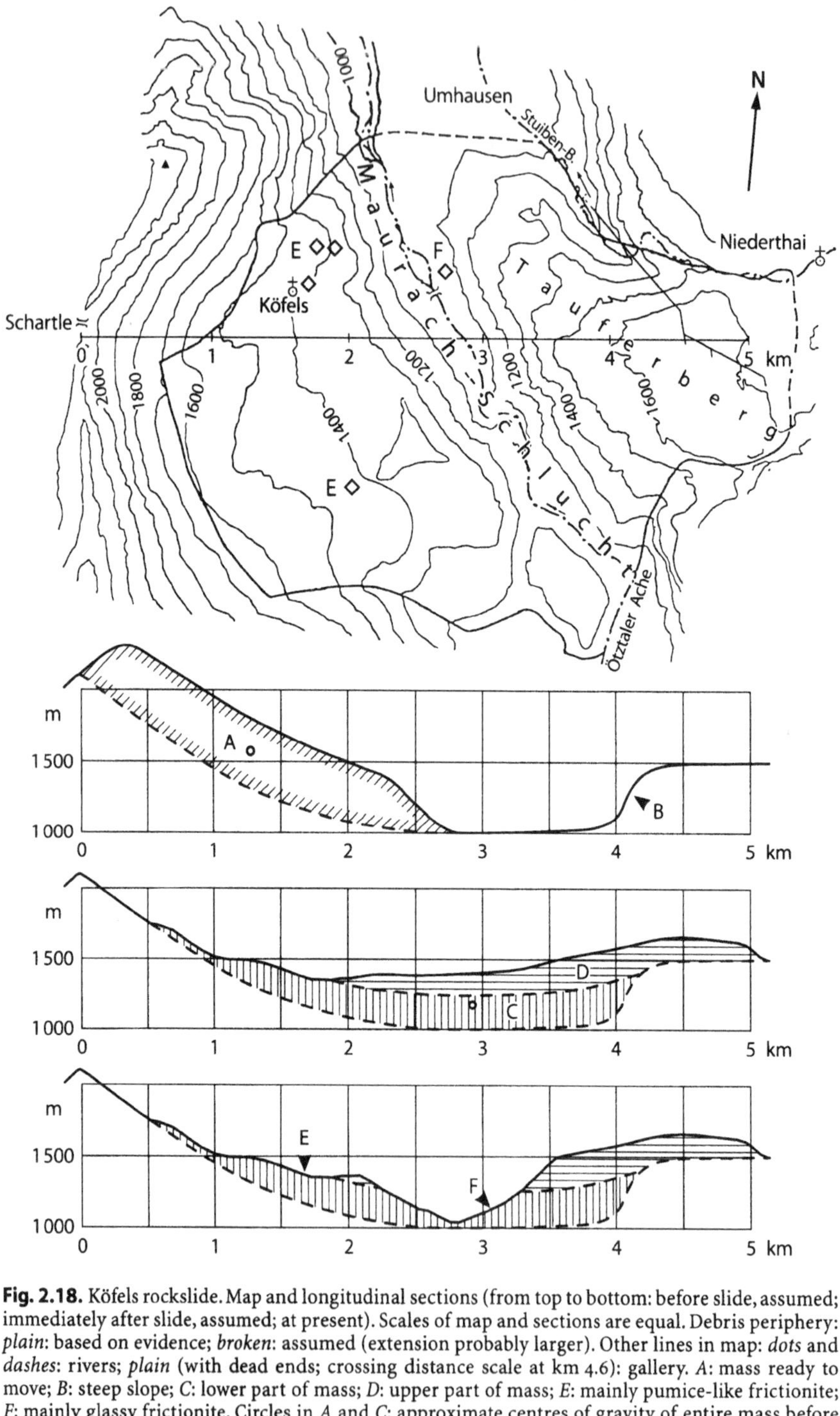

Fig. 2.18. Köfels rockslide. Map and longitudinal sections (from top to bottom: before slide, assumed; immediately after slide, assumed; at present). Scales of map and sections are equal. Debris periphery: *plain*: based on evidence; *broken*: assumed (extension probably larger). Other lines in map: *dots* and *dashes*: rivers; *plain* (with dead ends; crossing distance scale at km 4.6): gallery. *A*: mass ready to move; *B*: steep slope; *C*: lower part of mass; *D*: upper part of mass; *E*: mainly pumice-like frictionite; *F*: mainly glassy frictionite. Circles in *A* and *C*: approximate centres of gravity of entire mass before and after slide respectively (sketch by Erismann)

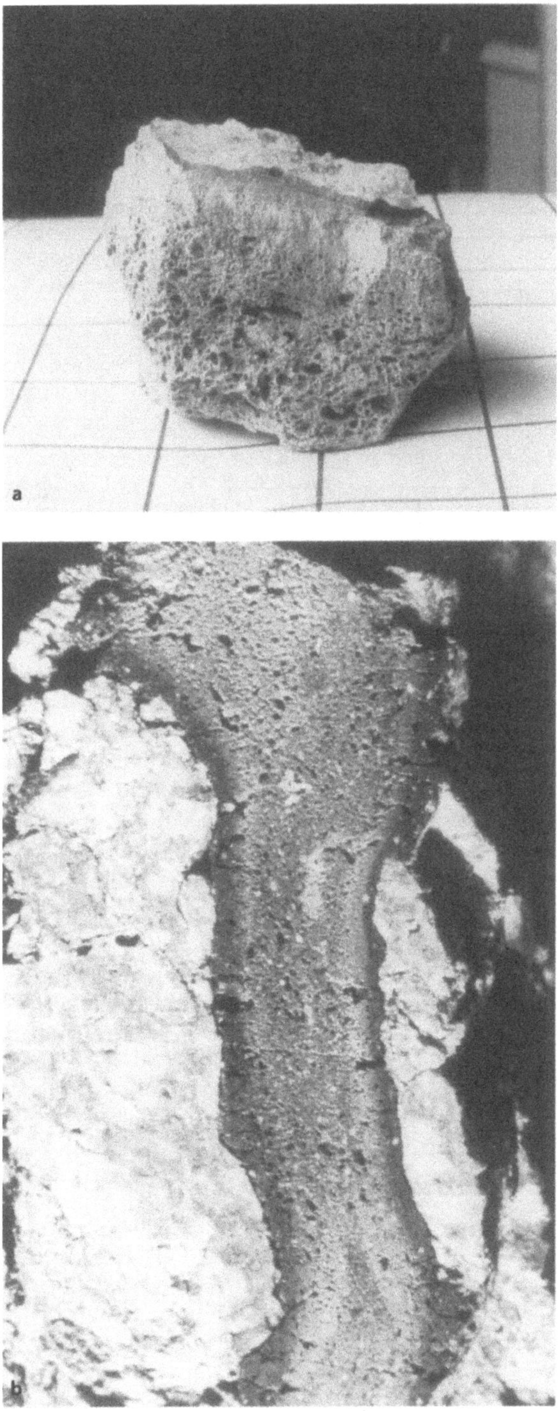

Fig. 2.19. Köfels rockslide. Fused rocks (frictionites). **a** typical porous, pumice-like piece (squares are 50 mm); **b** section through specimen showing low viscosity of melt (thickness of gap is about 10 mm) (by courtesy of E. Preuss)

To obtain a complete view of the problem of frictional melting, it was, in addition, necessary to go into tribological details and, in particular, to treat the differential equations of heat transfer and mechanical motion. So an even more *interdisciplinary team* was formed by incorporating a mechanical engineer with the required experience. After both in situ and theoretical investigations, this team treated various aspects of the fused rock in an extensive study (Erismann et al. 1977). The following sketchy description of the reconstructed case history is nothing but a compressed synopsis of this article. It is well understood, however, that a substantial part of the information was taken from the abundant literature dealing with the matter (a list of references, as far as not mentioned in the text, is presented at the end of this section).

*

The motion of the mass, directed approximately from west to east, was initiated at the Schartle Pass (Fig. 2.17, 2.18) where the crest, at the time, probably was substantially higher than nowadays. The mechanism which triggered the *release* must be considered as unknown. Among other more or less speculative possibilities, the vibrations due to the impact of a small meteorite are not excluded in this context although the generation of fused rock is attributed to frictional heat. In this context the aspects raised by Surenian (1988, 1989) and contested by Leroux and Doukhan (1993) and Lyons et al. (1993) are noted with interest without being discussed in further detail as not being relevant in the present context.

In *descending to the valley floor* (situated at a lower level than at present, perhaps even lower than assumed in Fig. 2.18, a hypothesis no longer excluded after geoseismic investigations: Heuberger and Brückl 1993; Brückl 1998), the mass obviously did not encounter massive obstacles and thus remained in a more or less coherent state of giant blocks. After a travel of somewhat more than one kilometre, having acquired a velocity of about 50 m s^{-1} (Erismann et al. 1977, 93–94), it collided with the opposite slope of the valley, almost vertical and consisting of massive rock, as could be concluded from the neighbouring topography and was confirmed by a gallery driven into the debris in connection with a project for a hydroelectric power station (Ascher 1952).

This *collision*, as shown in Fig. 2.18, sheared the mass into two main parts, the upper continuing its motion in a partially still coherent state, the lower being arrested. This lower part was disintegrated by two mechanisms. On the one hand the momentary locus of collision, initially situated at the bottom of the mass, was displaced higher and higher until it reached the top of the arrested part. So a continuously displaced shearing stress divided the material into many approximately horizontal slices (Fig. 2.20). On the other hand, as the leading portions of each slice were hammered by the following mass against the anvil of the slope, they were smashed to small pieces, ranging in size between few millimetres and fractions of a metre. In spite of this radical disintegration the clasts remained remarkably well "in shape", especially in their vertical sequence. A good example is the diabase dike in an outcrop of the Maurach gorge shown in Fig. 2.21a. The phenomenon is, by the way, not confined to crystalline rock as demonstrated, under quite similar topographic conditions, by a calcite vein in the carbonate debris of the giant Flims rockslide, Fig. 2.21b. In either case, from release to disintegration, the mass had remained essentially coherent.

After this dramatic phase, the kinetic energy stored in the upper part of the mass was dissipated in a further motion of almost one kilometre, including a run-up of about 100 m and the formation of the Tauferberg. It is even probable (and evidence in a quarry

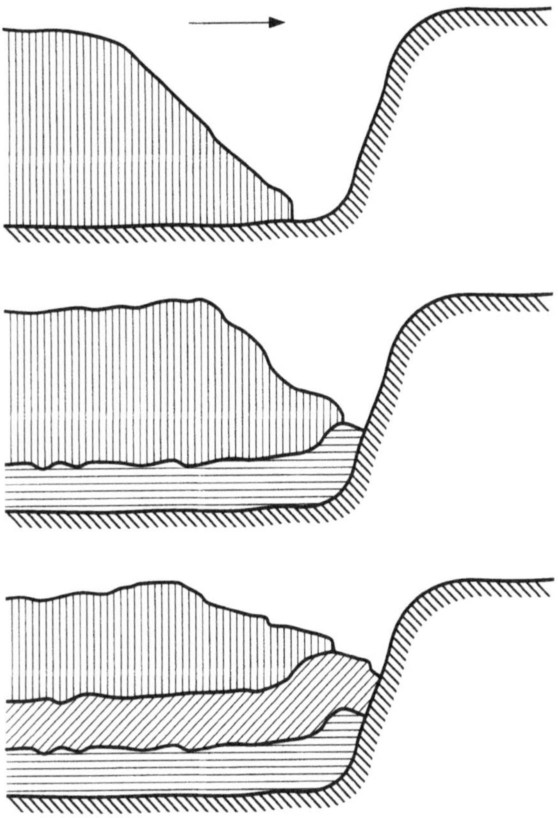

Fig. 2.20. Köfels rockslide. Collision of moving mass with the eastern slope of the Ötztal. Schematic representation of the mass being dissected into horizontal "slices". Very probably the real process was far more complicated owing to its three-dimensional character and a larger number of irregularly sized slices. Sliding surfaces could develop almost in any direction. *Arrow* shows direction of mass displacement (sketch by Erismann)

east of the Tauferberg backs up this assumption; Abele 1991b, 33) that valley fill has been pushed in front of the debris (as represented in Fig. 2.7) so that the actual area involved was larger than shown in Fig. 2.18. The same applies to laterally displaced material after the collision with the opposite slope. For instance, the accumulation of debris "Lärchbühel" (Fig. 2.15), about two kilometres north of the deposit might be a last rest of such a transverse motion (although its origin from a neighbouring slide cannot be excluded).

*

Anticipating the investigations discussed hereafter, it should be observed that, the more the results confirmed the hypothesis of frictional heat, the more it became questionable to use existing terms for the fused rock of Köfels. "Pumice", as employed by the first investigators, would point to a volcanic origin; "Köfelsit", proposed by Suess, had been linked to an impactite; "Hyalomylonite" would not allow for a distinction from material of endogenous origin. So the *new term "Frictionite"* was introduced and defined as "*any at least partially melted products generated by friction near the surface and allowing to be put down to exogenous processes exclusively depending on terrestrial gravitation...*" (Erismann et al. 1977, 81). It goes without saying that in case of a non-terrestrial rockslide the gravitation of the respective celestial body has to be used.

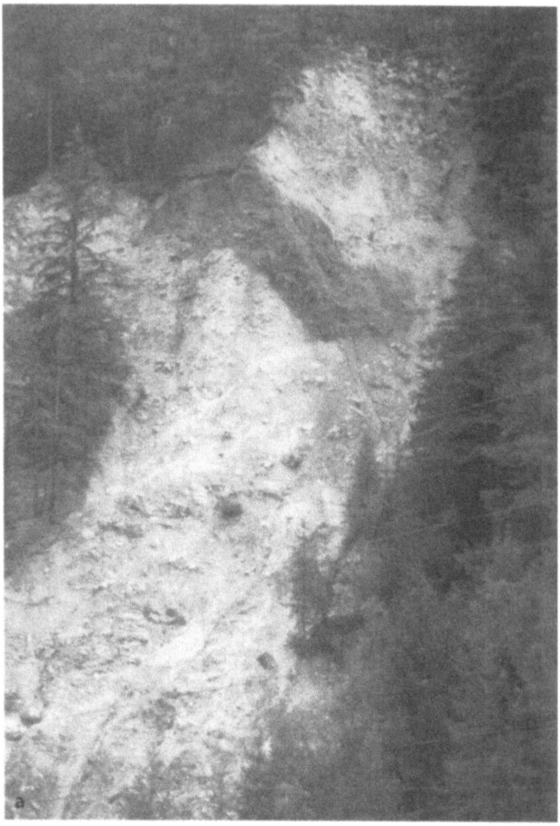

Fig. 2.21. a Köfels and **b** Flims rockslides. Deposits, severely disintegrated after collisions with solid obstacles, yet containing "in shape" dikes of diabase (**a**) and calcite (**b**) respectively (photos by Erismann)

Contrarily to melting processes due to earthquakes which can be deduced from pseudotachylyte occurrences in gneissose areas (Mckenzie and Brune 1972), no previous scientifically relevant field evidence was known about rocks being fused by frictional heat in rockslides. Muir (1912, 59) had spoken of falling boulders "*...luminous from friction...*" in a nightly rockslide; Heim (1932, 140) knew about luminous phenomena and proposed various explanations, the nearest to frictional heat being impact heat; and the mention, in an old Tyrolean legend (Wolff 1977, 656–657), of light emitted by large "landslides" (or whatever the German word "Mure" may mean in the particular context) is most probably based on experience, but certainly not in a scientific sense. So, in parallel to the work in situ, the problem was checked by *extensive calculations* (Erismann et al. 1977, 84–104). It is perhaps worthwhile to outline in short also this part of the investigation. In view of a nearly circular topography of the sliding surfaces the motion was calculated as that of a pendulum with Coulombian friction as damping mechanism (pp. 84–85, 91–94). A constant coefficient of friction was assumed. As will be demonstrated under Heading 6.2, the results of calculations on this basis give useful, though not perfectly correct approximations of the actual velocities.

The *amount of energy* thus becoming available was deduced from the estimates of volume, density, and vertical displacement of the mass (pp. 79–80). The result, about 16 500 TJ, was illustrated in various manners, for instance by the explosion of a H-bomb of 4.5 megatons, the burning of 400 kilotons of oil, or – closer to the reality in hand – heating of 9.3 megatons of Köfels rock (70% feldspar, 30% quartz) to a critical temperature of 1700 °C at which – even in the limiting case of a complete non-equilibrium melting of the individual phases of the host rocks – all feldspar is fused and quartz begins to melt. It may be remarked that from this mass of rock a crystalline Cheops Pyramid might be erected...

Subsidiary calculations of non-Coulombian resistance mechanisms were aimed at the question of how far energy dissipation was concentrated near the sliding surfaces (Erismann et al. 1977, 86–91). It was made evident that energy losses by aerodynamic drag were negligible, that a "roller bearing" effect had to be excluded owing to immediate and complete crushing of the "rollers" even in case of an ideal geometry (Sect. 5.3), and, eventually, that the available kinetic energy was amply sufficient not only to separate the mass into an upper and a lower part but also to cause the radical disintegration of the distal portion of the lower part (pp. 88–90, 104, estimated on the basis of fracture mechanical data: Hillemeier 1976). As its main result, this thermodynamic modeling of the event yielded the conclusion that the lion's share of the total energy dissipated before the collision was concentrated in the immediate vicinity of the sliding surface in the form of frictional heat and thus had to be considered as a potential means to generate frictionite.

Thus both the abundance of energy and its locally concentrated transformation into heat were in agreement with the frictionite hypothesis. It had, however, to be ascertained whether this energy remained near the critical zone, thus heating the material; in fact, it might have escaped fast enough by conduction to make heating ineffective (convection could be excluded in a coherently moving mass, and radiation was possible only at its surface). So the unidimensional partial differential *equation of heat conduction* in large coherent blocks was integrated in a numerical process, using finite time intervals and a vertical succession of finite-thickness layers (Erismann et al.

1977, 94–99). The result was impressive: for a mass 500 m thick, as probable for the travel before the collision, a layer of 10 mm thickness would be heated to 1 700 °C after less than 50 m travel, and even in the secondary motion after the collision (material thickness about 125 m) the same effect would be obtained after less than 200 m. Initiation of feldspar melting (at about 1 200 °C if no eutectic phenomena are taken into account) was obtained for distances reduced by almost 40%, and, of course, calculation with thinner layers would have yielded melting after an accordingly shorter travel.

It has to be observed that these calculations were based on average values of compressive stress upon the sliding surface. They may, therefore, be considered as conservative owing to the necessarily nonhomogeneous force distribution in the real event: *locally far higher rates of heat generation* have to be expected so that, at the points of highest compression, melting very probably took place after substantially shorter travel.

*

After having yielded such promising results, the *calculations were cross-checked by a simple experiment.* As the use of existing machines (for instance friction welding sets as experimented with by Spray 1987) was not considered as sufficiently near to reality, a flywheel-driven, servo-hydraulically operated, and completely instrumented device was conceived in which a rotating ring-shaped sliding surface of Köfels rock could rapidly (in several ms) be pressed against an equally shaped surface remaining at rest (Fig. 2.22). The thickness of the overburden mass was simulated by the hydraulic force exerted upon this "braking device". In spite of a rather low working velocity of about 10 m s^{-1}, the products of abrasion thus obtained from perfectly crystalline specimens were partially amorphous and of porous appearance (Fig. 2.23). As could be expected, the percentage of such artificial frictionites rapidly increased with the simulated thickness.

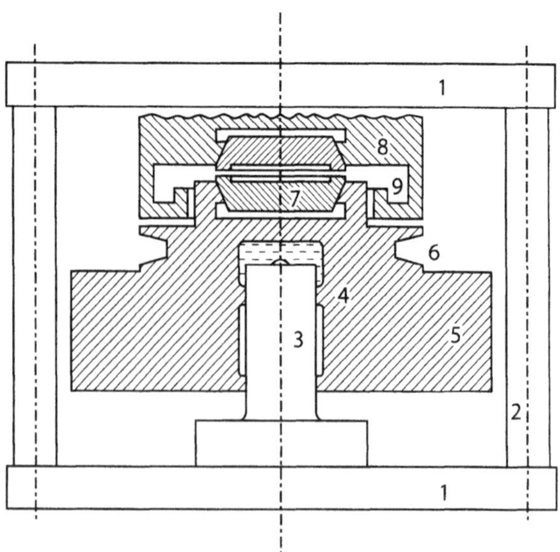

Fig. 2.22. Section through device for generating artificial frictionite: *1*: platens; *2*: columns; *3*: immobile servo-hydraulic piston; *4*: rotating servo-hydraulic actuator (able to apply programmed load in few milliseconds); *5*: flywheel; *6*: groove for driving belt; *7*: specimens (gneiss from Köfels); *8*: gripping device for immobile specimen; *9*: groove for retaining debris (sketch by Erismann)

Fig. 2.23. Artificial frictionites, produced from Köfels augengneiss using device shown in Fig. 2.22. **a** Specimen after test, with frictionites and gripping device; **b** examples of frictionites (photos by EMPA, Swiss Federal Laboratory for Testing Materials and Research, Dübendorf)

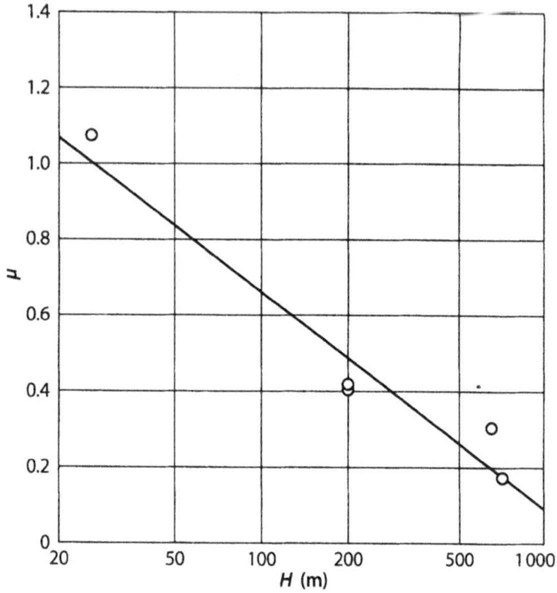

Fig. 2.24. Coefficient of friction μ in function of simulated thickness H obtained with device shown in Fig. 2.22. *Circles* show results of tests. *Straight line* is obtained by regression (sketch by Erismann)

As an initially surprising result of the tests, the coefficient of friction was found to decrease substantially with increasing simulated overburden thickness (Fig. 2.24). This fact gave the investigation a new direction: it was speculated that a powerful *thickness-dependent mechanism of self-lubrication* was active (i.e. lubrication by local transformation into a lubricant of at least one of the materials undergoing friction; s. also Sect. 5.5). Further it was reasoned that a lubrication of this kind might be particularly effective owing to the fact that the lubricant, a viscous fluid, was produced right at the locations where it was most needed, namely at the points of highest friction.

And, of course, the causal connection between thickness and friction was considered as particularly important. This was, at last, a mechanism clearly favouring large events with respect to the economy of locomotion and thus promising an explanation for the old question of *size effect* already mentioned under Heading 2.2 (Sect. 6.3; Heim 1932, 121–130; Scheidegger 1973). In a study aiming at a further evaluation of the results obtained from Köfels (Erismann 1979, 36–43), the question was theoretically treated for a disintegrated mass (Chap. 4; Sect. 5.1). The author of this paper was aware of the fact that the results, though reasonably founded as concerning the generation of lubricant, have to be considered with a certain reserve: the assumptions required in connection with the escape of lubricant – and in particular the granulometric parameters – are very difficult to estimate (p. 40, Item f).

*

This is true not only for fused rock but also for another mechanism discussed in the last-mentioned publication. This second mechanism was postulated in spite of a complete lack of field evidence: *self-lubrication by thermal generation of carbonic acid* (so to say, a hovercraft effect). It was concluded that, in carbonate as well as in crystalline rock, frictional heat may transform the material near the sliding surfaces. The possibility of melting, however, was left aside: the conditions required (contrarily to eutectic circumstances sometimes encountered deeper in the earth and known for a long time, Wyllie and Tuttle 1960; Keller 1981), are not likely to be present in a rockslide mass of current size. So it was assumed that carbonates, when heated to a temperature in the range between 900 and 1050 °C (depending on the pressure), dissociate according to the reaction $CaCO_3 \longrightarrow CaO + CO_2$ (or another similar endothermic process, depending on the particular material). The amounts of energy required to dissociate a mass unit of carbonate rock and to melt a mass unit of gneissic rock were compared for the two largest Alpine rockslides (Flims and Köfels). They turned out to be practically equal (Erismann 1979, 35). CaO is produced as a powder that, under the influence of atmospheric agents, is recycled to carbonate by the reactions $CaO + H_2O \longrightarrow Ca(OH)_2$ and $Ca(OH)_2 + CO_2 \longrightarrow CaCO_3 + H_2O$. So after a short time all traces of a possible transformation by heat are practically annihilated: the particular frictionite is unstable.

Only once, to the knowledge of the authors, dissociation in a rockslide could be observed in the field. Hewitt (1988, 66), in describing three large consecutive slides in the Karakoram, reported that his skin was *burnt by touching unslaked lime*. He also observed that "*...naturally generated limes in large landslides may easily go unreported...*". This remark is of high value as it raises the probability of self-lubrication by a gas – in principle, though not in the abundance of field evidence and without experimental background – to a similar level as self-lubrication by a liquid: in both cases lubrication is granted by the generation of sufficient lubricant; its effect, however, is difficult to quantify as the escape of lubricant depends on the permeability of the moving mass, in particular of its bottom zone (and thus is, probably, best in case of coherent motion, s. Sect. 5.1).

*

But even in the case of crystalline material the *detection of fused rock may be difficult*. As can be seen from Fig. 2.18, the hitherto mentioned findings of frictionite do not belong to the main sliding surface. And the probability of getting at such material without exorbitant digging is very limited: most of it is buried under a thick layer of debris. And in case of being exposed to weathering on open sliding surfaces, porous pumice-like rocks are subjected to fast erosion or hidden under the loose material that inevitably follows a rockslide. So in the case of Köfels the existence of a secondary sliding surface offering partial protection against such mechanisms must be considered as particularly good luck.

Fortunately a *main sliding surface was found* in another rockslide of probably even larger size. In the valley of Langtang (Himalaya, Nepal, Fig. 2.1, 2.25) an extended layer of glassy material was investigated by Scott and Drever (1953). Uncovered by a ravine over a length of about 170 m, it follows a slope dipping about 26°, and its *straightness* makes the impression of being generated by a machine tool. The thickness of the layer is in the range of 20 mm, and various transverse cracks are also filled with the same amorphous material. Scott and Drever assumed displacement along a thrust – that would be uphill – and considered the possibility of frictional heat having melted the rock "*...in a short, quick and continuous slide...*" (p. 125). In addition, they speculated about the ability of the fluid material to have acted as a solvent, thus reducing the melting temperature by the eutectic power of dissolved water under a pressure corresponding to an overburden material thickness of more than 4000 m. Masch and Preuss (1977), after a visit to the site, concluded that the overburden thickness must have been less than 1500 m and that the movement was downhill. Their statements, though for-

Fig. 2.25. Langtang rockslide. Main sliding surface with layer of glassy, predominantly compact frictionite. *Arrows* show transverse crack filled with frictionite even at a considerable distance from the sliding surface. Note the almost perfect straightness of the surface. Division of stick is 0.1 m (by courtesy of H. Heuberger)

mulated in a cautious manner, might be interpreted as early doubts in the tectonic origin of the melt. Later (Masch and Preuss 1977; Masch et al. 1985; Heuberger et al. 1984) it became clear that the event of Langtang had been a very large rockslide and that most of the debris had later been removed by a glacier.

Most of the fused rock found in Langtang is glassy with only few bubble-shaped voids. By the end of the seventies the major part of the corresponding material known from Köfels was porous (some pieces able to float in water). Obviously the difference came from the different circumstances of solidification: rapidity (due to the ambient temperature) and pressure (due to the overburden thickness). Thus it was concluded that, as the mentioned secondary sliding in the Maurach gorge necessarily had taken place under heavy overburden, *glassy material with low porosity* might be found in the outcrops shown in Fig. 2.16. A closer inspection confirmed this expectation. So it was made clear that, as observed in the experiment, even relatively short displacements sufficed for the generation of frictionite (Fig. 2.26a).

More than that: the sliding surfaces found in the Maurach gorge showed the same *striking straightness* in the presumed direction of displacement as those of Langtang, thus pointing to a more general tendency of sliding masses to generate straight surfaces if moving more than very short distances under heavy overburden. Besides showing a characteristic example, Fig. 2.26b reveals another interesting aspect: on the particular surface, sliding seems to have been arrested just at the moment when temperature had approached the lowest melting level. So the surface, instead of frictionite, is covered with very fine sand, a few particles of which are at least partly amorphous.

Fig. 2.26. Köfels rockslide. Secondary sliding surfaces in the Maurach gorge (s. Fig. 2.18). **a** Glassy, compact frictionite. Displacement probably was across the plane of the image (s. legend to Fig. 2.20); **b** sliding surface, probably represented along its direction of motion and thus perfectly straight. In the particular case only a "sand bed" (preceding fusion) was generated (photos by Erismann)

Obviously protruding asperities, before being melted, had been sheared off and ground to a breccia of finer and finer grain. (Erismann et al. 1977, 95–96).

Of course, it would be most interesting to know how far frictional heating is able to transform material in *smaller events* than Langtang or Köfels. Unfortunately, additional difficulties make such investigations almost hopeless. If only sand-sized particles show traces of melting and no protection (as, for instance, encountered in the Maurach gorge) is active, the next rain may wash away most traces of frictionite. In addition, the inaccessibility of promising locations may exclude an effective search. The authors and three other scientists (Cancelli, Crosta, Masch) had to acknowledge this fact in the course of two excursions to the Val Pola slide. For the slide of Saleshan (1983, Dongxian, China) traces pointing to temperatures exceeding 800 °C are reported by Yin Kunlong (1990, 1769–1773). He asserts, on the basis of calculations, the existence of fused material below the deposits. A longitudinal section (p. 1773) shows an initial thickness far below those assumed for Köfels and Langtang (about 100 m).

*

The work of Preuss and his team was, by the way, not the first in the field of thermally generated lubricating mechanisms. Habib (1967) already had pointed out that, if water is present, *vapour lubrication* may take place at the best-suited points (p. 151), that the travel necessary to make the mechanism operational may be short (pp. 152–153), and that the required velocity of displacement is far below that of a rockslide. Later he reiterated and complemented his ideas with a special side-look to the catastrophe of Vaiont (Habib 1975, 193, s. also Sect. 2.6). Goguel, in a study on the role of water and heat in tectonic phenomena (1969), mentioned frictional heat as energy source for both vaporisation of water and *dehydration* of materials like gypsum or serpentine (pp. 157–158). He also drew the reader's attention to the critical pressure of 22.1 MPa for vaporisation, corresponding to a thickness of about 850 m for coherent rock, and thus irrelevant for most rockslides (p. 160). Two remarks may be useful in this context: if ample water is present, the thermal effectiveness of the mechanisms discussed by Habib and Goguel may be reduced by cooling; and as a rule these mechanisms do not produce easy-to-perceive traces like frictionite so that often they may pass unreported like dissociated rock.

*

As announced, this section is closed by a list of further articles connected with the "detective story" of Köfels. Although containing valuable (and in some exceptional cases even fantastic) information, their importance for the particular scope of the present book is limited (s. also list of references): Deutsch et al. 1994; Hammer and Reithofer 1936; Heissel 1938; Heuberger 1996; Klebelsberg 1935, 1949; Kranz 1938; Penck 1925, Purtscheller 1971; Tollmann and Tollmann 1994. For readers interested in the laborious life of an Alpine scientist-clergyman in the 19th century Fliri's report about Trientl (1817–1897) is most edifying (1992 in German).

2.5 Val Pola

The key events hitherto presented are not suspected to have done any *harm to human life* or property. Blackhawk and Köfels occurred in prehistoric times when population in the respective regions – if any – was extremely sparse; and the mountain ranges surrounding Pandemonium Creek, in spite of their beauty, still are remote enough to remain uninhabited and inaccessible to mass tourism. Contrarily thereto, the three following slides – Val Pola, Vaiont, and Huascarán – represent a dramatic crescendo of killed persons and destroyed dwellings. Obviously this tragic score is connected with the fact that, on a world-wide scale, places of residence, under the pressure of a growing population, more and more expand into mountainous areas. How far other reasons may have contributed, will be discussed in Sect. 3.1.

In the context thus evoked geomorphological evidence has to be complemented by a review of the *actions taken by persons* directly or indirectly confronted with dangerous situations.

In the Alps the time between July 15 and 28, 1987, was a period of particularly *heavy precipitation*. In addition, the high temperature (freezing level above 3 500 m) entailed an abnormal runoff from the Alpine glaciers. A great number of flood disasters and gravitational mass movements were consequences of this meteorological coincidence. One of the areas most severely struck was the valley Valtellina (in German publications: "Veltlin") in northern Italy, near the Swiss border (Notarpietro 1990). To illustrate the general situation, only two of many reports may be cited: "*… the storm of July… that… triggered more than 500 mass movements in three left tributary basins of the Adda…*" (Crosta 1990, 247). "*About 18 people were killed in this region during this time; 10 lives were lost in Tartano when a hotel was struck by a debris flow…*" (Costa 1991, 19).

A certain *ambiguity of a dense population* – on the one hand an increased number of threatened persons, on the other hand the existence of a powerful infrastructure assisted by an efficient geological survey – had already shown its character in the mentioned week: no less than about 3 500 people had been evacuated, and there is no doubt that the number of lives thus saved exceeds by far that of victims.

This situation came to a climax when, on July 18 and 19, the possibility of a far larger event had to be faced. It "*… was located on the east slope of Mt. Zandila… at the head of Val Pola, a small steep torrent tributary to the Adda River on the west side of the Valtellina. The entire northern mountainside is part of a large prehistoric landslide at the intersection of two major joint sets dipping 45° and 80° into the Adda River valley on an average hillside slope of 32°. Bedrock consists of northward-dipping isoclinal folds of highly fractured and jointed gneiss intruded by gabbro and diorite… Heavy rainfall… caused flooding and debris flows in the Val Pola that eroded the north side of the prehistoric landslide, and formed a debris fan… that dammed the Adda River. A… lake with an estimated volume of 50 000 cubic metres and a maximum depth of 5 metres formed… The lake was partly drained by excavation of an outlet channel…, but new sediment was continually added to the debris fan. On July 25, geologists reported a 600 m long crack at the 2 200 m elevation on the slope of Mt. Zandila adjacent to Val Pola. Over the next two days, the crack increased in width, and rockfalls from the unstable mountainside became more frequent. Between 500 and 600 residences… near the base of the unstable slope were evacuated…*" (Costa 1991).

Fig. 2.27. Val Pola rockslide, view of descent from the east ("Plaz"). Photograph taken eight years after event shows marks of erosion, especially in large fan on bottom (accessibility of certain sites has distinctly changed in three years). Rhombic area on top is the scar. At right thereof, unstable protrusion (s. Fig. 2.33). Note essentially channel-like shape of slope (photo by Erismann)

This was the situation which, on the morning of July 28, was followed by the *rockslide of Val Pola* (Fig. 2.1, 2.27, 2.28, 2.29, 2.30; Cancelli et al. 1990) with an estimated volume of 0.032–0.040 km^3 and a total vertical extension of about 1 250 metres. In spite of the mentioned successful evacuations in the threatened area, a toll of 27 lives was taken. Seven of these were workers engaged in the excavation of the outlet channel – an extremely risky (not to say heroic) task under the described conditions. This risk had been taken to prevent worse. The other victims, however, could have been saved if an adequate forecast of the slide's motion and its effect upon the lake had been available in time. This is nothing but a statement of facts and by no means intended to blame anybody for not having done the best possible to prevent damage: the persons involved, responsible for hundreds of human lives, had to do their work under extreme stress. Who could imagine, in such instances, that a small accumulation of water, apparently negligible in comparison with the immense mass of rock expected to be detached a kilometre above the valley, would play the most disastrous role in the drama? And there was a tragic irony in the fact that just those courageous workers engaged in the reduction of this accumulation of water were first to perish.

The fatality consisted in the fact that the mass of debris hit the lake at a sufficiently high velocity to entail a *flood wave* that rapidly travelled upstream. Some general information about waves excited by rockslides will be given in Sect. 7.1. It must, however, be pointed out (and was confirmed by an oral communication of Prof. Vischer, ETH, Zürich) that algorithms for the calculation of waves presented by various authors (e.g. Noda 1970; Vischer 1986) are valid, as a rule, for conditions not fulfilled in the

▲
Fig. 2.28. Val Pola rockslide, general aerial view from the east, few days after event. Involved area stained bright by debris, mud, and water. Note resemblance to Fig. 2.30 and characteristic details: protrusion "Plaz" facing track; newly formed lake at right; intact strip of forest at left. Behind the scar Monte Zandila (2 936 m). Low on its slope the shadow of the opposite range that the helicopter is crossing in the light of the raising sun (photo by unknown helicopter pilot having taken part in rescue; by courtesy of the former Swiss Federal Office for Military Airfields)

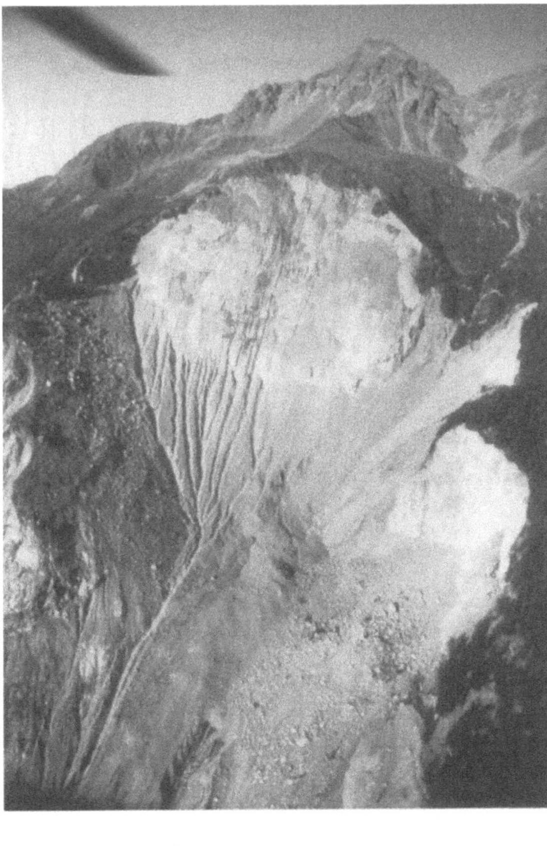

Fig. 2.29. Val Pola rockslide, aerial view of scar with Monte Zandila in the background. Photo taken few days after event. Note characteristic details: channelling geometry of lateral surfaces, particularly well-developed on the left; beginning erosion by two torrents; the one at right is Pola (photo by unknown helicopter pilot having taken part in rescue; by courtesy of the former Swiss Federal Office for Military Airfields) ▶

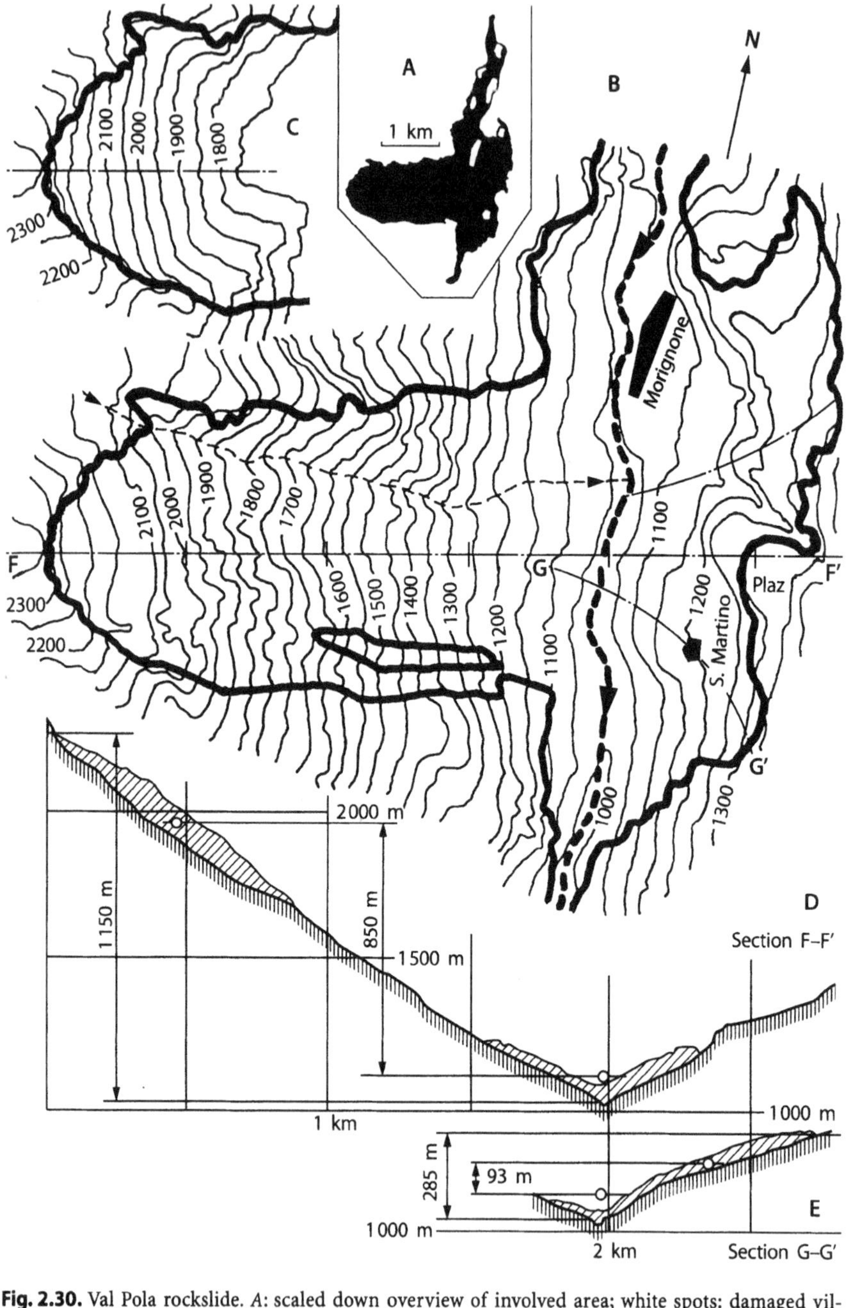

Fig. 2.30. Val Pola rockslide. *A*: scaled down overview of involved area; white spots: damaged villages and hamlets, from north to south: Aquilone, Tirindrè, S. Antonio Morignone, Poz, S. Bartolomeo, Morignone, S. Martino Serravalle). Maps (*B* before, *C* after event) and longitudinal sections (*D*, *E*) along *dots-and-dashes lines F-F'* and *G-G'*. Scales of *B*, *C*, *D*, and *E* are equal. Lines in maps: *plain*: periphery of involved area; *broken*: River Adda, torrent Pola. Hatched areas in sections: mass at start; in crossing valley; in ultimate run-up position (circles indicate approximate locations of centre of gravity) (sketch by Erismann)

Fig. 2.31. Val Pola rockslide. Eastern slope of Valtellina with massive protrusion "Plaz" which separated the debris into a northern lobe at left and a southern one at right. The entire foreground is rearranged to warrant safe passage of Adda river and road traffic (photo by Erismann)

case of Val Pola. In fact, one of their main assumptions is that the volume of the rock mass be small in comparison with that of the basin into which it plunges. Exactly the opposite was the case: parts of a rock mass hundreds of times larger than the lake literally swept away the water, mixing it with mud and gravel.

Perhaps this unexpected effect would not have happened in such a dramatic manner if the mass had been free to run out in ascending a gentle slope on the eastern side of the valley. But instead there was the massive bedrock protrusion of "Plaz" that, situated in a generally steep slope and right in front of the central axis of the moving mass (Fig. 2.31), acted like the bow of a ship and *cut the debris into two lobes.* Each lobe ran up on its side of Plaz and fell back, overrunning once more the valley floor and coming to a rest with its distal elements on the western slope again. By the way, as suggested in Fig. 2.32, the particular geometry made the elements turn to run back more or less simultaneously and then move rather side by side than in line. Anyhow, it is probable that the main hit upon the lake occurred when the northern lobe slid back westward and was focused to the lake and the village of Morignone by the concave shape of the adjacent slope.

Two geomorphological facts can be put forward in favour of this hypothesis. On the one hand, the mass, at the beginning of its main descent, was definitely well-channelled by lateral confinements (s. Fig. 2.27, 2.29, 2.30c, 2.33). Thus lateral spreading could not develop on a large scale until it was forced upon the mass by the action of Plaz. In other words, it is not excluded that the mass, in rushing downhill, did not (or only to a small extent) hit the lake. On the other hand, clear marks of a *secondary run-up* can be observed on the western slope at a height exceeding 1150 m (s. in Fig. 2.30b: protrusion of involved area pointing to the silhouette A). This is exactly the location which

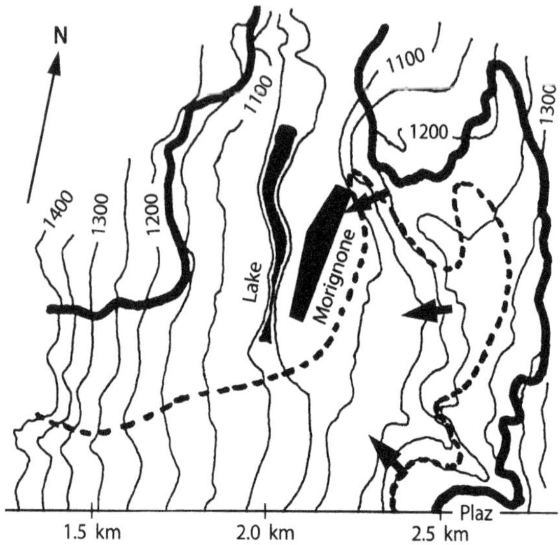

Fig. 2.32. Val Pola rockslide, northern lobe (bottom of figure is section *F–F'*, Fig. 2.30). *Dotted line*: core of debris in assumed ultimate run-up position (only poor deposits beyond dotted line). *Arrows*: effective run-back on steep slope portions. Note focusing upon lake (location and shape are hypothetical) and Morignone, as possible cause of secondary run-up below contour line 1 200 m (sketch by Erismann)

must be expected when considering the mechanism suggested in Fig. 2.32. It is conspicuous that the respective protrusion of the southern lobe, in spite of an approximately equal run-up height of about 1 350 m, is far less prominent. This fact points to a substantial difference which very probably was in first instance one of displaced material: speaking in terms of energy, water moves at a lower cost than rock.

Anyhow, in spite of the relatively low velocity of running back there was ample energy available to get as far as can be concluded from erosion and mud marks on the western slope of the valley: initially the *wave* must have been about 100 m high, and after a distance of 1.3 km some 15–20 m (Govi 1989). And even after a total travel of 2.1 km, when it reached the village Aquilone (not evacuated because of its considerable distance from the slide), it still had the power to damage houses and to kill persons.

*

The destructive power and extreme mobility of a water-and-mud mixture, even if mobilised by a relatively slow mass, is the first lesson to be learned from Val Pola. The second one refers to the determination of the *velocity acquired by the mass* of a rockslide.

In Sect. 1.1 the importance of reach (position of the distal elements) and velocity already have been stressed. Both depend in first instance on the *available potential energy*. This energy is expressed by $m\Delta zg$ where m is the mass, Δz its vertical displacement, and g gravitational acceleration.

Obviously the available energy also can be written in terms of *kinetic energy*. The respective transformation,

$$v = \sqrt{(2g\Delta z)} \ , \qquad (2.1)$$

yields the velocity v acquired in a free fall over Δz. In this formula, resulting directly from Newton's fundamental equations, the mass does not appear since both energies, potential and kinetic, are proportional thereto.

Fig. 2.33. Val Pola rockslide, main descending slope, view from south-east. Note lateral confinements, crowned in the background by protrusion of dubious stability (a carefully monitored potential source of danger). Some debris of the small lateral lobe in the forest at left emerges below the trees it was unable to chop down (photo by Erismann)

It is a trivial fact that in a rockslide or a rockfall only part of the available energy can assume kinetic form. The rest is transformed – directly or via intermediate states – into heat. In other words: *heat is the final effect of all forms of resistance*, be it by friction, impact, irreversible deformation or fracturing of material, aerodynamic or hydrodynamic drag, or what else. Thus kinetic energy is nothing but energy not dissipated by resistance, and at the point where the available energy is completely "eaten" by resistance so that no energy can subsist in kinetic form, motion is arrested. This description, though somewhat simplified in details (for instance in case of "recycling" potential energy in running up and down a non-arresting obstacle), gives a clear idea of what essentially happens in a gravitational mass movement. Similar considerations actually yield the energetic basis for Chap. 5 and 6. In the present section the crucial point consists in the fact that motion (and in particular its velocity) directly depends on the energy available beyond that needed to overcome resistance.

In the context of endangering human life easy-to-apply *methods for predicting velocity*, derived from post-eventum analysis, are a powerful basis for decision-making. And the geomorphological circumstances in the case of Val Pola (Fig. 2.27, 2.28, 2.30), in first instance the topography of its longitudinal section, may be considered as a showpiece for how such analyses should be approached (and perhaps even more how they not should be approached). In addition, as compared with most other (even recent) events, Val Pola is significant for the existence of maps both for the ante-eventum and the post-eventum states. So the influence of geometric errors is relatively small. It is, however, not the scope of a section aimed at geomorphological aspects to describe methods of this kind in detail. So, as a working hypothesis, the physical background of the method discussed hereafter will be taken for granted (it is presented and critically analysed in Sect. 6.2). Here only scarce remarks are made in this respect.

According to Francis and Baker (1977) the *maximum velocity* v_{max} attained by a mass between a descent (loss of altitude Δz_1) and an arresting run-up (gain of altitude Δz_2) is obtained from Eq. 2.1 by using

$$\Delta z = \sqrt{(\Delta z_1 \, \Delta z_2)} \tag{2.2}$$

to calculate the potential energy remaining after overcoming all effects of resistance. Anticipating some of the results given under Heading 6.2, this much should be said about the physical basis of this remarkably elegant (though easy to misuse) method: Eq. 2.2 is correct if (1) the percentage of potential energy transformed into kinetic energy in descending is equal to the percentage of kinetic energy re-transformed into potential energy in running up and if (2) no other losses of energy have to be taken into account (e.g. losses in horizontal motion between descent and run-up).

The event of Val Pola, with only little near-to-horizontal displacement between descent and run-up, seems to fulfil condition of (2) fairly well. And as there is no stringent reason to assume substantially different mechanisms of resistance on the two sides of the valley (disintegration took place in an early stage so that most of the travel occurred in a more or less unchanged state), also condition of (1) appears to be rather unproblematic. So it is plausible that several scientists used Eq. 2.1 and 2.2 as a basis for an estimate of the highest velocity attained at the bottom of the valley (Völk 1989; Costa 1991). The calculation of Δz_1 and Δz_2 was based, however, on the elevations of the three following points: the bottom of the valley and, respectively, the highest points of the head scar and the run-up. In doing so, a fundamental physical condition was violated: any energetic calculation on a coherent body has to be made with respect to *the body's centre of gravity*. And at least in two of its relevant positions the mass, in spite of disintegration and lateral spreading, had to be treated like a coherent body: when starting, it actually was more or less coherent; and when crossing the valley, it was longitudinally held together by the driving forces of its (descending) proximal portion and the braking forces of the (ascending) distal portion. Only in running up along curved paths on the eastern slope the elements of the mass could have lost contact with each other. But even then its centre of gravity obviously could not reach an altitude differing from that given by the kinetic energy when crossing the valley minus the sum of all resisting (potential or heat-generating) energies subsequently acting upon its elements.

Now the centre of gravity of the mass never was as high as the top of the head scar, it never was as low as the sole of the valley (not only because of the thickness of the mass but also because of the distal and proximal ends being higher than the centre), and, of course, it never could reach the highest position attained by the topmost elements. So, as can be seen from Fig. 2.30, the *differences of altitude are cut down* drastically. Instead of $\Delta z_1 = 1150$ m and $\Delta z_2 = 285$ m the respective values are $\Delta z_1 = 850$ m and $\Delta z_2 = 93$ m, thus reducing the maximum velocity from 106.0 m s^{-1} to 74.3 m s^{-1}. Once more anticipating the results of Sect. 6.2, it can be confirmed that this reduction, in spite of the inherent weak points of the used algorithm, brings down the estimate of velocity to a far more realistic level: Val Pola, without any doubt, was a very fast rockslide, yet there is no evidence in favour of a velocity in the range of 100 m s^{-1} or even 400 km h^{-1} (111 m s^{-1}) as claimed by some investigators.

The problem might pass as trivial if there were not a general *tendency to overestimate velocity*. And the destructive power of an event is, as a rule, proportional to the square of velocity. The fact that the prediction of a catastrophe thus will be rather on the safe side (i.e. an overestimate), is a poor consolation; and it even may turn to the wrong side in calculations used to establish a set of standard parameters as a basis for a more general use. This is by no means an alarmist's fantasy: this is, as reported under Heading 6.2, exactly what happened more than 60 years ago when the first approaches were made to a physically plausible method for the calculation of velocity! And ever since, the wrong results of these early attempts have been re-copied so many times... It is a great satisfaction for an old rockslide-fan to observe how, step by step, a new generation of scientists becomes aware of the difference between the periphery of a mass and its physical centre (Evans et al. 1989, 442, though confined to Körner's model, s. Sect. 6.2; Crosta 1991, 104).

There is a third lesson, somewhat hidden behind the extensive discussion that followed the event of Val Pola. It demonstrates how carefully *quantitative information taken from the literature* should be considered before using it as an argument in con-

Fig. 2.34. Val Pola rockslide, secondary run-up of water on western slope of Valtellina seen from present valley floor, about 40 m above engulfed Morignone (Fig. 2.30, 2.31, 2.32). Peaks of run-up reach almost another 100 m higher. Though taken eight years after event, photograph clearly shows traces of completely annihilated vegetation (photo by Erismann)

nection with important parameters. It also shows, by the way, that the author of a book about rockslides is not eo ipso exempt from committing such errors.

*

The rockslide of Val Pola was one of the few events which, as having occurred in the very heart of Europe, were recorded by a considerable number of more or less nearby seismographs. The fact was used by some investigators (e.g. Costa 1991, 25) to back up their assumptions about the attained velocity by referring the distance covered to the *duration of a seismic signal* and thus obtaining an estimate for the average velocity. In a paper orally presented at the Deutscher Geographentag 1991 and subsequently published, Erismann (1992, 13–15; Kienholz et al. 1993, 306–309) used this information for a somewhat more differentiated analysis, in particular aimed at the necessity of taking into account the centre of gravity and its displacement. Later, when analysing the seismographic records of various Italian (De Simoni et al. 1990) and Swiss laboratories (obtained by courtesy of the Swiss Earthquake Survey at ETH Zürich), some doubts came up with respect to the obtainable accuracy. These doubts were confirmed in an oral discussion with Dr. D. Mayer-Rosa, section leader in the mentioned survey: in spite of the fact that one of the Swiss seismographs is only at a distance of about 30 km from Val Pola, there are at least two problematic points in the interpretation of the records. It is (1) not known how far the mass had been displaced when the first oscillation able to produce an answer in the recorder was emitted; and (2) even

Fig. 2.35. Val Pola rockslide lake formed after the event, seen from the north (near Aquilone), several months after the slide. At this time it still was far larger than that due to rockfalls preceding the main catastrophe. Sliding surfaces scarcely visible in the background. At right, near the basin separated from the main lake by a tongue, the bright traces of the secondary run-up (Fig. 2.34). In the foreground at left a pumping station provided to lower the water's level (photo by Abele)

over so short a distance the signals, owing to reflections, are transmitted on more than one single path, and the time required to get to the seismograph may strongly depend not only on the length of the path but also on the material. In other words, neither the beginning nor the end of the signal can, at least without an extended mathematical analysis (probably by autocorrelation techniques), unequivocally be correlated with the beginning and the end of the event.

*

On the whole the Val Pola rockslide, in spite of its tragic aspect, must be considered as an event in which one of the heaviest catastrophes of the century could be avoided by a remarkably efficient *hazard management* (Costa 1991, 26–36). And this is true not only for the work done under the stress of immediate danger: also in the following time the necessary actions for controlling the massively dammed river (Fig. 2.35), restoring traffic, etc. have been undertaken with admirable vision, skill, and energy.

Not to forget the actual stability of the slope: an extensive, mainly telemetric *monitoring network* has been installed with microseismic stations, surface and vertical extensometers, piezometers, inclinometers, and, of course, the required processing equipment (I.S.M.E.S. 1990, 4–11). In particular a large protrusion besides the head scar, as being of dubious stability and thus signalling the possibility of a new (though definitely smaller) slide, is observed with due care (Fig. 2.33).

2.6
Vaiont

The catastrophe of Vaiont, like any rockslide that takes a death toll, was a human tragedy. It was, however, not only a tragedy for the victims and their families: it also was a *tragedy for the experts and engineers* in charge. Not many large rockslides have been the object of such extended ante eventum studies, and, post eventum, probably no one so mercilessly gave the lie to the authors of such studies.

In principle it is not the scope of this book to incriminate experts for not having estimated an impending danger realistically, especially if, as in the particular case of Vaiont, they had to deal with unexpected phenomena. Under the stress of eminent commercial demands, decisions of great import had to be taken in connection with a large mass of rock threatening the valley of Vaiont River (an eastern tributary of the Piave in Valle Serpentine, Northern Italy, Fig. 2.1), and *no adequate measures were taken* to protect thousands of persons. However, with future damage prevention in mind, in spite of the just-mentioned principle, the question is inevitable how this could happen. And this question necessarily gives rise to a comparison of (1) what the experts had expected ante eventum, checked against (2) the state of the art at the critical time and, above all, against (3) the real course of the event.

*

Since the first decades of the century it was clear that the valley of Vaiont presented excellent topographic and hydrologic conditions for the installation of a *hydro-electric power station*. In fact, it is rich in water, sufficiently voluminous and elevated to hold a reasonably large storage basin, and it descends to Val Serpentine through a narrow and steep gorge, predestinate for the erection of a dam (Fig. 2.36).

The stability of the slopes surrounding the basin had been examined by eminent geologists long before the project could be started. First studies were conducted by Dal Piaz in 1928. In spite of the care invested in such studies in the course of about 30 years, and the reassuring results initially obtained, the first tentative filling of the basin, carried out from March to November 1960, alarmed experts and engineers: a *creeping motion of a large mass* (at the time estimated to have a volume of about 0.2 km^3, but later found to be even larger) was observed (Fig. 2.37). This mass originated from two cracks (partly on top of the scarps of the later event) in the steep northern slope of Monte Toc and descended at a softer angle to the gorge of the Vaiont immediately upstream from the dam. Shortly after this first warning, several small and rather slow slides came down in the lower part of the mentioned slope. However, as soon as the water level was substantially lowered, the velocity of creeping at a characteristic measuring point (one of many which were observed), after having reached a maximum of almost 40 mm day^{-1} (up to 80 mm day^{-1} at other points), fell to slightly varying values below 1 mm day^{-1}. This situation remained stable in 1961. For the major part of that year, the water was held at a low level (about 600 m) and no further filling took place: warned by the spectacular observations made in 1960, the engineers of ENEL (Ente Nazionale per l'Energia Elettrica) had decided to bypass the critical portion of the valley by a tunnel and thus avoid an inundation in case of the normal outlet of the basin being blocked by a slide. The reduction of creeping velocity gave the reassuring impression of a limited motion that had more or less found a new equilibrium of driving and resisting forces.

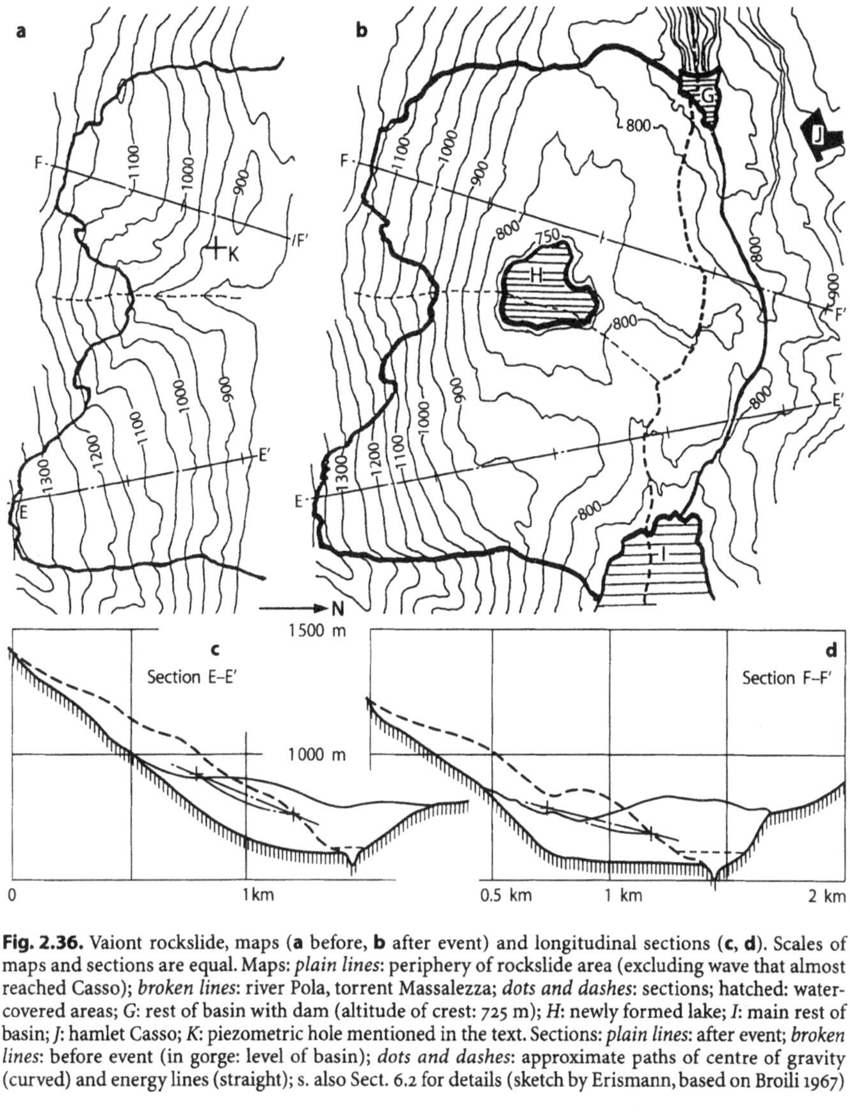

Fig. 2.36. Vaiont rockslide, maps (**a** before, **b** after event) and longitudinal sections (**c**, **d**). Scales of maps and sections are equal. Maps: *plain lines*: periphery of rockslide area (excluding wave that almost reached Casso); *broken lines*: river Pola, torrent Massalezza; *dots and dashes*: sections; hatched: water-covered areas; *G*: rest of basin with dam (altitude of crest: 725 m); *H*: newly formed lake; *I*: main rest of basin; *J*: hamlet Casso; *K*: piezometric hole mentioned in the text. Sections: *plain lines*: after event; *broken lines*: before event (in gorge: level of basin); *dots and dashes*: approximate paths of centre of gravity (curved) and energy lines (straight); s. also Sect. 6.2 for details (sketch by Erismann, based on Broili 1967)

In such instances it was a fatal circumstance that, in 1962, a markedly higher further tentative filling (water level at 700 m against 650 m in 1960) resulted in a comparatively *low second peak of creeping velocity* (about 10 mm day^{-1}, i.e. almost four times less than in 1960). So the impression of a process of stabilisation in course was cemented, and the decision was taken to continue in a progressive filling cycle in 1963: first the level was intended to be lowered to 650 m, then raised to 710 m. In case of encouraging results, further cycles were provided until the elevation of the dam's crest at 725 m would approximately be reached. A cautious step-by-step approach with slow raising and lowering phases in combination with draining was recommended by at least one of the experts as the best means to gain control of the situation (Müller 1964, 166).

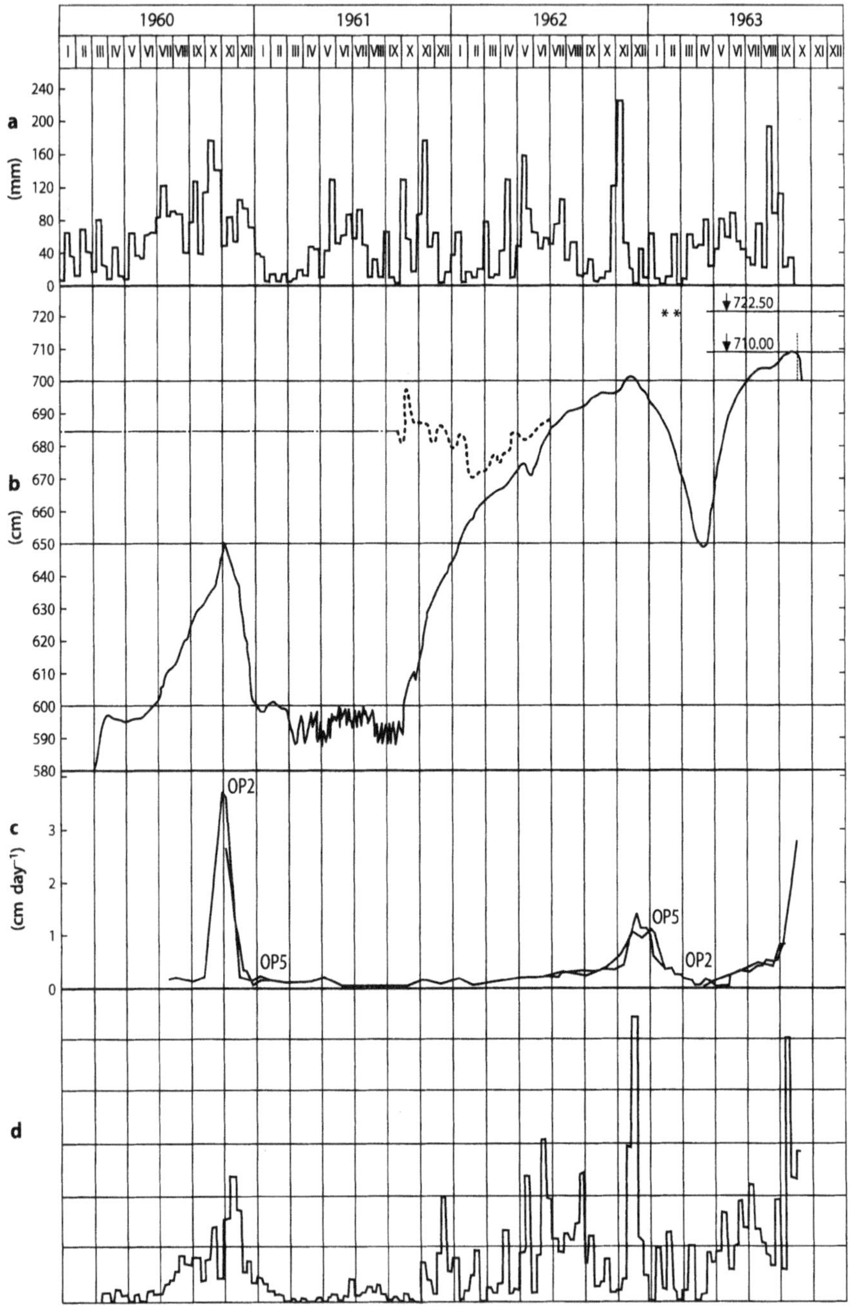

Fig. 2.37. Vaiont rockslide, creeping phase preceding the slide proper. Diagrams: **a** precipitation; **b** water level in basin; **c** creeping velocity in two observed points; **d** product of precipitation (shifted forward by 30 days) and water level in basin. *Dotted line* in **b** level in piezometric hole (for location s. Fig. 2.36, Item *K*), suggesting buoyancy effect at higher level than in the basin (important for filling in 1960). Note correlation between **c** and **d** (sketch by Erismann, partly based on Müller 1964)

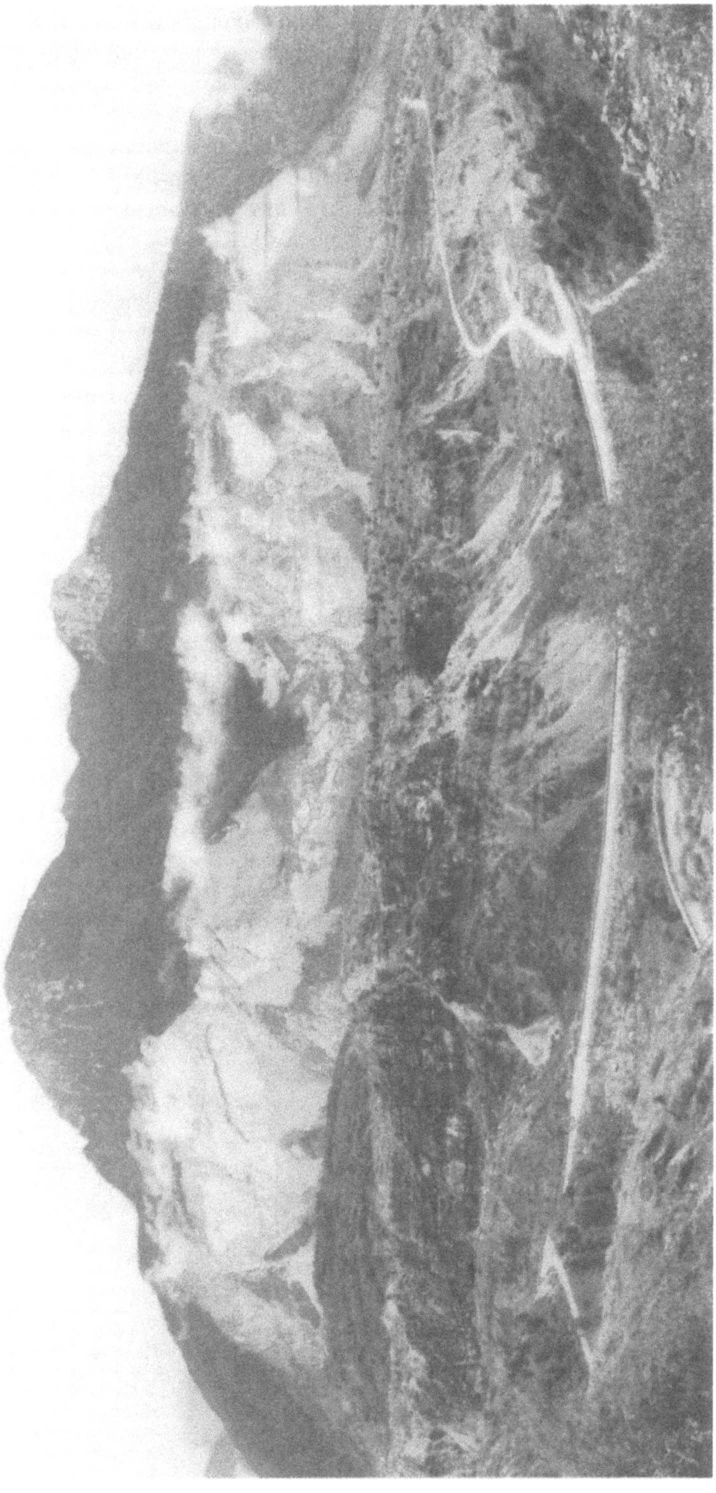

Fig. 2.38. Vaiont rockslide, general view from north, taken in 1984 slightly below Casso (Fig. 2.36, Item *J*). Distal limit of deposits is approximately marked by the main road in the foreground. Besides the disintegrated material at their front, deposits show their pre-event character (including forest commented in detail under Fig. 2.41). In the background Monte Toc with the twin scars, partly separated by the forested ditch of torrent Massalezza (and by a patch of fog) (panoramic photo by Erismann)

In seeming accordance with the hypothesis of stabilisation the velocity of creep, after the water's level had been lowered to 650 m, returned to almost negligible values. Strangely enough, the confidence thus acquired was not lost by the warning that accompanied the rising of the water's level: a sudden "jump" to 3 mm day^{-1} by end of May, 1963, followed by a continuous acceleration to 5 mm day^{-1} in the course of three months. And even in September, when the acceleration at least equalled that of 1960, the filling process was not interrupted. Obviously nobody was aware of the possibility that the *velocity of locomotion might change* drastically.

*

Then, on October 9, late in the evening, a mass totalling about 0.3 km^3 of rock slid down from both scars to the bottom of the valley (Fig. 2.38). Its velocity reached a maximum more than *seven powers of ten faster than the fastest creeping* previously observed (more than 20 m s^{-1} against 80 mm day^{-1}), and it changed completely the topography of the valley. As far as nothing but displacement of rock was concerned, the event took about half a minute's time. Owing to the steep slope on the opposite side and the moderate velocity, only low (if any) true run-up resulted (details with respect to this last statement will follow hereafter). So the centres of gravity of the two half-slides (Fig. 2.36, Sect. E–E' and F–F') accomplished short travels of some 400–450 m in the horizontal and 110–160 m in the vertical direction. As a result of the steep slopes on both sides of the valley the mass was somewhat compressed in the direction of displacement: at the distal end the horizontal displacement was shorter (240–360 m), at the proximal end accordingly longer (500–550 m). Especially in the eastern half-slide this fact resulted in a certain run-up or overthrust within the distal portion of the mass (though on a far smaller scale than in the case of Köfels, Sect. 2.4).

The slide came down in an *exceptionally coherent* state (Fig. 2.39, 2.40). Particularly spectacular in this context was the fact that the gorge of Vaiont River, about 80 m wide at the respective location, was perfectly bridged without large portions of the mass falling to its bottom. Another noteworthy detail was the soil which, in spite of the dramatic downhill ride, remained on top of the displaced mass, still bearing trees in an upright position (Fig. 2.41), though slightly tilted as a consequence of the change in slope.

*

All this sounds interesting, perhaps also impressing, but by no means tragic. As a matter of fact, the rockslide proper, though it completed a tremendous work of destruction in making the Vaiont power station useless, *did neither kill nor injure any persons*. And still, in the course of this half-minute a mechanism was initiated which, for another few minutes, transformed the neighbouring part of Valle Serpentine into an inferno of collapsing dwellings and drowning inhabitants: *one of the heaviest catastrophes* ever caused in historical times by gravity-driven masses had become reality.

Nature and reach of the fatal *mechanism behind this cataclysm* were clearly perceptible from its effects. The rockslide had brutally displaced the water of the basin into which it had been partly immersed before the event and then had plunged right from the start of fast motion. "Plunged", however, is perhaps the wrong word as the volume of the rockslide was almost by a factor of eight larger than that of the water involved, thus yielding a water-to-rock mass ratio in the range of 1/20. So *the water rather was swept away than plunged into*, and, as stated for the case of Val Pola

Fig. 2.39. Vaiont rockslide from the north-east. Foreground: deposits of eastern half-slide (run-up and gorge of Vaiont River at right). Background: scars (eastern with Monte Toc at left, western at right). Photograph approximately corresponds to section *E–E′* (Fig. 2.36c) and illustrates chair-like shape acquired by this section post eventum. The remark made in the legend of Fig. 2.41 with respect to humus and vegetation, though valid, is somewhat less evident for reasons of scale (panoramic photo by Erismann)

Fig. 2.40. Vaiont rockslide. Detail of distal deposits belonging to the eastern half-slide (near section *E–E'*, Fig. 2.36c). The low degree of disintegration is clearly perceptible (photo by Erismann)

(Sect. 2.5), the equations otherwise used for the calculation of wave dimensions would be useless (for details s. Sect. 7.1).

Anyhow, once launched, the water raced up the northern slope, annihilating almost the entire vegetation on its track, but sparing most of the hamlet Casso, well 250 m above the basin (Fig. 2.42). Then, as in the east two barriers impeded lateral spreading (Fig. 2.36b, contour lines 850 and 900 on both sides of section E–E'), the major part of the water descended to the western part of the newly-shaped valley, ran up along the western limit of the scar, crossed the dam as a wave of more than 100 m height without seriously damaging it (Fig. 2.41), and finally rushed down to Val Serpentine (river Piave at 2 km from the dam and about 250 m below its crest) where it almost completely *destroyed the large village of Longarone* and the smaller rural communes of Fae, Pirago, Rivalta and Villanova, thus extinguishing about 1900 lives. The disaster came too fast to allow the slightest chance of escape in the darkness of an October night at about 22:40 hours.

To imagine the *destructive power of the water* two considerations may be useful. If the assumption is made that 0.04 km^3 of water were displaced (Müller 1964, 199) and that 75 percent thereof (0.03 km^3) reached Valle Serpentine, this would mean that the

Fig. 2.41. Vaiont rockslide. Practically undamaged dam of the basin. In the background at left the western scar. Photograph, taken in 1984, shows trees of a certain size having subsisted in riding downhill, and younger vegetation witnessing for the humus having remained on top of the mass. Area devastated by water is marked by horizontal periphery reaching from bottom right corner of scar almost to the horizon, then turning downward to disappear behind right end of dam (photo by Erismann)

Fig. 2.42. Vaiont rockslide, map of area devastated by flood wave (black silhouette). *A*: Longarone (altitude 478 m); *B*: Val Serpentine (*arrows* signal that water moved further, though with reduced destructive power); *C*: rockslide area (corresponding to Fig. 2.36b); *D*: Casso; *E*: area in Vaiont Valley (slightly exceeding the periphery of the basin) (sketch by Erismann)

neighbourhood of Longarone (area between arrows in Fig. 2.42, approximately 2.5 km^2) theoretically could be covered by water about 12 m high. So it is not surprising that under run-up conditions far higher traces of water are reported (70 to 90 m according to Müller 1964, 199). And the water jet at the bottom of a wave of 100 m, in passing the crest of the dam, must have assumed a velocity in the range of 45 m s^{-1}, not to speak of possible further acceleration in descending another 250 m to the bottom of the valley.

A smaller portion of the water inundated Vaiont Valley more than three kilometres upstream without producing particularly spectacular damage (Fig. 2.42).

*

One of the essential intentions of this abridged case history (mainly compiled from the particularly complete studies by Müller 1964, 1968) was to make obvious the *discrepancies between expectations and reality*. In fact,

11. nobody had expected so rapid a *transition from slow to fast* motion that no time would be left for appropriate measures. Even for the case of a dramatic acceleration,
12. nobody had expected a *velocity exceeding 2–8 m s^{-1}* (Müller 1964, 199). As a consequence,
13. nobody had expected the cataclysm of a *flood wave.*

It goes without explanation that the mere possibility of an event beginning as a rapid transition from creeping to sliding and followed by an acceleration up to more than

20 m s^{-1}, inevitably would have implicated the threat of a flood wave which, in its turn, would have evoked a discussion about effective measures. And such measures, as all conceivable means of arresting or well-controlled release of the unstable rocks would have been either beyond human bounds or too risky with a mass of such dimensions (Müller 1964, 166), hardly could have had other aims than large-scale evacuations.

The following discussion of the state of the art at the critical time cannot be approached without going into certain further details of both the expectations and the case history. So this discussion rather will be a comparison within the *triangle expectations/state of the art/reality* than a mere confrontation of two aspects. This comparison will be conducted according to the order of the three aspects established above (rapid acceleration – reach and maximum velocity – flood wave) and closed by some considerations on creeping.

The assumption of a gentle *transition from slow to fast motion* mainly was based on the idea of a glacier-like creeping mechanism with visco-plastically deformed joints, filled with clay or some other material having an essentially velocity-dependent shearing resistance (Müller 1964, 185). This assumption was questionable both with respect to facts and to the state of the art. In a detailed study Broili (1967) came to the following conclusions: "*...intercalations will... be of subordinate importance, because... we must take into account also the roughness...*" (p. 46), "*...the sequence involved in the slide does not include any clay layers or any of those rocks that are usually considered 'clayey'...*" (p. 49), and "*...the technical characters of the rocks which form the Vaiont sequence can usually be considered more than satisfactory...*" (p. 49). The remark on p. 46 is made in connection with the statement that clayey interbeds sporadically found in joints regularly had a thickness far below the height of asperities on the respective surfaces. In such instances even a lubricant of high quality would remain more or less ineffective. The consequence with respect to the material considered as a basis for a creeping displacement is trivial: the expected material was non-existent so that *the expected sliding mechanism could not occur.*

But even if the rock of the slide had been rich in clayey layers, the question of transition from slow to fast motion would have merited more attention in the reading of the *literature existing at the time*, in particular of Heim's standard work (1932) which, in spite of its age of more than 30 years, still was one of the best references concerning rockslides (even another 35 years later it has remained worth consultation). In connection with the famous Goldau event, a slide that was released on a remarkably soft slope of about 20° and had travelled mainly on a clayey track, Heim describes the critical transition as follows: "*...motion in the scar region begins slowly, crack openings grow, the large rock slabs... slide very slowly, then faster and faster. In Goldau many eyewitnesses have seen the slow sliding. Some of them could save themselves in jumping across the lateral shearing crack...*" (p. 61, translation; detailed description of the event pp. 71–74). Especially the last statement is illustrative: motion must have been barely perceptible if the mentioned persons dared to stand on the creeping mass; yet the acceleration must have been so rapid that they had to jump, not to step for salvation. Obviously no one of the experts was aware of this sentence when expecting a gentle transition from creep to rush.

Once engaged in a more careful reading of Heim's work (1932), the experts also would have found precious information to deal with the *question of reach* in a sufficiently accurate way to know that a heavy catastrophe could not be excluded. As a matter of fact, Heim gives perhaps the first list of slides with physically relevant figures (pp. 114–119), in particular volume and what he calls "Fahrböschung" (i.e. the overall slope, defined as the inclination between the top of the scar and the foremost distal deposits, measured along the main track of the mass; for details s. Sect. 6.2, 6.3). A plot of Fahrböschung against volume (as already used for Pandemonium Creek, Fig. 2.5) is a powerful means for the comparison of various rockslides.

*

So far for post-eventum analysis. The crucial question, however, is whether an expert well acquainted with Heim's work would have been able to apply this knowledge to *prediction*. The required line of thought in determining range probably would run as follows:

21. recognise that danger is not excluded;
22. plot Fahrböschung against volume from Heim's table and – by estimation or calculation – determine an approximate value for the expected volume of 0.2 km^3 (Fig. 2.43);

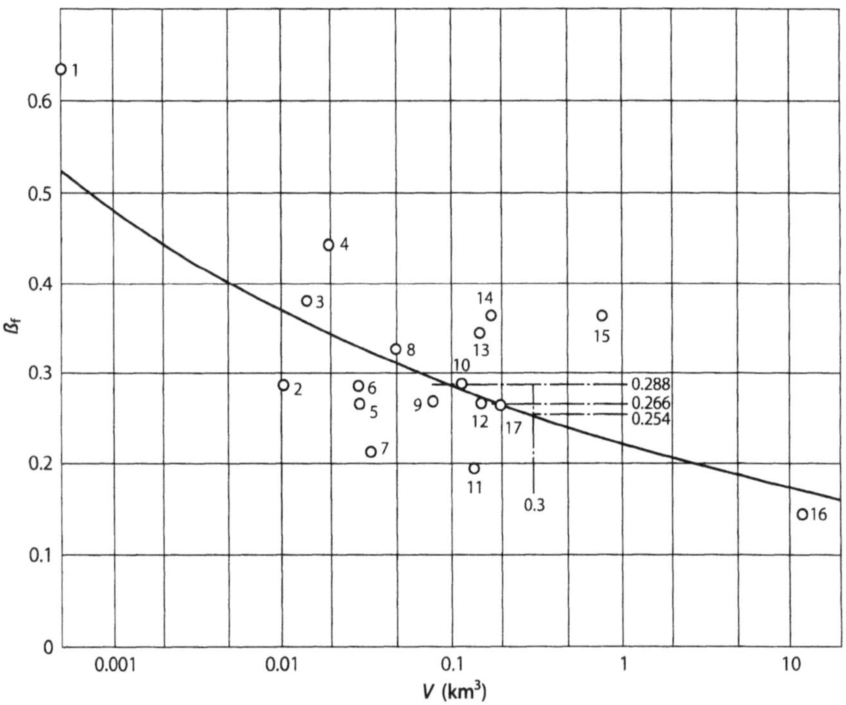

Fig. 2.43. Plot of "Fahrböschung" β_f against volume V according to the table established by Heim (1932, 114–119). Rockslides: *1*: Airolo; *2*: Elm; *3*: Disentis; *4*: Cima di Dosdè; *5*: Voralpsee; *6*: Frank; *7*: Goldau; *8*: Diablerets; *9*: Cima di Saoseo; *10*: Obersee; *11*: Kandertal; *12*: Parpan; *13*: Poschiavo; *14*: Bórmio; *15*: Glärnisch-Guppen; *16*: Flims. Besides Cima di Dosdè, Bórmio (both Italy), and Frank (Canada) all events are in Switzerland. The interpolation line is a Scheidegger function (s. 6.3). Vaiont rockslide (*17*) is given only by its volume of 0.2 km^3 (complemented by 0.1 and 0.3) as assumed ante eventum. The entire information used was available in 1963 (sketch by Erismann)

23. establish a representative longitudinal section for the considered half-slide (Fig. 2.44) and use the obtained value of Fahrböschung to find out how completely the water would be dislocated.

A reasonable prediction of velocity (as mentioned hereafter) would probably be difficult as neither the duration of the slide nor the shape of the sliding surface (not an easy guess under 200 m of rock) were previously known with due accuracy.

Figure 2.43 is nothing but a visual representation of Fahrböschung against volume. The figures are plotted according to Heim's table (1932, 114–119) without any corrections, thus showing the information as it was available in 1963. Few parameters in the table are expressed by maxima and minima instead of one single value. In such cases the arithmetic mean values are used. To obtain a mathematically defined interpolation, a log-log regression curve was calculated (s. Sect. 6.2; Scheidegger 1973). It is obvious, however, that the result (Fahrböschung = 0.266 for a mass of 0.2 km^3) cannot be far from what would be obtained by any other reasonable method, mere visual estimation not being excluded. The real volume and the volume when considering the half-slide as independent event (0.3 and 0.1 km^3) are within less than ±10 percent (Fahrböschung 0.254 and 0.288 respectively). Introduction of the angle thus obtained in longitudinal sections of the half-slides (Fig. 2.44) yields the reach. The result is clear: in both cases practically the entire volume of water within the width covered by rock necessarily would be swept away. *A flood wave was inevitable.* The comparison with the real situation post eventum shows that for the eastern half-slide the predicted reach would have been overestimated by about 250 m; for the western half-slide it would have been underestimated by less than 20 m. In other words: the mechanisms of the Vaiont slide were, as concerning *reach, not far from those of other large rockslides.*

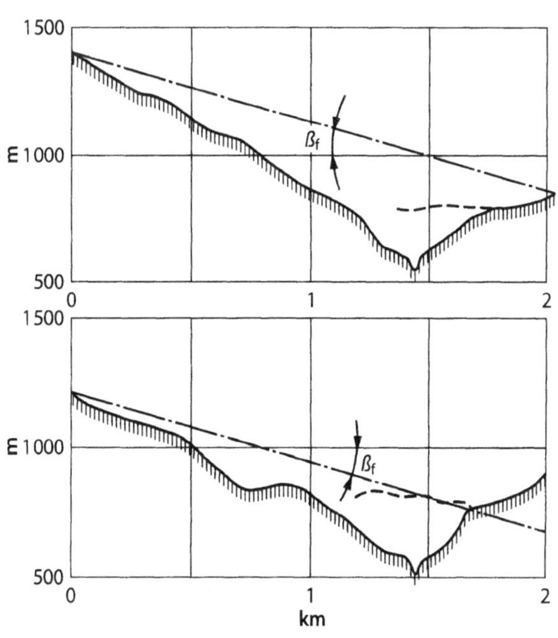

Fig. 2.44. Vaiont rockslide, Sect. *E–E'* and *F–F'* of Fig. 2.36 in ante eventum state. The angle $\beta_f = 14.9°$ ($\tan\beta_f = 0.266$) of Fahrböschung, as determined in Fig. 2.43, is inserted from the top of either scar (cracks were known at the time). Broken lines: distal elements of the slide post eventum. The figure shows that on the basis of Heim's book (1932) the reach would have been rather overestimated than underestimated (sketch by Erismann)

Once the dislocation of the entire water in front of the moving mass being certain, the question of a *flood wave excited by the slide* needs little comment. In fact, the bare idea of a volume, large enough to transform an area of 2.5 km^2 into a lake more than ten metres deep, rushing upon Longarone in a cascade 250 m high, was in itself a sufficiently serious threat to evoke activity. In addition, two events, recent at the time, were appropriate to make aware the responsible persons of the impending danger: the catastrophe of Fréjus in Southern France (1959) where the bursting of a dam had taken more than 400 lives; and the publication by Miller (1960, s. also Sect. 7.1) about giant waves in Lituya Bay (Alaska), excited by a rockslide and having left their traces of destruction up to an altitude of no less than 530 m (1740 ft). In spite of not being directly comparable with the rockslide of Vaiont, these events gave an impressing demonstration of the destructive capacity residing in a large quantity of rapidly displaced water.

*

In addition to his considerations on size and reach, Heim (1932) presents *two methods for the estimate of velocity post eventum.* One (pp. 92–94) is based on the determination of the time elapsing between two well-observed positions of the mass. On this basis, and considering in first instance reports from the famous events of Elm and Goldau, Heim came to the general statement that in rockslides velocities of 50 to 150 m s^{-1} must be the rule. This was an overestimation as Heim assumed that the maximum velocity used to be twice the average. This relation is true for a constant-acceleration-constant-deceleration cycle, a somewhat too primitive schedule. So a relation 3/2 and a velocity of 30 to 100 m s^{-1} may be considered as more realistic. The second method (pp. 143–152; for details s. Sect. 6.2), a result of Heim's co-operation with Müller-Bernet (not the expert involved in Vaiont), makes use of the potential energy not absorbed by overcoming resistance and thus transformed into kinetic form. For application in a particular case a vertical section is laid along the path of the centre of gravity of the mass ("CoG path"). In this section besides the CoG path the constant-slope line between start and end of the path (the "energy line") is considered. Provided that the coefficient of friction is constant, the kinetic energy is proportional to the vertical distance Δz between any point of the CoG path and the respective point of the energy line. The velocity is obtained from Eq. 2.1. It is $v = \sqrt{(2g\Delta z)}$ where g is gravitational acceleration.

Of course the absence of a well-founded prediction of velocity reduced the estimate of destructive capacity to the formula: "a large volume of water descending from considerable altitude". Post eventum, however, this gap easily can be closed. In Fig. 2.36c and 2.36d the required CoG paths and energy lines are inserted as well as possible. The resulting maximal values of Δz, around 27.5 m for the eastern and 22.5 m for the western half-slide, yield velocities of 23.2 and 21.0 m s^{-1} respectively. Owing to the low vertical travel of the centre of gravity Vaiont was a comparatively slow slide. And its *velocity needs no extraordinary mechanisms* to be explained.

Obviously there is an open contradiction concerning the *assessment of velocity*. And it is difficult to understand how Müller, who must have known Heim's minimum of 50 m s^{-1}, after an extensive study of the literature, had come to the persuasion that "...rock-*slides... varied between 0.8 and 20 m/sec...*" (1964, 198) and 40 to 70 m s^{-1} were reserved for rock-falls (whatever the terms "slides" and "falls" were standing for in the particular cases). So it is regrettable that the possibility of estimating the prospective velocity by Heim's and Müller-Bernet's method was at the time barred by insufficient knowledge of the geometry. In itself, the ante eventum application of a method originally conceived for post eventum use hardly could be considered as an innovation requiring the mental powers of a genius.

It is with good reasons that both for reach and velocity the possibility of explanations within the normal range of other rockslides and without extraordinary mechanisms was stressed. In fact, Müller's frequently repeated opinion that the velocity was extraordinary made him suspect the existence of a particularly effective, *hitherto unknown sliding mechanism.* He presented ideas of Weiss (1964), Mencl (1966), Haefeli (1967), and Falconnier (oral, s. Müller 1968, 83), as well as his own hypothesis of a "mass thixothropy" or "quasi-thixothropy" (1964, 208, 1968, 85). By exception of Falconnier who tried to establish an analogy with large subaquatic slides, the considerations of all mentioned authors were based on the fact that friction can be reduced by shocks (as applied when knocking at certain mechanical measuring instruments). Sources of shocks were suspected either in numerous ante eventum earthquakes or in the collapse of voids within the mass and transmission by shock waves either in the rock or in the interstitial water. Later a similar (though not identical) mechanism was quantitatively investigated by Melosh (1979, 1983) under the name of "acoustic fluidisation". This idea will be discussed under Heading 5.6 so that here the bare results will suffice.

Provided that the physical circumstances (above all geometry and elasticity) are favourable, such a mechanism can, as a matter of fact, reduce friction to a remarkably high degree. Yet this effect works only if the relative velocity between two bodies (or two layers of a single body) is substantially lower than that observed between the coherently moving mass and the ground in the slide of Vaiont. So, if at all, *such a mechanism rather might have influenced creeping* than fast motion. In light of the considerations discussed here-above this result could be expected: in the presented optics, velocity, though far higher than expected, was quite moderate and, in any case, far from abnormal. There might, however, be some unknown mechanism hidden in the phenomena observed before the stroke of the catastrophic descent.

*

In such instances a closer look to the physically relevant circumstances observed between 1960 and 1963 is opportune. Possibly it may at least answer the question concerning the occurrence of *creeping without intervention of an unknown mechanism.* The topographic features of the western half-slide are particularly suited for an investigation of this kind.

As Fig. 2.45 shows, a longitudinal section corresponding to F–F' in Fig. 2.36 has, roughly speaking, a *chair-like shape* with a steep "back" and an almost horizontal "seat". This geometry (assumed already before the slide, s. Müller 1964, 169) allows for a separate quantitative analysis of the known parameters, the "back" (area A_1 in the figure) being the essential driving part, unaffected by the water in the storage basin, while the "seat" has in first instance to be considered as driven and consisting of a non-immersed portion (A_2) and an immersed one (A_3), according to the water's level. To simplify things, no special portion was assigned to the intermediate bend between "back" and "seat" (triangles in Fig. 2.45b–d). In case of submersion the error committed thereby is small as on the one hand friction, on the other hand drive is reduced.

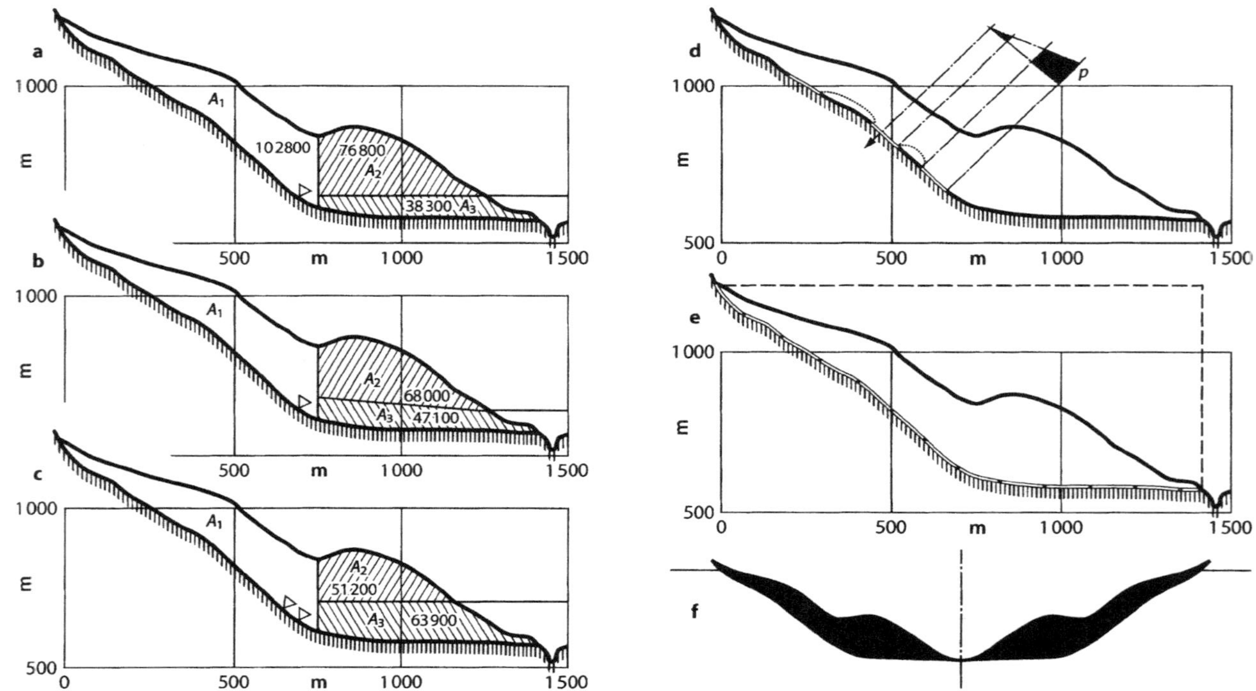

Fig. 2.45. Vaiont rockslide, water pressure acting on western half-slide (Fig. 2.36, Sect. *F–F'*). **a**, **b**, **c** Buoyancy of A_3 reduces resistance. A_1, A_2, A_3 are expressed in m^2. *Triangles:* zone of neglected buoyancy (s. text for explanation). **a** Basin and ground water at 650 m; **b** basin at 650 m, ground water plane oblique; **c** basin and ground water at 710 m. **d** water-filled voids (white portions in sliding surface) can increase drive by reducing friction. Central void, hydraulically connected (*dotted lines*) with two other voids, is partly active owing to local water escape (*arrow*); so top void is inactive, bottom void strongly active (pressure *p*; force, proportional to stained area, at right angles to surface). **e**, **f** "total" buoyancy (required supply/escape ratio is unrealistic). Nearly entire surface water-pressurised. For a moment mass might float. Resemblance with cross-section of ship-hull shown by mirror-imaging in **f** (water-lines indicate floating condition) (sketch by Erismann)

According to Coulomb's rule (for details s. Sect. 5.4), a body on an inclined plane undergoes a gravitational force component $mg\sin\beta$ in the direction of dip, antagonised by a frictional force $\mu mg\cos\beta$. Therein m is the mass of the body, g gravitational acceleration, β the slope angle, and μ the total coefficient of friction (no matter, whether acting between mass and bedrock or being due to internal resistance within the mass, s. Sect. 5.4). For the further analysis it is assumed that the section is representative for a volume of breadth B, large enough to be dominant as far as release of motion is considered. If the density of the material is ∂ and i the ordinal index (1,2,3) of the respective portion, the mass can be expressed by $A_i B \partial_i$ so that for portion i

$$F_i = A_i B \partial_i g(\sin\beta_i - \mu_i \cos\beta_i) \tag{2.3}$$

is the total force generated by A_i in the direction of dip. In order to be correct in adding up the three forces F_1, F_2, F_3 so as to obtain critical values of μ_i, several details of the compound mechanism need closer consideration.

31. There are *three coefficients of friction* in the system, μ_1 in the "back", μ_3 in the "seat", and μ_{13} in the bend between "back" and "seat". By no means are their values necessarily equal. For instance, there may be pressurised water in the joint between "back" and bedrock, a mechanism reducing the friction-generating component of weight and thus resulting in an apparent reduction of the coefficient μ_1. And the rock subjected to μ_{13} is more jointed than the rest of the mass.
32. In determining F_3 it has to be taken into account that, as a result of buoyancy (Fig. 2.45b–d), the acting density in a *submersed volume* is the difference $\Delta\partial = \partial_r - \partial_w$ between ∂_r and ∂_w, the respective densities of rock and water.
33. *Friction in the bend between "back" and "seat"* is generated by the forces required to deviate force F_1. Such forces exist even if the rock is jointed before the initiation of creep.

In a *differential portion of the bend* the angular difference $d\beta$ between input and output forces (F and $F - dF$ respectively) produces a differential transverse force $dF_t = F d\beta$ which, in its turn, gives rise to a frictional reduction of F_1 by $dF = -F_1 \mu d\beta$. The solution of this differential equation yields the force F_{13} acting as a drive upon A_2 and A_3. This force is

$$F_{13} = F_1 e^{(-\mu_{13}\Delta\beta)} \tag{2.4}$$

where $\Delta\beta = \beta_1 - \beta_3$ is the angle between "back" and "seat". Of course this angle has to be expressed in radian units (multiply by $\pi / 180 = 0.01745$ if angle is given in degrees).

As there are three coefficients of friction, a complete parametric presentation of the conditions of equilibrium would lack clarity. Hence the weight of each coefficient was assessed, and μ_{13} *was found of moderate influence* for $\Delta\beta = 33.5°$ as observed in Vaiont: within the range $0.3 < \mu_{13} < 0.5$ force F_{13} varies within about 12 percent ($0.7465 < F_{13} / F_1 < 0.8391$). So a constant value of $F_{13} = 0.7915 F_1$ was obtained by assuming $\mu_{13} = 0.4$ (rather a high figure for a predetermined sliding surface as encountered in the western half-slide), and an uncertainty of ±6% had to be tolerated for F_{13}.

After such preliminary considerations the *labile equilibrium condition* could be formulated with the variables μ_1 and μ_3 and taking into account Eq. 2.3, 2.4, and

$$F_{13} + F_2 + F_3 = 0 \tag{2.5}$$

as basic condition. The critical value μ_{3c} for a given value of μ_1,

$$\mu_{3c} = \tan\beta_3 + \frac{A_1 e^{-\mu_{13}\Delta\beta}}{A_2 + A_3 \Delta\partial / \partial_r} \frac{\sin\beta_1 - \mu_1 \cos\beta_1}{\cos\beta_3} \tag{2.6}$$

describes said equilibrium of the system between standstill and motion. With the values given for the western half-slide ($\beta_1 = 37.5°$; $\beta_3 = 4°$; $\Delta\beta = 33.5° = 0.5847$; $\Delta\partial / \partial_r = 0.6$) this lengthy equation obtains a handy form,

$$\mu_{3c} = 0.070 + A_1 \frac{0.483 - 0.629\mu_1}{A_2 - 0.6A_3}$$

and thus yields a clear answer to the question initially asked (Fig. 2.46, whereby the auxiliary lines marking the differences between μ_1 and μ_3 are helpful). For instance, it is immediately perceivable that, according to the water level in the basin, equal values of both coefficients would lie between 0.32 and 0.36. Now for the western half-slide the slope of the energy line was about 0.20 (Fig. 2.36d) and the estimated Fahrböschung 0.266. In such instances there is nothing exceptional in the fact that transition from creeping to sliding with a water level at 710 m took place with a coefficient of friction about 80 percent above that of the energy line (and it might even be asked why the catastrophe did not happen in an earlier stage). Like that of reach, also the question of *slow-to-fast transition needs no ad hoc mechanisms* and can be explained on the simple basis of Coulomb's rule.

*

Here it is, perhaps, opportune to have – by way of parenthesis – a closer look at the weak side of the most valuable source of information existing in connection with Vaiont. The respective deficiencies can be subsumed under a slogan: extraordinary mastery of a scientific field does not automatically include infallibility in the art of *transposing knowledge from qualitative to quantitative form* (i.e. formulation of algorithms). A striking example – besides the mentioned problem of velocity – is Müller's (1968, 82–83) somewhat cryptic equation for the energy balance in labile equilibrium: what has the integral of weight over the distance covered (in the used notation: $\partial Q\, ds$; in other words: the energy required to lift the mass vertically as high as it moves obliquely on the slope), to do with far smaller energies as involved in the sliding process? It appears hopeless to re-enact what really was meant. Probably – so far a guess seems founded – the main idea was to separate Coulombian friction between mass and bedrock from energy losses within the mass by deformation and fracture of its layers. If so, two points were neglected.

41. Post eventum knowledge, as far as obtained from the geometry of displacement (e.g. Fahrböschung, slope of energy line etc.), always refers to the *sum of both forms of energy dissipation* so that analogous considerations would, according to the par-

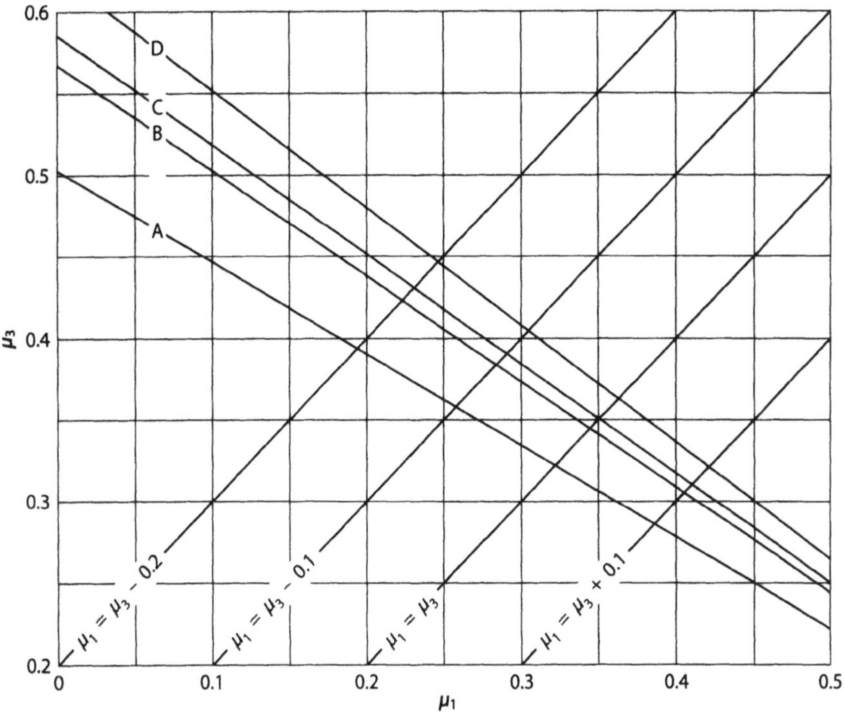

Fig. 2.46. Vaiont rockslide, coefficients of friction μ_1 ("back") and μ_3 ("seat") resulting in labile equilibrium according to Eq. 2.6. Coefficient μ_{13} (bend between "back" and "seat") is assumed to be 0.4. Curves *B*, *C*, *D* correspond to respective sections in Fig. 2.45. *A*: No buoyancy; *B*: basin and ground water at 650 m; *C*: basin at 650 m, ground water plane oblique; *D*: basin and ground water at 710 m. Note auxiliary lines allowing to correlate μ_1 and μ_3 (sketch by Erismann)

ticular circumstances, be applicable to any other rockslide used for comparison (and represented in Fig. 2.43).

42. It was assumed that the *mass was heavily fractured* (1964, 185); and this assumption appears justified if, in the course of creeping, the layers of rock first were delaminated and then fractured in the bend (s. Sect. 3.1). Where from, in such instances, could come a significant resistance of this material against deformation in the bend between "back" and "seat"?

Another example in this context is connected with the stop of the mass on the right bank of Vaiont Valley. On the one hand, maximum velocities of 25 to 30 m s^{-1} (1964, 179, 198, 1968, 83) are stipulated, on the other hand a run-up height of 140 m (1964, 188) is claimed. This statement might be understood as if there had been a true run-up, with corresponding vertical displacement of the centre of gravity, requiring (even under frictionless conditions) a velocity of at least 52 m s^{-1}. The real mechanism probably was a far less dramatic one: the layers of rock, separated beforehand by delamination, may have run out in relative displacement to each other as schematically shown in Fig. 2.20 for the case of Köfels.

Closing the parenthesis, it should be observed that the conclusions about the slow-to-fast transition do not refer to the creeping mechanism proper. It would not be opportune to extend this long section even more by going into all details of creeping. Nevertheless, Vaiont is remarkably well suited to demonstrate some important circumstances of rock-on-rock creep. Therefore several thoughts should be anticipated from the more careful analysis in Sect. 3.2. Most imaginable mechanisms, once launched, show a tendency to accelerate irreversibly. So the central problem is the modeling of a more or less *stable creep of rock on rock*. Coulomb's rule alone demands acceleration as soon as the increased friction of standstill is overcome, and single crack propagation as well as progressive failure, when occurring under constant force, necessarily lead into complete failure. So, in order to remain stable, rock-on-rock creep requires "auxiliary" means. And the most probable quality of such means is to act in a periodic (though not necessarily regular) manner. The mechanism suspected to have worked in Vaiont is the precipitation-dependent joint-water pressure.

To exemplify the *effect of joint-water pressure* in a drastic, though notoriously unrealistic manner, Fig. 2.45f shows the section of the western half-slide with so dense a system of voids that practically the entire bottom surface of both "back" and "seat" is contacted by water. If, in addition (and to bring unrealism to the limit), all voids were hydraulically interconnected and the water supply (on top) able to compensate for any existing leakage, the entire mass of the slide would be subjected to buoyancy. And as rock would fill less than 35 percent of the volume indicated by broken lines in the figure, the entire mass would become waterborne! Figure 2.45g makes this seemingly absurd statement plausible by mirror-imaging the section and thus giving it the aspect of a somewhat queer-looking cross-section of a ship-hull. In continuing this mental experiment and turning back to the single-sided version, one could ask what would happen to the mass if it remained coherent and undistorted in spite of having been lifted. The answer would be a trivial one. First of all, owing to the lateral component of pressure, especially at the "back", the mass would be lifted obliquely instead of vertically. And then, the passages available for escaping water having dramatically exceeded the capacity of any conceivable supply, the entire mass would fall back after a ridiculous little jump ...

What obviously is nothing but a fantastic mental game for a mass of 0.3 km^3, may perhaps have functioned locally on a drastically reduced scale. In various manners (Sect. 3.2; Hubbert and Ruby 1959; Ruby and Hubbert 1959 – once more a study available in 1963!) water pressure may have displaced comparatively small slabs, thus introducing seemingly insignificant changes in the distribution of forces. And local water pressure, the cause of each minute displacement, immediately would have been annihilated by the escape passages it had created in generating motion. In its vicinity, any such incremental event would have either reduced the ability to support overthrust weight, or it would have increased the weight supported by elements located below (Hutchinson and Bhandari 1971). Eventually, many small changes would have resulted in a perceptible displacement of the entire mass. So it becomes plausible to consider *creep as a step-by-step motion*, not as a continuous one, whereby mainly the "back" would have played the part of a driving motor according to the balance of water supply and escape.

In the steep slope of Monte Toc the supply of water to the "back" obviously was in first instance *a function of precipitation*. So it cannot be excluded that, while buoyancy reduced the resistance of the "seat", precipitation, by "pressure-lubricating" the "back", increased its push for the short duration preceding local pressure collapse. And it is perhaps more than pure coincidence that all three phases of creep in Fig. 2.37 not only are correlated with peaks of water level in the basin but that they regularly follow peaks of precipitation. In fact, a step-by-step calculation of the product of precipitation and water level (the former with 30 days time lag) yields impressive peaks correlated with the periods of creep (Fig. 2.37d).

In this context a remark should be made concerning *buoyancy in* 1960. The respective peak of creep velocity looks undersized in comparison with those of 1962 and 1963. Of course, this discrepancy may be explained by the random character of the assumed process of locomotion: creeping velocity must be considered as the result of a complicated interaction between numerous local parameters. It was, however, observed that in 1961 the ground water level sloped so as to hold at least 680 m in an appropriately located piezometric hole (Fig. 2.36, 2.37b) while the level of the basin was below 650 m. It is not excluded that this phenomenon (which disappeared in 1962, the level in the respective hole henceforth following more or less exactly that of the basin) already had existed in 1960 so that not alone the level of the basin but also that of the ground water had determined buoyancy. If so, the peaks for 1960 in Fig. 2.37 (b, d) would be higher, thus backing up the hypothesis of interaction of buoyancy in the "seat" and pressure lubrication in the "back".

*

In closing this section it is essential to point out *what has been proved and what not.*

51. On the one hand, it has been proved on the exclusive basis of *ante-eventum information* that neither creep, nor slow-to-fast transition, nor reach need exotic mechanisms to be explained.
52. In addition, it has been proved on the basis of *post-eventum information* that also fast motion as observed could have occurred without exotic mechanisms.
53. On the other hand, it has *not been proved* that none of the mechanisms proposed in the literature could have contributed to the final result.

However, it should perhaps be recalled that, as a general rule, an *explanation by mechanisms known* as being valid from a large multitude of observations has a better chance of being correct than one by mechanisms more or less created ad hoc.

The *literature* dealing with Vaiont is abundant, and here-above only the studies directly interesting in the present context have been mentioned. For digging deeper the following publications might be useful: De Nardi (1965); Ente Nazionale del'Energia Elettrica (1964); Kenney (1967); Kiersch (1964); Schnitter and Weber (1964); Selli and Trevisan (1964a, 1964b); Semenza (1966–1967). Unfortunately some important reports have remained unpublished, in particular the geological investigations by Dal Piaz (1948, 1953, 1960) and seismic tests by Caloi and Spadea (1960a, 1960b, 1961, 1962). To a certain extent, the results thereof are reflected in Müller's (1964, 1968) and Broili's (1967) studies.

2.7 Huascarán

In human history, if volcanic phenomena are excluded, no catastrophe due to the downhill motion of gravity-driven masses can be compared with the event of Huascarán. And it is definitely improbable that in prehistoric epochs one single mass displacement ever could have taken more than 18 000 lives. Where did exist, in those early times, a sufficiently dense population of homo sapiens to provide the victims for a disaster that took its origin at an elevation of, roughly speaking, 6 000 metres?

The involved mass, though by no means exorbitant in volume, moved, at least in certain phases, at an *unparalleled velocity*. And this fact gives the possibility of scientific investigation "in extremis", a field of particular interest from the physical point of view and, as can be seen hereafter, also seducing to speculation by the thrill of previously unknown dimensions.

Were it not for this particular fact, the mention of the Huascarán event might be considered as questionable in the frame of a book aimed at rockslides and rockfalls. It is with good reason that in the literature a reluctance to use the wording "Huascarán rockslide" can be observed (various combinations with "avalanche" prevail in English, in German also the word "Mure" has been used, pointing to a large amount of water being involved). What was it really? Rocks, of course, took part in the motion and probably constituted a major part of the mass; but there also was water (be it in liquid state, be it as ready-to-melt snow or ice) in sufficient quantity to form mud with rock powder and soil. So, looked at from the viewpoint of displaced material, the event was a *combined phenomenon* and its character is not easy to characterise by one single word.

On the other hand, the Huascarán event not only offers an excellent possibility to study mechanisms connected with high velocity: it also makes available certain clues for a quantitative approach. So there is a basis for investigating the *interaction between velocity and other phenomena* in a well-founded manner, and the mentioned reserve has to be disregarded in view of opportunities that exist only in this particular case.

Yet, before providing a short description serving as a background for further considerations, it has to be pointed out that the catastrophe of May 31, 1970, was not an event that changed once for ever a situation, thus excluding further danger. On the contrary, it was *preceded by at least two similar events* at the same location. One of them, so to say a reduced-scale forerunner, came down on January 10, 1962. In spite of being eclipsed in practically every respect eight years later, it has to be regarded as a heavy disaster of its own right; and if it is not discussed hereafter in more detail, this is exclusively due to the fact that the most interesting mechanisms were more clearly developed (and left more conclusive traces) in 1970. The deposits of the other (prehistoric or, at least, pre-Columbian) predecessor, however, spread wider and reached farther than those of both modern events (Plafker and Ericksen 1978, 310–311; Stadelmann 1983, 66–69)!

*

Motion took its origin at the west face of Nevado Huascarán Pico Norte (6 654 m, Fig. 2.1, 2.47, 2.48), the lower summit of the highest mountain in the Cordillera Blanca (Huascarán Pico Sur is 6 768 m high) where a hanging glacier reaches an extremely steep wall (about 60° over a difference in altitude of 700 m). The occurrence of more

Fig. 2.47. Huascarán event, panoramic view taken from the Cordillera Negra (south-west of Cordillera Blanca) soon after May 31, 1970. By comparison with Fig. 2.48 the phases of descent can be followed from top to bottom: Huascarán, with Pico Norte, 6 654 m, at left (Pico Sur, 6 768 m, at right), hanging glacier, and dark steep wall; snow-covered Glacier 511; "funnel"; wide area of giant bounces and splashes; "Shacsha Narrows" with marks of superelevation; ridge of Aira with bifurcation into two lobes: smaller Yungay lobe at left with the town hidden under debris and the cemetery hill separating two tongues; wide-fanned Matacoto lobe at right; Rio Santa Valley in the foreground; parts of Matacoto at the bottom right (by courtesy of W. Welsch)

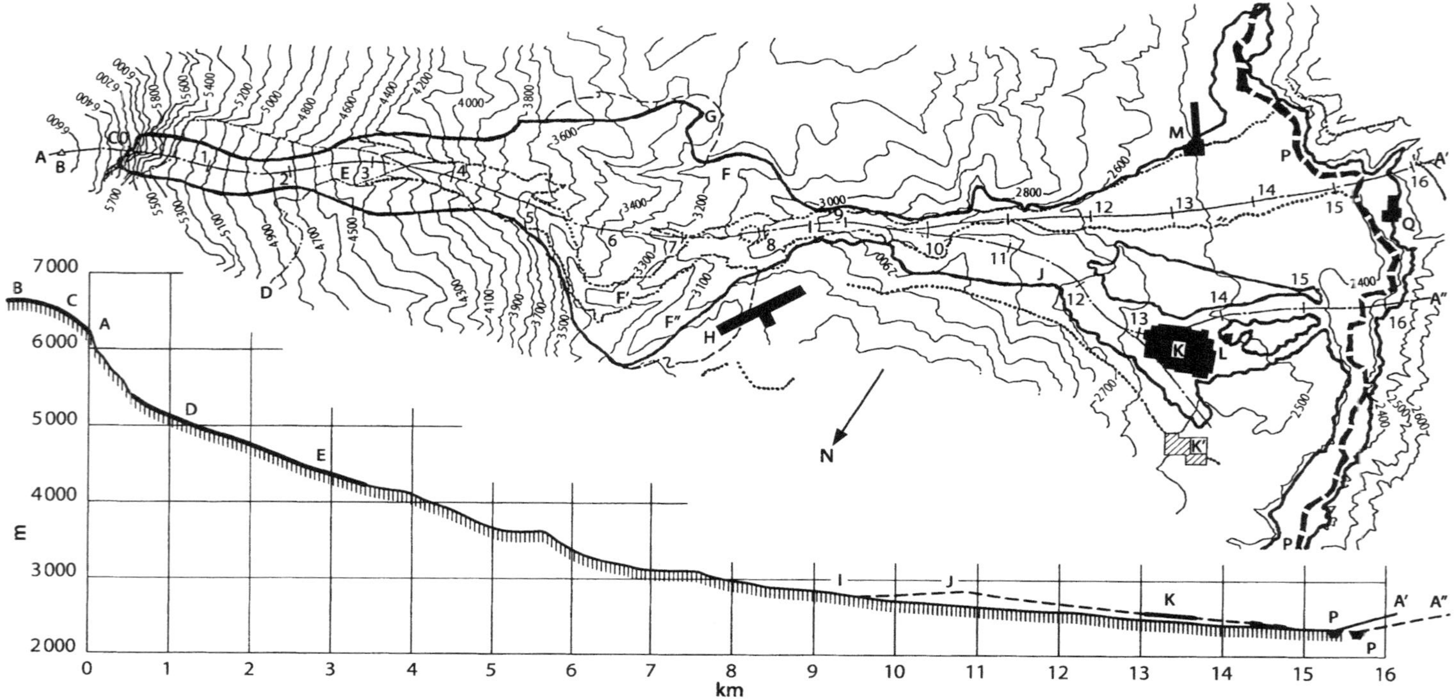

Fig. 2.48. Huascarán event, map and longitudinal sections (*A–A'*: main track and Matacoto lobe, *I–A"*: Yungay lobe). Scales of map and sections are equal. Lines in map: *plain:* periphery of event (1970); *broken:* Rio Santa, splashing zones; *dotted:* peripheries of previous events (1962 within, prehistoric beyond those of 1970); *dots and dashes:* sections. Lines in sections: *plain: A–A'*; *broken: I–A"*. *B:* Huascarán, Pico Norte, 6 654 m; *C:* hanging glacier; *D:* Glacier 511; *E:* "funnel"; *F, F', F"*: Quebradas: Incayoc, Armapampa, Shacsha; *G:* splashing zone; *H:* Huashau with splashing zone; *I:* "Shacsha Narrows"; *J:* Aira; *K, K'*: Yungay, Nueva Yungay (within prehistoric debris!); *L:* cemetery hill; *M:* Ranrahirca; *P:* Rio Santa; *Q:* Matacoto (sketch by Erismann)

Fig. 2.49. Huascarán event, headscar and steep wall. *Broken line* shows approximately the periphery of the displaced material. The photo, taken soon after the event, clearly illustrates the difficulties encountered in estimating the initial volume of the mass (by courtesy of W. Welsch)

or less regular icefalls is nothing but normal in such instances. More than that: the strata of the wall, dipping west and not free from approximately vertical joints, favour the casually *simultaneous release of ice and rock* (Fig. 2.49). Normally, however, such events remain on a comparatively small scale so that no harm is done to inhabitable regions. To generate the catastrophic event of 1970, a particularly powerful trigger was required, able to mobilise material that, in the ordinary course of things, would have been detached over a period of dozens or hundreds of years. This trigger was the heaviest *earthquake* known in the history of South America (s. Sect. 3.2; Körner 1983, 78–79).

The mass was released in a zone of *elevation between 5 600 and 6 200 m* approximately. These figures are a compromise between the results obtained, after careful observation, by Welsch (1983, 42–43, and map, scale 1 : 25 000) with a rather narrow range of elevations, and Plafker and Ericksen (1978, 284–285) with a wider one. The uncertainty thus signalled is due to the lack of photographs allowing for sufficiently accurate photogrammetric comparison of contours before and after the event: terrestrial pre-event photos, taken from Cordillera Negra, are handicapped by the shortness of the available basis. The same uncertainty is valid for the *volume of material released* at the start: for rock the estimations (in cubic kilometres) vary between 0.007–0.008 (Lliboutry 1975, 358) and 0.050–0.100 (Plafker et al. 1971, 550); the respective figures for ice are 0.001 (Lliboutry) and 0.005 (Plafker et al.).

If the accessible solid constituents of the debris are considered (according to Plafker and Ericksen 1978, 297–298, the material deposited was "*... mainly mud and boulders...*"; "*... boulders and angular blocks of rocks... appear to constitute 10–50% of the* *total volume...*"), an initial volume of 0.050 km^3, mainly consisting of rock, is more plausible than the idea of Ghiglino Antunez (1970, 87), suggesting a small mass at the start (rock and ice in a proportion of 0.005 to 0.009 km^3), later subjected to certain losses and a tremendous growth, finally resulting in 0.053 km^3 of rock, moraine, ice, snow, and soil. It is, in fact, not very probable that, in an event moving at a top velocity exceeding 100 m s^{-1}, a small initial mass might have been able to accelerate so large a secondary mass without a substantial loss of kinetic energy (i.e. velocity). In addition, Plafker et al. (1971, 553) showed that a moderate percentage of water was sufficient to form mud with the fine-grained material. So hereafter a *starting volume in the range of 0.050 km^3* will be assumed as a working hypothesis, not excluding a priori the possibility of a smaller (or even a larger) mass.

*

After having been detached from its original position, the mass descended along the mentioned wall as a *rock-and-ice fall*. If its centre of gravity (CoG) initially was at 5 900 m, and at 5 325 m when hitting the so-called "Glacier 511" (Fig. 2.48D), and if a horizontal travel of 400 m and a coefficient of friction of 0.25–0.5 are assumed, the velocity at the foot of the wall must have been something between 86 and 96.5 m s^{-1} (s. Sect. 5.3 and mark the characteristic of a very steep slope: moderate influence of friction upon the velocity attained). To complete the information: free fall over an equal differential altitude would have yielded 106 m s^{-1} (Eq. 2.1).

For the further development of motion the fact was essential that, after the almost vertical fall, the *transition to Glacier 511* was continuous enough to preserve a substantial part of the kinetic energy of the mass. Still the impact must have been a massive one: after the event, on the surface of the glacier concentric circular striations were observed, for the origin of which Welsch (1970, 42) offers two possible explanations, namely erosion by the moving debris or shock-waves having caused a plastic deformation of the ice. If the last-mentioned hypothesis is correct (a third one might be rebounding of debris after the impact), this would – besides the cloud seen and the bang heard by many witnesses – be strong evidence for a powerful impact. Anyhow, the main part of the mass slid down on the moderately steep glacier (average slope angle about 20°), and very probably under its weight ice and/or snow melted by friction, thus providing a self-lubrication in which the weak point of water as a lubricant, its fast escaping due to low viscosity (Sect. 5.5), was counterbalanced by immediate generation of new lubricant exactly at the locations of highest friction. The process was enhanced by the weather: it was a particularly hot day with temperatures in the range of +15 °C at an altitude of almost 4 000 m (Stadelmann 1983, 57). So it is probable that the three kilometres on Glacier 511 (Fig. 2.48, section A–A', km 0.6–3.4) resulted in acceleration and, perhaps, made up for the kinetic energy lost in the impact.

A seducing circumstance is given by the slight curve to the left that the mass described in passing over Glacier 511. Indeed, *curves may offer possibilities for determining velocity* as mentioned in the case of Pandemonium Creek (Sect. 2.2) and treated later in more detail (s. "Shacsha Narrows" hereafter; Sect. 5.7; McSaveney 1978, 227; Evans et al. 1989, 436–443). Körner (1983, 96–97) presented a calculation of this kind, based on the equation developed by McSaveney for a mass moving in a direction between strike and dip on an inclined plane surface. The result, if sufficiently reliable, would offer a useful hint to the energy loss suffered in hitting the glacier. Unfortunately, a closer inspection of the contour lines of Fig. 2.48 (and the map, scale 1 : 25 000, of the book containing Welsch's and Körner's articles) shows that the actual direction of displacement did not significantly deviate from that of dip so that Körner's approach, in spite of yielding a plausible result ($\geq$91 m s^{-1}), is revealed as not sufficiently well founded.

Anyhow, the motion on Glacier 511 not only must be regarded as one characterised by low resistance: the large amount of water (no matter how the respective snow had been melted, be it by impact, by friction, or – at least partially – by solar heat), necessarily swept along by the mass, in presence of fine-grained material obviously started the formation of mud. So it was on the glacier that parts of the mass assumed certain *properties normally attributed to mud flows*, and also in its further course low resistance could be expected.

*

The lowest tongue of Glacier 511 and the following track were confined by high and steep moraines forming a *funnel* (Fig. 2.50), wide at the entrance (almost 700 m at 4 500 m, km 2.5 in Fig. 2.48) and relatively narrow at the end (not even 300 m at 3 700 m, km 4.1). This funnel has given rise to the idea that it locally had accelerated the mass or parts thereof to a velocity exceeding 1 000 or even 1 150 km h^{-1} (almost 280 and 320 m s^{-1} respectively), the last-mentioned being approximately sonic (Plafker and Ericksen 1978, 304–305). To obtain a founded judgement concerning such an acceleration, with good reason described as puzzling by the cited authors (the kinetic energy at 300 m s^{-1} corresponds to that acquired in a free fall of almost 4 600 m in vacuum!), it is useful to discuss the possibilities of the Venturi effect.

When passing through a bi-conical tube, a quasi-incompressible gas or a liquid can reach in the locus of smallest cross-section a velocity exceeding by several times that at the wider entrance (Fig. 2.51). The extra energy required to materialise this phenomenon, the so-called *Venturi effect*, is taken from the pressure in the fluid. In other words: the pressure in the narrow section necessarily is lower than that at the entrance, and the fluid would be unable to overcome the surrounding pressure, if it were not re-pressurised in an appropriate manner after having passed trough the narrow section (otherwise the possibility of a first order perpetuum mobile would be demonstrated...). As the figure shows, in technical applications (increased velocity in wind tunnels and low-fall water turbines; increased power for driving gyroscopes in aircraft) this problem is mainly solved by a very gentle widening of the outlet section. The efficiency of such equipment depends to a high degree on the care applied in designing this "diffuser" (widening part of the device). The effect can be generated also in other than spherical sections, for instance between two wing-shaped bodies. But then it fades out near the wing tips and a reasonable acceleration only can be obtained if the "wing span" from root to tip is sufficiently large in comparison with the gap between the wings.

When *applying to the Huascarán event* these physically based and technically well-tried considerations, it becomes obvious that none of the enumerated conditions is fulfilled to a degree promising any hope for massive acceleration. First of all, the hydrostatic pressure in the material has to be compared with the energy required for acceleration, expressed in potential form, i.e. about 4 600 m of vertical travel. Now, on the one hand, from the start to the narrow passage at the end of the funnel the CoG of the mass had made a vertical travel not exceeding substantially 1 800 m so that – even in a perfectly frictionless ride – a deficiency of around 2 800 m would remain. On the

Fig. 2.50. Huascarán event, upper half (s. also Fig. 2.47). From top to bottom: traces in the almost vertical wall; funnel-like shape of the moraines below snow-covered Glacier 511; complex topography of the following area of (from the left to the right) Huashau, Quebrada Shacsha (half hidden), Quebrada Armapampa (with sharp curve at left), and Quebrada Incayoc; gorge of "Shacsha Narrows" with superelevation marks mainly on the right (orographically left) side; at the bottom, begin of bifurcation marked by excessive superelevation on the left and (less relevant) on the right (by courtesy of W. Welsch)

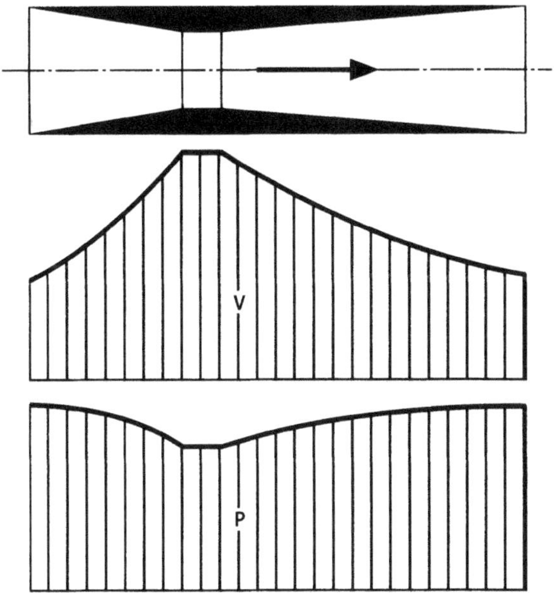

Fig. 2.51. Venturi pipe. In passing through the inlet conus (*arrow* shows direction), the moving fluid gains velocity v. The required energy is obtained by a reduction of the fluid's pressure p. In the gently widening outlet conus (diffuser) the inverse process (loss of velocity, gain of pressure) is accomplished with low energy loss. Efficiency depends on careful shaping (sketch by Erismann)

other hand, the hydrostatic pressure in the material cannot exceed the figure given by its thickness of, say, 100 m. The comparison of this moderate value with 2 800 m makes any comment superfluous: the water in the mass would have been evaporated long before a non-negligible acceleration had taken place. And even if the thickness had been substantially higher than assumed, the geometry would be far from meeting the conditions of a Venturi effect that might count: the lateral moraines are low in comparison with the distance in between, and not even a rudimentary diffuser can be observed. So there is only one possible conclusion: the Venturi effect, if any, was definitely moderate, and at the end of the funnel the *velocity was far from attaining that of free fall* (not to speak of equalling the velocity of sound). To conclude this line of thought, it should be noted that an accelerated motion in the funnel is not at all excluded: both the channelling effect (as discussed under Heading 5.1) and the above-mentioned consequences of ample water supply for lubrication and/or mud generation are strong arguments in favour of a low resistance. Still, in spite of remaining moderate, this resistance was opposed to acceleration and not increasing it...

*

The physically founded limitation thus resulting for the maximum velocity has more than a purely academic significance. It helps to answer the question of where the boulders came from which, after having crossed Quebrada (valley) Shacsha in *prodigious bounces*, bombarded the hamlet Huashau and killed an unknown number of persons. Perhaps it would be more correct to say that this limitation, by excluding certain take-off locations, answers the question where-from these boulders and the mud accompanying them could not have come: they did not come, as assumed by Plafker and Ericksen (1978, 288–292), from the moraines confining the lower part of the funnel on its north-western side. They simply had not the kinetic energy required for a bounce in the range of at least 3 500 metres. Readers desiring to dig deeper are referred to

Sect. 5.3 where they will find the necessary mathematical tools and the take-off velocities calculated for various presumable "jumping hills". In anticipation of such details it should be mentioned that Körner (1983, 96–99), in citing the study of Plafker and Ericksen, must have had a good (though unmentioned) reason to speak only of bouncing distances up to 1 000 m and a velocity of 119 m s^{-1}. If the remaining uncertainties are taken into account, these figures are in reasonable accordance with those presented in Sect. 5.3.

In considering the mechanisms of giant bounces a further interesting fact should be mentioned. According to the statements of witnesses, more or less regularly *a splash of mud preceded the landing of blocks*. From the mud-free "shadow" generated by obstacles of known height, Plafker and Ericksen (1978, 288; 291) found out that the angle of incidence of this mud was lower than that of the boulders (8–13° against 15–30°). The conclusion drawn therefrom was that the mud had accomplished remarkably flat trajectories at very high velocity. This conclusion cannot pass uncommented. In Sect. 5.3 it will be shown that boulders larger than a certain size (say, 1 m diameter) bounce in air almost as far as in vacuum. For small and relatively light particles like drops of mud this is not the case, and aerodynamic forces have to be taken into account. And as the motion of large blocks – especially if simultaneous bounces of several blocks in a dense cluster are considered – necessarily must have excited mighty gusts, also the effect of such phenomena has to be accounted for. Although a detailed analysis of this marginal mechanism would exceed the limits of the present book, it has to be pointed out that the splashing of mud essentially was not the result of self-contained bouncing but rather one of being dragged along by air which, in its turn, was mobilised by bouncing blocks. To quote a clue: in Fig. 5.10 trajectories of stones from 0.01 to 1.0 m diameter are compared. How far would a drop of mud get without external drive coming from a more powerful partner? And it is needless to say that, in the process of landing, a boulder may displace air both in the sense of preceding it and of straightening the apparent trajectory of a comparatively light object (Fig. 2.52).

In showing that bounces of several kilometres were excluded by physically stringent circumstances, the necessary "jumping hills" must be found below the funnel. So the attention inevitably is focused upon the *central portion of the track* (Fig. 2.48, km 4–8; Fig. 2.50) where the mass spread to the extraordinary width of approximately 4 km with splashing zones on both sides. As Fig. 5.10 shows, opportunities of launching blocks were present at so many locations that in one single straight section more than one thereof can be presented. This was due to the high volume displaced in 1970: the

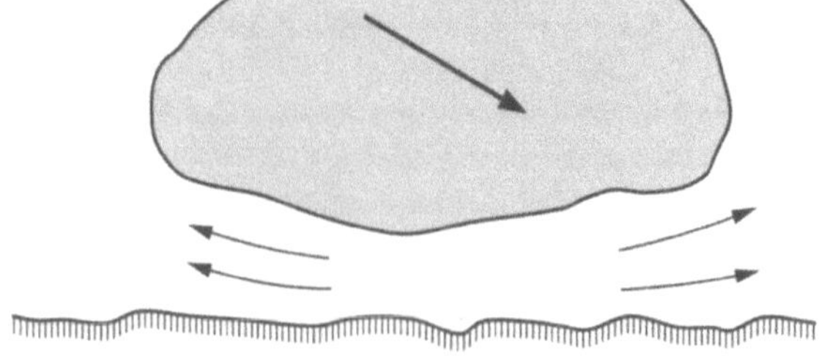

Fig. 2.52. Landing of boulder after giant bounce (*bold arrow* shows direction of displacement). Accompanying air ("bow wave" gust, boundary layer) containing fine drops of mud is blown away when the gap between boulder and ground dramatically decreases (*fine arrows* show approximate directions). Part of these particles necessarily overtake and thus precede the boulder (sketch by Erismann)

event of 1962 remained within the narrow limits of Quebrada Armapampa (Fig. 2.48F') while in 1970 the debris, overflowing the banks of this valley and the adjacent uneven area, eventually reached the slopes beyond the steeper Quebrada Incayoc (Fig. 2.48F). This fact throws some light on the general shape of the descending mass: if it had been 3 or 4 km long and accordingly slender when leaving the funnel, it might have been split up into an Armapampa and an Incayoc arm, but it not would have "flooded" an area of 7 or 8 km^2 (a rudimentary motion in this sense was arrested after some 400 m in 1962, s. Fig. 2.48, near km 5). Leaving aside considerations about dynamics, Quebrada Armapampa was sufficiently voluminous to accommodate almost 0.1 km^3 at once. So it is very probable that in 1970 the mass came down in a more bulky shape (this hypothesis is backed up by the overflow on the orographically right side of the funnel). Furthermore it is probable that still a large portion of the debris followed the course of Quebrada Armapampa and that, as a whole, a considerable percentage of the kinetic energy was lost, be it in the sharp right-angle curve of this valley, be it in the general consecution of ups and downs, be it, finally, by collisions of formerly separated portions and/or jamming before entering Quebrada Shacsha (Fig. 2.50).

*

This relatively narrow passage, in the following referred to as "Shacsha Narrows" (an unofficial denomination used exclusively in the present study), will turn out as being of particular importance: *here the fate of Yungay was sealed* by an intricate configuration of mechanisms which enabled part of the debris to overrun the ridge of Aira, thus bifurcating the mass into a smaller "Yungay lobe" and a larger "Matacoto lobe". Once more it will make sense, in view of such ill-fated importance, to anticipate some results of more systematic deductions presented later in this book (Sect. 5.7).

To yield its disastrous issue, a certain portion of the mass (i.e. the prospective Yungay lobe) had to be brought somehow into a markedly superelevated position with respect to the rest. The kinetic energy thereby spent had to be small enough to keep said portion able to overrun, in a second phase, the slope separating it from the crest of the ridge. So the very key to the subsequent phenomena lies in the topographic course of Shacsha Narrows. Slightly meandering, as it is, it necessarily implies *centrifugal forces* acting to and fro. As a consequence, the mass, forced into a motion superposed at right angles to the general course of descent, behaved, so to say, like a pendulum riding on a descending vehicle and suspended to oscillate laterally. Mechanisms of this kind may accomplish spectacular (and sometimes unexpected) motions if excited so as to boost amplitudes by the interaction between the succession of exciting forces and the natural frequency of the pendulum. More physical details and some easy-to-follow examples are presented in Sect. 5.7.

Quantitative analysis is, besides the difficulties enumerated hereafter, complicated by an annoying fact: *no reliable estimates of velocity* are available in the literature. There is, of course, Körner's remarkable pioneer study (1983, 97; 99) resulting in a value of 66 m s^{-1} near km 10.4. Körner actually was first to trace systematically all available possibilities of determining velocity and thus to insert a speedometric diagram into a set of comparatively reliable points of reference. This endeavour deserves high esteem despite the fact that, in a particular case, the chosen approach may have been questionable. In fact, to obtain a velocity-relevant difference of elevations at the end of Shacsha Narrows (Eq. 2.1), Körner subtracted the approximate altitude of the surface of the mass at km 10.4 (altitude 2 800 m, valley floor 2 700 m, s. Fig. 2.48) from

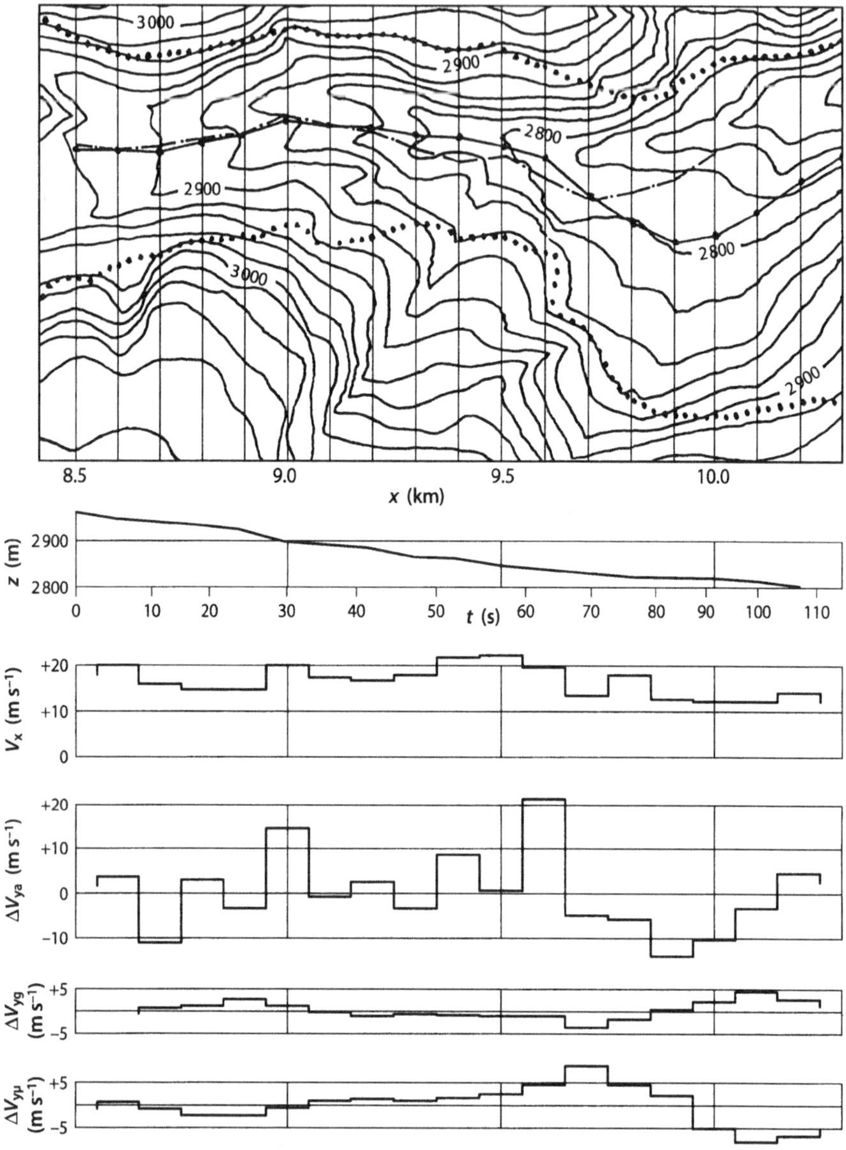

Fig. 2.53. Huascarán event, "Shacsha Narrows" (unofficial denomination used exclusively in this book). Enlarged map and, on equal longitudinal scale, main parameters. Map: *x:* distance in abscissa direction; *vertical grid:* cross-sections, partly shown in Fig. 2.54; *bold dotted lines:* periphery of debris; *plain polygonal line with dots at intersections with grid:* actual path of cross-section's centre of gravity (CoG); *corresponding dots-and-dashes line:* virtual "zero velocity" path of CoG (distance between the two CoG lines signals superelevations). Note excessive superelevation near km 10.0. Graph *z:* elevation of CoG (note practically continuous descent); scale *t:* time scale for 20 m s^{-1} basic velocity at km 9.0 (chosen extremely low with a good reason). Last four graphs are discontinuous owing to evaluation in cross-sections only. v_x: velocity (conditions as for *t* scale); Δv_{ya}: lateral velocity increment due to centrifugal force; Δv_{yg}: lateral velocity increment due to gravitation; $\Delta v_{y\mu}$: lateral velocity increment due to Coulombian friction. Note dominance of Δv_{ya} revealing necessity of further force (most probably internal resistance in mass) to obtain balance (sketch by Erismann)

that of the highest point reached by the north-western edge of the mass (2 945 m), increased by 70 m to account for a local velocity of 37 m s^{-1} (calculated from the geometry of a nearby bounce). Thus a difference of 215 m and the reported velocity were deduced. In doing so, various errors were committed: the surface of the mass (and even an edge thereof, situated on a slope) was considered as if it had been its CoG (Sect. 2.2, 6.2); the fact was disregarded that separation of the mass into two lobes was a rather complex process that had been initiated earlier than at km 10.4; the silent implication of equal physical conditions for the two lobes was made despite its questionable evidence; and, last but not least, the implications of relevant centrifugal forces were not taken into account.

Thus it was compulsive, for the present study, to consider *velocity as a free parameter*, and analysis was aimed at clues helping to understand how part of the mass was enabled to cross the ridge of Aira. Figures 2.53 and 2.54 illustrate the results thus obtained.

*

First of all, between km 8 500 and 10 500 *cross-sections* at distances of 100 m (Fig. 2.53, 2.54) were evaluated from a map (Welsch 1983) showing the traces of the slide and thus allowing to reconstruct superelevation, i.e. the difference in elevation between the ends of each cross-section. As a trivial result, the evaluation yielded sectional areas and thus relative variations of velocity, provided that the uncertainty of Item 1 hereafter is not too serious. Then the respective positions of the centres of gravity could be obtained both for the really observed (superelevated) cross-sections and their virtual zero-velocity counterparts with equal areas but horizontal surfaces (s. Fig. 2.54). As a compromise, the surface of the mass in the cross-sections was assumed as having been straight (for comments thereto s. hereafter and Sect. 5.7). Thus an approximate quantitative connection between superelevation and respective horizontal deviation of the CoG could be established.

Even with the reduced ambition of a non-simulating approach some *uncertainties* should not be dissimulated with respect to the assumptions made and the parameters used.

1. Both superelevation and velocity measurements are correct only if the observed borders of traces sufficiently well express the situation of maximum mass transport and if this maximum was sufficiently constant during the descent through Shacsha Narrows. Thus variations of mass transport, time shifts between the maxima on the right and the left border, and local phenomena like waves are neglected.
2. The assumption of a straight surface in superelevated state is contested by the considerations made under Heading 5.7. As a priori neither concave nor convex surfaces are excluded and only in extreme circumstances the resulting curvatures are considerable, the use of a straight line may be accepted as a first approximation. In connection with the cross-section at km 10.4 an exception will be discussed hereafter.
3. The motion of a quasi-pendulum would be well represented by cross-sections shaped like circular segments with continuous changes along the track and homogeneous distribution of density. Obviously these conditions cannot be fulfilled exactly. So determining the natural period (as required hereafter) necessarily is inaccurate.

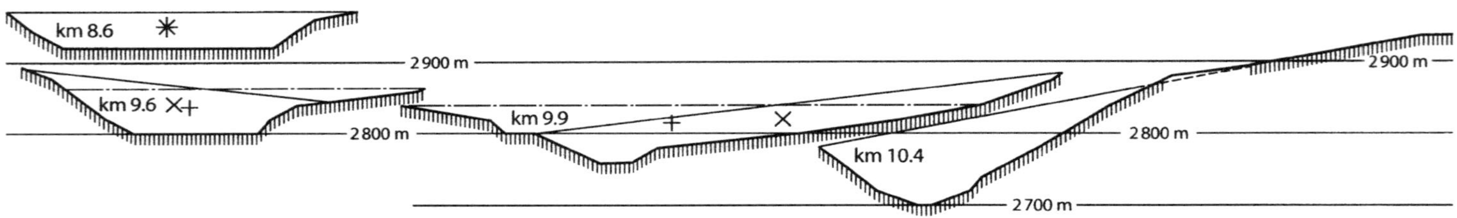

Fig. 2.54. Huascarán event, cross-sections at km 8.6, 9.6, 9.9, and 10.4 along "Shacsha Narrows". For locations, and comments s. text and Fig. 2.53. Reference lines are elevations. As far as differences are sufficient, both actual (superelevated) and "zero velocity" (horizontal) surfaces are shown (*plain lines* and *dots-and-dashes*, respectively). *Oblique crosses:* actual (superelevated) centres of gravity (CoG); *upright crosses:* virtual "zero velocity" CoG. Note continuous descent of CoG in spite of massive superelevations (sketch by Erismann)

4. The neighbourhood of a cross-section (so to say, a slice 100 m thick) certainly cannot be considered as an independent pendulum. In fact, it is damped not only by (Coulombian or other) friction against the ground but also by internal, so to say inter-sectional, coherence within the mass. In first instance such forces are damping the swing of the pendulum.

In view of such difficulties the following *quantitative analysis* will, as far as possible, be conducted so as to yield useful results even on the basis of roughly approximated figures. In addition, a simulating approach (i.e. the complete reconstruction of the motion's differential equation) will be replaced by an open-loop consideration of the main forces acting in a cross-section. This simplified approach is acceptable in view of the short step length of 100 m. Despite its obvious imperfections it will allow to draw important conclusions about the motion of the mass and, in particular, its capability to form the Yungay lobe.

Before going into further details of Fig. 2.53, a glance at Fig. 2.54 gives first complements to the map. Four *characteristic sections* are shown, including their positions in the grid of the map (km 8.6, 9.6, 9.9, 10.4). CoG locations of superelevated cross-sections are marked by oblique crosses and their surfaces by plain lines; the respective marks for the virtual zero-velocity cross-sections are upright crosses and dots and dashes. The figure is arranged to show directly the elevation of each cross-section (s. levels 2700, 2800, 2900 m; the horizontal arrangement is irrelevant). So it is immediately perceivable that a continuous descent of the CoG takes place, even between the slightly superelevated section at km 9.6 and the highly superelevated one at km 9.9.

As elevations are directly proportional to the respective potential energies, this statement bears an important *physical meaning*: the energy input by descending is, all in all, far higher than the – remarkably low – energy required to attain the highest observed superelevations (9.6 is actually the highest one on the orographically left side, 9.9 belongs to the extreme right-side displacement of the CoG). The important qualitative message thereby is that in the highly energetic process of descending even a small irregularity may suffice to evoke a substantial transversal motion. By the way, the mentioned continuity of descent becomes evident also in the diagram z (i.e. CoG elevation) of Fig. 2.53: only a ridiculously low run-up of 2 m exists between km 9.7 and km 9.8.

*

The cross-section at km 10.4, located slightly beyond the area shown in the map of Fig. 2.53, has a particular importance of its own. It makes directly visible the irreversible *separation between the two lobes* of the event. The long, moderately steep slope on its right side, ending in a short horizontal portion, belongs to the ridge of Aira. And the entire extension of this impressing "outrigger", more than 350 m in the horizontal and more than 50 in the vertical, is covered by traces of having been overrun. The straight line between the lateral ends of such traces, however, passes below the shoulder that marks the transition from steep to moderate slope! Two conclusions are suggested by this fact. On the one hand, in this extreme situation assumption of Item 2 no longer can be sustained: a convex surface has to be assumed, signalling rather a constant velocity than a constant angular velocity within the cross-section (for details s. Sect. 5.7). On the other hand, the mentioned shoulder materially divides the mass into two parts so that the germ of irreversible bifurcation becomes manifest. In such instances it makes little sense to consider the mass as an integral whole even at a somewhat earlier stage. So the map was cut at km 10.3 and the zero-velocity CoG path at km 10.0.

*

On the same longitudinal scale as the map, Fig. 2.53 shows, between km 8.6 and 10.2, some *easily accessible parameters.*

On the x-axis of the already-mentioned diagram of CoG elevation z a *scale of time t* is attached. It is valid for the neglect of uncertainties 1 and 2 and a "basic" velocity of 20 m s^{-1} through the cross-section at km 9.0. The reason for the choice of so low a standard will emerge soon.

The next diagram shows the *velocity v_x* in the direction of x. For reasons of uniformity with the parameters following below it is represented as if the value determined in a given section were constant over the respective slice (as defined here-above). In this context it should be observed that the direction of x and those of actual CoG displacements, though approximately identical over the total length of the map, locally may differ, especially in the large curve beginning at km 9.6. Accordingly, the resulting local velocities are slightly higher (with a maximum of 27% between km 9.6 and 9.7).

The last diagrams of the figure show, in principle, three important *forces acting upon a slice* of 100 m thickness. Each force, assumed to be constant within a step of 100 m length, is multiplied by the time required to cover this length (thus obtaining the respective momentum) and divided by the mass of the slice. The final result assumes the dimension m s^{-1} and gives the increment of transversal velocity Δv_y induced to the slice by the considered force. This result, as being calculated in an open loop, does not account for the feedback upon the exciting force. The neglected error however, is necessarily small as the value taken from the respective cross-section cannot dramatically differ from a figure correctly averaged over a distance of 100 m. This has to do with the fact that the exact solution of the differential equation is expressed in the (known) path of the CoG.

Δv_{ya} is the transversal velocity *increment induced by centrifugal force* (key word: "acceleration"). Again a basic velocity $v_{x9} = 20$ m s^{-1} is assumed in the cross-section at km 9.0. Of course, the centrifugal force has to be compatible with the well-known relation v^2 / R. As in the map a polygonal representation of the CoG path is used, the question might arise, how the theoretically infinitesimal radius R can be dealt with. The answer is simple: the acceleration in the corner is infinite, but its time of action is infinitely short. And as both are inversely proportional to each other, their product (and therewith the resulting momentum) turns out to be constant so that the momentum only depends on mass, velocity, and differential angle. In other words, the effect equals that of a perfectly elastic collision with an infinite mass in adequate position. The influence of v_{x9} upon Δv_y is easy to determine: on the one hand, the centrifugal force is proportional to the square of velocity, on the other hand the time of action is inversely proportional to velocity. So Δv_{ya} is proportional to v_{x9}.

Δv_{yg} is the transversal velocity *increment induced by gravitation* in the superelevated state. The determination is essentially trivial. Only the fact has to be accounted for that the slice, simultaneously with its pendulating motion, descends in x-direction whereby part of the potential energy of descent is converted into kinetic energy of the pendulum. The respective trigonometric function is given by a sin(atan) relation.

The sign simply depends on the cross-section's CoG being on the right or the left side with respect to its deepest (i.e. the zero-velocity) position. Obviously the force exerted is independent of velocity, but the time of its action is inversely proportional thereto. So Δv_{yg} is inversely proportional to v_{x9}.

Finally $\Delta v_{y\mu}$ is the transversal velocity *increment induced by friction* between mass and ground. To make things simple, Coulombian friction is assumed so that, once more, determination is easy. A low coefficient of friction $\mu = 0.1$ was suggested by the long travel on a track of approximately equal slope: after having lost much of its kinetic energy between km 5 and km 8, the mass, with a higher resistance, would have been arrested rather quickly on so soft a slope. As friction obviously is opposed to the vector of total velocity, once more a sin(atan) function is involved (for details s. Sect. 5.7). And once more, the force being independent of velocity, while its time of action is proportional thereto, $\Delta v_{y\mu}$ is inversely proportional to v_{x9}.

In considering the three diagrams (with a side-glance to the relations of the respective effects with velocity) the *choice of the basic velocity* – $v_{x9} = 20$ m s^{-1} is below the realistic range – becomes plausible. With, say, 40 m s^{-1} the dominance of centrifugal forces would be increased by a factor of four (Δv_{ya} being doubled, Δv_{yg} and $\Delta v_{y\mu}$ halved), and the two smaller increments almost would require a magnifying glass to be discerned...

Now the physical content of the present discussion certainly is not exhausted by the mere comparison of three forces acting upon a single body. The real goal consists in finding a realistic configuration granting a permanent *balance of external and inertial forces*. And a look at the three diagrams makes it evident that gravitational and frictional forces show similar profiles and thus easily might be combined to an almost perfectly balanced sum. As a first approximation, it would suffice to multiply μ by 0.5 (not a priori unrealistic, s. Sect. 2.2) to obtain a result nowhere exceeding a value of 1.27 m s^{-1}, with a mean square root of 0.66 m s^{-1}. The profile of the inertial force, however, is so different from the two others that no adjustment of the two free parameters v_{x9} and μ can yield a reasonably satisfactory result. An exception would, of course, be the brutal suppression of inertial forces by reducing v_{x9} below 10 m s^{-1}, a completely unacceptable proposition.

The dilemma probably would be solved by taking into account the uncertainty of Item 4. The three forces so far considered are certainly an incomplete set, and an effective counterpart to centrifugal force could be given by the existence of an inter-slices (or better: mass-internal) *resistance against distortions*, i.e. lateral relative displacements within the mass. Even a speculation on the consistence of the mass might make sense: couldn't the amount of water – at least in the most bulky portion of the moving mass – have been lower than might be assumed at first sight? It is not intended to push speculation further without additional knowledge about the involved parameters.

*

A last fundamental argument refers to the possibility of energy restitution or – to speak in terms of oscillating systems – *resonance*. A first hint into this direction was given by Jätzold (1971, 113, s. also Stadelmann 1983, 62) who proposed, on the basis of aerial photos, an effect of restitution resulting in boosting motion from the left to the right slope of Shacsha Valley. A certain additional effect – increasing or reducing oscillation – might come from the curve at km 9.0. Increase, however, only can occur in such a configuration if the driving force of the second curve (at km 9.6) acts while the oscillation due to the first curve (at km 9.0) moves in the direction of said force. In other words: as far as the two force pulses are separated by less than an entire period, the time elapsing between them must be below one half of the natural period of the system, the optimum lying at one quarter. In a cross-section the approximate length of the pendulum can be determined from the horizontal displacement of the CoG and the angle of superelevation. For example, between km 9.0 and 9.6 it ranges from 230 to 380 m, yielding an average period of 35.4 s. Accordingly, to obtain resonance, a time below 18 s should elapse in covering 600 m. Accounting for the varying cross-sections a basic velocity v_{x9} of at least 34 m s^{-1} would offer resonance. So a value around 40 m s^{-1}, as postulated later in this book (Sect. 5.7), might make sense.

Thus the basis for the last *assault upon the ridge of Aira* really was optimal: the right wing of the mass, by favourable circumstances (including permanent energy input and resonance) highly superelevated and no longer linked to the rest of the mass, raced almost frontally against the ridge (s. right bottom corner of map), and its velocity, even if it did not exceed 40 m s^{-1}, was amply sufficient for a last climb of 20 or 30 m.

*

It is considered useful to *summarise the results* of this lengthy qualitative, but quantitatively based excursion. Five short statements will suffice. The uncertainties (1.–4.) are not repeated at this occasion but should be kept in mind.

11. Probably in Shacsha Narrows the descending velocity was lower than hitherto assumed: about 40 m s^{-1} or somewhat more instead of 65 m s^{-1}.
12. Owing to a surprisingly low demand for energy, substantial superelevations may be fed by minute fractions of the energy of descent.
13. At least in certain superelevations the surface of cross-sections apparently was convex, thus pointing to low inequalities of velocity (Sect. 5.7).
14. Separation in two lobes mainly was due to a high superelevation, boosted by energy from the preceding opposite superelevation, very probably in its turn boosted by resonance.
15. The balance of forces due to gravity, friction, and inertia being imperfect, a substantial contribution from mass distortions is highly probable.

*

For readers interested in the *physico-mathematical background* of the above considerations the equations are given hereafter – without further explanations in view of their quasi-triviality and the fact that some of them are commented under Heading 5.7.

The following notation is used: x_i, y_i, z_i: co-ordinates of CoG in superelevated cross-section i (Fig. 2.53, 2.54); $x_{i-1}, y_{i-1}, z_{i-1}, x_{i+1}, y_{i+1}, z_{i+1}$: co-ordinates of CoG in cross-sections preceding and following i; y_{i0}, z_{i0}: co-ordinates of CoG in zero-velocity cross-section i; A_i: area of cross-section i; A_9: reference area of cross-section at km 9.0; T_i: natural period of cross-section i; Δx: distance between cross-sections (100 m). The abbreviations $\Delta z_{i-} = z_i - z_{i-1}, \Delta z_{i+} = z_{i+1} - z_i, \Delta y_{i-} = y_i - y_{i-1}, \Delta y_{i+} = y_{i+1} - y_i, \Delta y_{i0} = y_i - y_{i0}$, $\Delta z_{i0} = z_i - z_{i0}$ serve to simplify the equations. For cross-section i the following values are obtained:

- velocity in x-direction:

$$v_{xi} = v_{x9} A_9 / A_i \tag{2.7}$$

- lateral velocity increment due to centrifugal force:

$$\Delta v_{yai} = v_{xi} \frac{\Delta y_{i-} - \Delta y_{i+}}{\Delta x_i} \tag{2.8}$$

- lateral velocity increment due to gravitation on slope:

$$\Delta v_{ygi} = \left(\Delta z_{i-} \sin \operatorname{atan} \frac{\Delta y_{i-}}{\Delta x} + \Delta z_{i+} \sin \operatorname{atan} \frac{\Delta y_{i+}}{\Delta x} \right) \frac{g}{v_{xi}} \tag{2.9}$$

- lateral velocity increment due to friction:

$$\Delta v_{y\mu i} = \left(\sin \operatorname{atan} \frac{\Delta y_{i-}}{\Delta x} + \sin \operatorname{atan} \frac{\Delta y_{i+}}{\Delta x} \right) \frac{g \mu \Delta x}{v_{xi}} \tag{2.10}$$

- natural period of oscillation:

$$T_i = 2\pi \sqrt{\frac{\Delta y_{i0}}{g \sin \operatorname{atan} \frac{\Delta z_{i0}}{\Delta y_{i0}}}} \tag{2.11}$$

In Eq. 2.8, 2.9, and 2.10 due attention should be given to the signs which may be somewhat confusing in certain cases. In Eq. 2.11 the influence of descending is disregarded as being not very important.

*

Anyhow, in 1970 the *Yungay lobe* was able to overrun the ridge which had protected the town in 1962. It raced down, covering at first most of the hamlet Aira and then a substantial part of the slightly sloping surroundings of Yungay, including its entire centre. Finally, the foremost elements, in two relatively narrow tongues, were joined by the far larger mass of the Matocoto lobe in the valley of Rio Santa. The death toll was appalling: in a few seconds almost the entire population (more than 15 000 inhabitants, an overwhelming majority of the 18 000 victims as estimated for the entire event) and an unknown number of market-day visitors were extinguished.

Nevertheless, there was not the absolute hopelessness as in the nightmare of Vaiont. *Escape was possible* for those who, alert by the earthquake, realised the approaching danger and, finding themselves in a favourable position, ran in the right direction. An acoustic alarm was given by the beginning descent: immediately after the end of the earthquake in Yungay a roar was heard from the Huascarán. This fact makes it probable that the main part of the mass was released by the first seismic waves: on the one hand, the earthquake lasted for 45 seconds; on the other hand, the quasi-vertical fall of the mass and the transmission to Yungay of the sound emitted in hitting Glacier 511 took an approximately equal space of time. This signal came early enough to give time for salvation in at least 93 cases (it is not excluded that further people could escape from other locations, running in other directions). This is the number of persons which survived on the cemetery hill, a kind of stepped pyramid (Fig. 2.55) which, like the bow of a ship, divided the Yungay lobe into the mentioned two tongues. The resulting bow-wave attained more than half its height.

From the scientific viewpoint it was good luck that two of the survivors were geoscientists: Casaverde and Patzelt (attention: the second-mentioned is a namesake of the geographer who edited the book containing the repeatedly quoted articles by Körner, Stadelmann and Welsch). The following *report of salvation* was given by Casaverde to Stadelmann (1983, 62–63): "*... The Patzelt couple and I were just leaving for Huaraz when the earthquake began. Our pickup truck several times was pushed vertically into the air by the shocks and it was difficult to keep it under control... along the street adobe houses were breaking down. Immediately after the end of the seismic waves we heard a deep roar from the Huascarán. In looking up we saw that the summit was hidden by a grey, rapidly darkening cloud. We realised that an avalanche was approaching. We immediately jumped down from the truck and ran to the cemetery hill, at a distance of 150 m. Other people also became aware of the danger, but they first searched the ruins for survivors or precious objects instead of saving their lives. The hill has five terraces containing arrays of niche graves. When reaching the first terrace, I looked back and saw the grey crest of a wave, about 50 m high, slopping over Cerro de Aira and rushing down to the town. No sooner had we, together with another couple and three children, reached the third terrace – in doing so we were bothered by an intense gust of wind –, as the avalanche surged with a bang against the third terrace, at a distance of hardly five metres from us. The ear-splitting noise was followed by a deathly silence...*" (translation by Erismann).

Fig. 2.55. Huascarán event, central part of Yungay lobe. In the left foreground the cemetery hill where 93 persons survived. The cluster of dots near the long road at right are remnants of the town's main square, including the tower of the church and the palms mentioned in the text. The mass moved, roughly speaking, from the right to the left (by courtesy of W. Welsch)

At first sight this report might appear useful as a means for a rather accurate estimation of the *average velocity* of the descent between the impact upon Glacier 511 and the moment of Yungay being engulfed. Yet in digging somewhat deeper one necessarily is confronted with the fact that only three circumstances can be quantified on a solid basis. The (21) time lag between the generation of a sound near the area of release and its being perceivable near Yungay was about 40 seconds. As Casaverde ascertained by repeating it, the (22) rush from the truck to the cemetery hill took another 100 seconds. And (23) the duration of the earthquake was approximately 45 seconds. All the rest must be guessed. Here some examples: Immediately after the end of the earthquake the scientists heard the roar. Is "immediately" within the next second or after 10 seconds? Came the roar from km 0.5 where the mass hit Glacier 511 or from the bouncing-and-splashing area around km 6? Had release taken place at one of the first strong waves (a highly probable but unproved possibility) or later? How much time elapsed in grasping the situation, stopping the truck, and leaving it? How many seconds were required to attain the third terrace of the hill in spite of being hindered by tremendous gusts of wind (note that during the climb from the first to the third terrace the distal end of the mass descended over a distance of more than 3 km from the ridge of Aira to the town)? Similar uncertainties are reflected in Stadelmann's study (1983, 64): a most instructive table shows estimations made by various authors on the basis of reports from witnesses. For the average velocity the reported figures vary between 250 and 470 km h^{-1} (69 and 131 m s^{-1}). Of course, the problem lies in the fact that even an eyewitness in a safe position hardly had the nerve to observe so dramatic a phenomenon quietly, with the stopwatch in hand (if at all he had a watch). So the only reasonable estimation can be given for the absolute minimum of duration referring to the displacement of the mass between hitting Glacier 511 (km 0.5) and appearing at the ridge of Aira (km 10.5):

40 s	(sound transfer from km 0.5 to Yungay)
+ 100 s	(rush from truck to cemetery hill)
+ 10 s	(stopping truck etc. and climbing first terrace)
= 150 s	

So the average velocity over the considered 10 kilometres of descent cannot have exceeded 67 m s^{-1} and probably was even lower. This fact implies a first additional argument in favour of a comparatively low velocity in and below Shacsha Narrows: if, for instance, velocity was braked down to 40 m s^{-1} as early as at km 7.5 (suggesting a distance of 3 km covered in 75 s), at least another 75 s remained for the upper 7 km, resulting in a high, though plausible average of 93 m s^{-1}.

Figure 2.55 faintly shows a further fact pointing into the same direction. Several palms on the main square of Yungay, in spite of having been engulfed with the surrounding buildings, remained upright after the catastrophe. The same is true for the tower of the nearby church. It is easy to estimate the hydrodynamic force acting upon such a building (s. Eq. 3.5). For an immersed sectional area $A = 5 \times 5 = 25$ m^2, a plausible coefficient $c = 1$, a density of the mass $\partial_l = 1250$ kg per m^3, and a velocity $v = 40$ m s^{-1} a force of 25 MN results, more than an entirely massive tower of said dimensions can weigh (about 15 MN). In such instances it is perfectly clear that, when overrunning Yungay, the mass moved at a *velocity below 40 m* s^{-1}. It has to be remarked that at least one of the above parameters deliberately was chosen at the lowest limit of plausibility, namely a density of the mass exceeding that of pure water only by 25 percent.

*

Meanwhile the *Matacoto lobe*, doing far less immediate harm (mainly in Ranrahirca, Fig. 2.48), formed an impressing fan before crossing Rio Santa and then had still the energy for a short run-up on its left side (Fig. 2.56), resulting in the death of 60 persons and the destruction of several houses in Matacoto. Without going into the details of this run-up (Plafker and Ericksen 1978, 297; Welsch 1983, 45), a short comment should be given to Körner's (1983, 99–100) *estimation of velocity* established on the basis of Eq. 2.1. As observed – in somewhat other terms – under the Headings 2.5 and 2.6, overestimation in first instance is due to the fact that distal elements often are pushed forward from behind far beyond the limits given by their own kinetic energy. There is no doubt that exactly this was the case in the considered run-up. Körner, in using the estimated difference between the surface of the approaching mass and the highest level attained, unconsciously made at least part of the necessary correction. Thus the figure obtained – 28 m s^{-1} –, according to the actual thickness of the mass, may be correct or slightly too high. Also this result fits into the idea of an essentially reduced velocity in the second half of the descent (the bare difference of 80 m between

Fig. 2.56. Huascarán event, traces of run-up on the left side of Rio Santa valley, view from the north-east. The highest points are slightly more than 80 m above the original valley floor. In the background the Cordillera Negra (by courtesy of W. Welsch)

Rio Santa and the highest points attained would yield at least 40 m s^{-1}, which would leave unanswered the question of the unavoidable deceleration in the rather flat fan).

The relatively low destructive effect of the Matacoto lobe should not be misinterpreted as being due to an immanent lack of destructive power. On the contrary, little failed, and a second, perhaps even more disastrous catastrophe would have followed that of Yungay. In fact, the mass not only formed the mentioned fan covering more than 4 km^2 with a layer about 10 m thick (after drying): it also *dammed Rio Santa*. And as the dam to a certain part consisted of mud and ice, it could not withstand the pressure of the growing lake for more than half an hour. Fortunately, an unmistakable signal preceded the inevitable rapid inundation: abruptly, the bed of the river dried out. So the inhabitants were warned and most of them could move to safe positions. The destruction over 150 kilometres – from Yungay to the sea – was immense: no bridge and no irrigation plant remained in working order. But the death toll was relatively moderate.

So the Matacoto lobe, in spite of having contributed many hundreds of victims to the catastrophic result (no accurate figures can be presented owing to the particular circumstances of the event), turns out as being of minor importance when looked at from an anthropocentric point of view.

2.8 Synopsis of Results

In the following the most essential information of the present chapter is summarised. For practical reasons the respective items are not presented in the order dictated by the succession of Sect. 2.2 to 2.7 but arrayed so as to comply with the *chronological course of an event*, beginning with the situation before release and ending with rest-hazard management post eventum.

Despite the care used in selecting the key events, it is needless to say that the following items are far from giving a complete enumeration of geomorphological aspects connected with rockslides and rockfalls. In the first instance, they are intended to serve as a mnemonic aid for *questions relevant in the frame of this book:* dynamics of displacement; to a certain extent mechanisms of release; and, as an important practical aspect, human activity in salvation, prevention and mitigation.

*

Ante-eventum items, as not being among the main subjects of the case histories, are not very numerous.

1. New information about the mechanisms of an event can influence the knowledge concerning its *geological past.* Example: in the case of Köfels, the frictionite hypothesis effectively contributed to replace previous hypotheses by setting up a concept that improves compatibility with geomorphological evidence (s. Item 33 hereafter).
2. For prediction, information about previous *similar phenomena at the same location* may be helpful. Example: Nueva Yungay, the town replacing the one engulfed by the Huascarán event, stands on potentially unsafe ground, partly within the deposits of a prehistoric catastrophe. And the situation in the zone of release points to future repetitions.
3. In uncertain ante-eventum situations *relevant publications* should be consulted with due care. In spite of not presenting the latest discoveries, old references may be of great use. Example: perhaps the tragic issue of Vaiont could have been prevented if this fact had been given heed to (s. Item 15).
4. *Creep of varying velocity* does not necessarily prove the existence of visco-plastic layers below the creeping mass. Example: erroneous assumption of experts before the event of Vaiont. And creep-to-rush transition can occur very quickly even on such layers. Example: Goldau (s. Items 3, 5).
5. *Rock-on-rock creep* is probably rather a sequence of increments – caused by at least one varying parameter – than a continuous process. Example: Vaiont where two such parameters are suggested, namely level-dependent buoyancy and precipitation-dependent joint-water pressure.
6. Release of masses is often correlated with heavy *precipitation.* Joint water pressure seems to be most important in this context. Examples: Vaiont; Val Pola and many smaller events in the same region.
7. It goes without special mention that *earthquakes* may be efficient triggers of rockslides and rockfalls and that their energy can influence the volume of the displaced mass. Example: Huascarán.

Once a mass is released, phenomena influencing its motion can, in principle, occur at any point of the track. In four of the six key events, however, an important phenomenon took place rather early: massive disintegration of the mass. In the context of this book some characteristics of *disintegrated motion* are of particular interest.

8. Only in the slide of Vaiont most of the mass accomplished its travel in an essentially *coherent state*. Köfels did as well, until the lower part of the mass, arrested by a solid wall of rock, was disintegrated.
9. Huascarán has to be considered as a special case. Early disintegration and plenty of water (initially ice and snow) favoured the *generation of mud*. So the event only to a certain extent was a slide.
10. Blackhawk and the disintegrated part of Köfels show the widespread phenomenon of moderate relative displacement between the particles, resulting in a well-maintained *sequential order* of components and a gradient of velocity concentrated near (or in) the sliding surface.
11. In two cases accessible sections through the zone of sliding (primary for Langtang, secondary for Köfels) display a remarkable *straightness of the sliding surfaces* in the direction of displacement, both on solid or on disintegrated ground. Similar observations are reported from other accessible sliding surfaces (e.g. Flims).
12. In many disintegrated rockslide deposits (e.g. Blackhawk) a coarse-on-top *gradation* is observed. If an adequate amount of water is present, low permeability results at the bottom of the mass, the passages between pieces of rock being clogged by compacted mud.

*

Velocimetry (s. Sect. 6.2) at a given point may be as difficult as it is important (the destructive capacity is proportional to the square of velocity!). Besides the methods listed hereafter, there is the calculation of velocity from the trajectory of a bouncing boulder (s. Item 22 and Sect. 5.3).

13. The simplest method of velocimetry uses the *time taken to cover a known distance*. Eye-witnesses (Huascarán, Elm) sometimes are handicapped by being obliged to run for their lives. Reconstruction seldom is feasible without partial estimation. Seismographic records (Val Pola) are difficult to correlate with a given position of the mass.
14. Velocimetry based on run-ups is a widespread source of overestimation due to the use of wrong reference altitudes (e.g. top of scar/floor of valley/top of run-up) instead of the respective *positions of the centre of gravity* of a mass (Pandemonium Creek, Val Pola, the latter with the claim for a maximum of 106 m s^{-1} instead of 73 m s^{-1}).
15. Normally the correlation between *velocity and destructive power* is positive. For Vaiont an underestimated prediction of velocity excluded the idea of a flood wave, thus paving the way for a catastrophe.
16. In deducing velocity from *superelevation* on a curved path, the inaccuracy due to dynamic pendulating motion must be remembered, especially in case of resonance (Huascarán). Still, precious results can be obtained in certain cases (Pandemonium Creek) (s. Item 17 and Sect. 5.7).

17. Velocimetry using the curvature resulting from *oblique motion on a dipping plane surface* depends on a favourable geometric configuration. A tentative application for Huascarán failed (s. Item 16).
18. Provided that the maximal mass displacement (in units of volume per unit of time) through several consecutive cross-sections is more or less constant, for all cross-sections the products of average velocity and sectional area are equal. So a *relative velocimetry* (comparison of velocities) between the cross-sections is possible as applied for Huascarán.
19. According to the velocimetric methods involved, the results are defined in somewhat different ways so that differences may occur in case of *two methods used in parallel.* The extent of such differences mainly depends on the physical correctness of the applied methods.

*

For a given velocity a radius of slope variation can be defined below which the centrifugal force exceeds the effect of gravity so that *bouncing* occurs. Calculation of the resulting trajectory may be possible and thus allow for ballistic velocimetry.

20. For sites with low atmospheric density (Huascarán; certain celestial bodies) and large boulders (diameter about 1 m or more on the Earth) *neglect of aerodynamic forces* is tolerable and vacuum trajectories can be used (s. Items 21, 22, 39).
21. Vacuum trajectories, as being parabolic, can be calculated from three relevant elements: points, angles, and velocity. In particular, "*aiming back*" is possible if parameters are better known for landing than for take off (Huascarán) (s. Items 20, 22).
22. According to Item 21 from three elements other than velocity (e.g. take off point, landing point, landing incidence angle), *velocity can be calculated* at any point, e.g. at take off. In the case of Huascarán certain assumed take off points were found to yield unrealistic velocities (s. Item 23).
23. If velocities obtained from bounces are suspected to be too high, the *velocimetric calculation can be inverted* and a plausible velocity used to find appropriate take off points (Huascarán) (s. Items 22, 27).

*

There are various *factors influencing the resistance to motion* and therewith also reach and velocity. Reduced resistance, as entailing the danger of long run-out, requires particular attention. Lubricating mechanisms will be presented later (s. Items 29–35).

24. The *size effect* (Köfels) expresses the farther reach of large masses in comparison with smaller ones. Its exclusive use for prediction is blocked by large scatter (Pandemonium Creek). This situation can be improved by knowledge of other effects (s. Items 26–35).
25. Motional *resistance is correlated with the geometry* of an event (Pandemonium Creek). This is a potential means to assess resistance.
26. Lateral confinement of the track, i.e. *channelling* (Pandemonium Creek, Val Pola, Huascarán), normally reduces resistance. A relation with the size effect is plausible if thickness of the mass is considered as important in the context of resistance (s. Item 24).

27. *Funnelling*, defined as motion through a narrowing channel, is known from Pandemonium Creek and Huascarán. Increasing velocity in a funnel, as postulated according to Item 18, means little more than jamming at the wide end. At any rate, velocities approaching that of free fall are suspect (s. Items 18, 26).
28. On *curved paths* energy dissipation can be very different. It is high if the mass, running back after a run-up, undergoes internal collisions. It is low in slightly meandering curves, especially if combined with channelling (Pandemonium Creek; Val Pola; Huascarán; s. Item 15).

*

Lubrication is a system of two solid bodies displaced with respect to each other, and a medium (the lubricant) able to exert upon said bodies a separating force in such a manner that the total resistance against displacement is smaller than without lubricant. Lubrication is at its best if the lubricant can separate the two bodies completely. Lubrication is a compromise between contradictory requirements: low resistance in the lubricant means low viscosity; but low viscosity means rapid escape and rapid fading of the lubricating effect.

29. *Hydrostatic pressure in joints* (Vaiont), in separating the lubricated bodies, normally increases the passages for escape of water, thus initiating motion, but not maintaining it at length (s. Item 5).
30. *Mud* (fine particles fluidised by water) can improve the poor quality of water as a lubricant by raising the viscosity. It may have played a certain part in the cases of Pandemonium Creek and Huascarán.
31. *Lubrication by pressurised water* is able to work over a long distance if new water is continually set under pressure by a mass moving over the water-saturated fill of a valley floor. Combination with channelling protects against undue escape of water. For Pandemonium Creek the probability of this mechanism is considered as high, and correlation with long run-out is high also for other slides in different parts of the world.
32. *Lubrication by compressed air*, as proposed for Blackhawk and other slides, besides being in contradiction with the size effect, lacks plausible mechanisms for compressing air and preventing its fast escape (be it lateral or by penetrating through the unstable mass under extensional stress; s. Item 24).
33. From geomorphological, theoretical, and experimental results (Köfels, Langtang) the effect of frictional heat upon crystalline rock and, as a consequence, *self-lubrication by melting rock* became evident. Strong points lie in the generation of lubricant right at the points of friction and in the compatibility with the size effect (s. Items 24, 34, 35).
34. *Self-lubrication by melting snow or ice* (or even by its evaporation) is mentioned in connection with Blackhawk and Köfels, and it is probable for motion on the glaciers of Pandemonium Creek and Huascarán. By lack of visible traces, detection in field evidence is difficult. Theoretical calculations point to efficient lubrication (s. Items 33, 35).
35. *Self-lubrication by chemical dissociation or dehydration* (carbonate or gypsum respectively), owing to fast recycling of the components to the original state, is also difficult to detect (Köfels). Theoretical calculations point to efficient lubrication (s. Items 33, 34).

The following items deal with *miscellaneous mechanisms*. On the one hand, secondary effects of rockslides refer to the last (but by no means the least) phase in the chronological order of an event. On the other hand, some rockslides on celestial bodies lie beyond the frame of individual chronology.

36. A *small mass of water*, hit by a far larger mass of rock, can reach a higher velocity than the rock and thus be very dangerous. The difference in death toll between two reported cases was due to human activity: the situation of Vaiont was misinterpreted, that of Val Pola almost perfectly understood, and thus most of the possible victims saved (s. Sect. 7.1).
37. The event of Huascarán could have turned out even more disastrous if Rio Santa had large tributaries below Yungay: it not would have disappeared, and the inhabitants would have suffered an unexpected flood wave after the collapse of the *dam consisting of debris*.
38. As reported for Val Pola, post eventum hazard management primarily means effective *monitoring* of potentially unstable masses. Mention of this activity leads back to ante eventum items.
39. In connection with Blackhawk, doubts were expressed against uncommented comparison of terrestrial with *extraterrestrial events*. In a particular case all relevant parameters have to be accounted for. Example: in case of lacking atmosphere, equal trajectories for large and small particles, if launched at equal velocity in equal direction.

Chapter 3

Comments on Mechanisms of Release

Then came such summer rains, as not had been known in the Hills for many seasons... Purun Bhagat heard the sound of something opening with a sigh, and saw two slabs of the floor draw away from each other... There was a sigh in the air that grew to a mutter, and a mutter that grew to a roar, and a roar that passed all sense of hearing...

Rudyard Kipling

3.1 Causes and Signals – a Pragmatic Approach

In a book mainly dealing with the dynamic mechanisms of moving masses, the release, i.e. the initiation of motion, cannot be discussed in equal detail as motion proper. Nevertheless some comments on the subject, as presented in this chapter, will be useful for different reasons. The most important thereof are the following. In the present section, a general discussion of the *conditions favouring release* of large masses (Sect. 3.1) will pave the way for a quantitative treatment of various impending problems. Some light will be thrown upon the mechanisms connected with the breakdown of cohesion and pre-release creeping; thereby methods hitherto unusual in connection with rockslides and rockfalls will be presented (Sect. 3.2). Eventually details of several particular mechanisms will illustrate applied quantification in cases ranging from everyday occurrences to exotica in the true sense (Sect. 3.3).

If considered more in detail, the descent of a rockslide or a rockfall normally turns out as having numerous causes. This fact suggests a *division into groups*, in first instance with respect to their physical basis and their logical connections.

*

One possible approach introduces the notions of *internal and external causes* (Howe 1909, 44–49; Terzaghi 1960, 88). Terzaghi's elegant definitions, based on the shearing stresses in the considered materials, are successfully used in soil mechanics and, to a certain extent, also for assessing the stability of rock masses (especially if extensively disintegrated before release). These definitions run as follows. "*External causes are those which produce an increase of the shearing stresses at unaltered shearing resistance of the material adjoining the slope... Internal causes are those which lead to a slide without any change in surface conditions and without the assistance of an earthquake shock.*"

Now many rockslides are released in a more or less coherent state, and the masses often are retained in place by mechanisms giving rise to high local stress concentrations and/or belonging to other modes than shearing. So Terzaghi's definitions would have to be substantially modified in order to fit into the frame of the present considerations. And the more general question arises how to assess the *value of a division into internal and external causes* in cases where cohesional strength is regarded as mainly depending on local stress concentrations due to cracks. By this approach it is, of course, not intended to minimise the importance of the phenomena near the prospective boundary between moved and unmoved material.

An argument against the use of internal and external causes is best illustrated by the extremely close interaction between both. Take a frequently occurring example. At the proximal end of a prospective rockslide mass a talus accumulates – an external cause. The process contributes to the increase of stresses in a cracked zone connecting the mass with stable rock. As a consequence thereof a crack grows and thus increases further the stress concentration at its front. So the material's resistance is reduced – an internal cause. In other words, the internal cause is caused by the external one, thus *blurring the difference between internal and external causes.*

A second argument is obtained from tracing back the causes of stress generation to gravitational acceleration, as the very cause of causes, and the difference in altitudes required to make acceleration do work when acting upon a mass. Now gravitation as being ubiquitous, does not ipso facto increase stresses: it creates them. So gravitation cannot be an external cause. And even less it fits into the definition of an internal cause. There simply is *no sense in attributing gravitation to this or that group.*

*

In such instances, the question of internal and external causes is left aside and another principle is considered: the *division into triggers and other causes.* The following definition describes the term "trigger" alone so that "other cause" is defined by exclusion. *The trigger of an event is the last cause which puts an end to the balance between driving and resisting forces and thus sets the mass moving.*

In spite of its being normally only one of many causes, the importance of a trigger may vary within extremely wide limits. In this context, a couple of *examples* will illustrate some characteristic configurations and raise some questions.

1. A coherent mass, otherwise ready to descend since millions of years, is retained in place by a solid, though *cracked, connection with stable rock.* One of the cracks, after having grown yearly by some fractions of a millimetre (e.g. owing to freeze-thaw cycles), attains a critical length, looses stability, and releases the mass. A "soft" release of similar kind is suspected for Pandemonium Creek (Sect. 2.2; Evans et al. 1989, 438).
2. In 1702, during the Spanish War of Succession, and in 1809, in their rebellion against Napoleon, Tyrolean warriors released considerable "*rock batteries*". Such artificial rockfalls, to quote a well-documented example, took place near Pontlatz, 9 km upstream from Landeck, Tyrol, Austria (Egger 1880, III, 673). Similar techniques have been used also in other mountainous regions (oral communication by Dr. H. R. Fuhrer, ETH Zürich). At present, prospective rockfalls threatening human lives and property sometimes are released artificially to keep the danger under control. In the old times triggering was nothing but knocking away wooden pillars or cutting ropes, nowadays it consists in firing an explosive charge.
3. Assume that the *earthquake of* 1970, obviously the trigger of the Huascarán event (Sect. 2.7), had been of moderate magnitude. Probably in such instances a smaller mass would have been released and the outcome might have been far less disastrous.
4. Probably the mass of the Vaiont event (Sect. 2.6) would have remained stable for a very long time *without human intervention.* So this intervention was the main cause of the rockslide, and the trigger was the unlucky coincidence of circumstances during the last filling of the basin. Now assume that the same mass, if untouched by man, would come down as early as in 2050. Would not, in such instances, the entire human activity be considered as trigger because it only would have been the last of many causes (whatever their nature might be) as postulated in the definition?
5. *Primary slides provoking secondary slides* have been reported or assumed by Russell (1927, 234), Abele (1991a, 31), and others. Now imagine that without the primary

slide no secondary slide would occur during millions of years. What would be the primary slide with respect to the secondary one – trigger alone (other causes having prepared the site) or essential cause and trigger in one?

From these examples it becomes evident that a trigger may be a minute increment, significant only by being, so to say, the last straw that breaks the camel's back (Items 1, 2). But it also may dramatically influence the character of the triggered event (Item 3). And sometimes a clear distinction between triggers and other causes may not be obvious (Items 4, 5). To compress the result into one single sentence: by its changes of aspect in accordance with the circumstances, the term *"trigger" recalls a chameleon* in more respects than might be assumed at first sight.

*

After all, the essential question in a particular case lies rather in the causal connection leading to an event than in the attribution of a mechanism to one group or another. The following considerations will be focused on this aspect, and they will be started by looking out for the *main parameters of release* in the case of rock-on-rock displacement as the most important configuration of rockslide and rockfall initiation. A well-known example will yield the necessary basis.

While, as a rule, rockslides – and all the more so rockfalls – are released on steep slopes so that the immediately subsequent motion may be a falling as well as a sliding one, there are cases in which rockslides have been released on rather *soft slopes*. Perhaps the best known example is the catastrophic event of Rossberg (Schwyz, Switzerland) that, in 1806, destroyed the village Goldau and killed 457 persons. The upper portion of the sliding surface, including that of the release region, has a remarkably constant slope angle of little more than 20°. Albert Heim (1932, 71–74) plausibly attributes the possibility of release on so mild a slope to the particular pre-existing circumstances: on the one hand, an almost perfectly plane surface, covered by a marl layer, was ready to serve as an ideal track with low friction; on the other hand deep, approximately vertical joints had reduced to a minimum the cohesion of the mass with adjacent stable rock. Eisbacher and Clague (1984, 156–159) add that said joints were water-filled and thus exerted an external force upon the ready-to-move mass. Of course, Rosserg/Goldau is not an unique example: similar conditions are reported, for instance, by Alden (1928, 347–349) for Gros Ventre, Wyoming, U.S.A., where, in riding on top of the mass that descended a gentle slope, trees remained upright.

Obviously three relevant circumstances – namely a combination of surface properties (shape and materials) granting low frictional resistance, the lack of cohesion between the prospective rockslide mass and stable surrounding rock, and the water pressure in the fissures – synergetically created the possibility to overcome the impeding effect of a fourth one, namely the low slope angle. So it is demonstrated that the *interaction of four parameters* is determining whether release will take place or not. In the particular case these parameters are:

11. the *slope angle* β of the underlying ground surface,
12. the *coefficient of friction* μ between the prospective rockslide mass and the ground,
13. the *cohesion* (in other words, the resistance against fracture) between the prospective rockslide mass and surrounding stable rock, and
14. *external forces* (the push of the hydraulic pressure in the joints).

Within the frame thus defined, a remarkably *wide field of applications* is opened. In fact, it is not easy (though not absolutely impossible) to imagine the release of a rockslide or a rockfall without the interaction of, at least, slope, friction and cohesion, in certain cases also external forces. In this context it is interesting that Terzaghi, in commenting the Rossberg/Goldau event (1960, 94), remarks: "*This catastrophe can be explained in at least three ways… the slope had gradually increased on account of tectonic movements… the resistance due to the* (cohesive) *bond was gradually reduced by progressive weathering…*" The third explanation, based on hydraulic arguments and somewhat too long for literal reproduction, describes a reduction of the coefficient of friction. So exactly the above-mentioned parameters are stipulated, namely an increase of slope (Item 11), due to external forces (Item 14), a reduction of cohesion (Item 13), and a reduction of friction (Item 12).

Of course, *more exotic circumstances* are not excluded a priori. Some of them will be treated in Sect. 3.3.

*

All this is, perhaps, perfectly plausible. It is, however, only one half of the truth. The other half is revealed when comparing the incidence of sites fulfilling the said conditions with the incidence of rockslides and rockfalls, especially large ones. To the author's knowledge a statistical investigation of this kind never has been made. But there is no doubt that it would yield a considerable discrepancy suggesting the existence of a causal connection. This fact was spotted in two of the earliest (and still of the best) references. Howe (1909, 27), before speaking of triggers (earthquakes, readjustment of stresses, water saturation), remarks: "*… at a good many other localities… precisely similar causal conditions now exist, and yet no landsliding has taken place…*". And in his comments on Rossberg and Goldau Heim (1932, 72) expresses his opinion as follows. "*A rockslide of this size can only be formed with a bedding slope angle below 30° because it requires long years of preparation. On a steeper slope, rock gradually falls in small portions, not waiting for the accumulation of a large event.*" (translation by Erismann). This formulation is, perhaps, somewhat too apodictic in postulating no other parameter but slope. Nevertheless the basic idea is correct. As a matter of fact, slope plus low resistance against motion plus lack of cohesion with adjacent stable rock, in the normal course of things, result in gradual erosion or, at most, in a sequence of small mass displacements. To put it the other way round: large events are exceptions. And in numerous cases their existence can be explained by, so to say, negative causes, i.e. mechanisms inhibiting gradual displacement. Thus, by being retained in place in combination with gradual alimentation of any kind, a considerable mass has the time to accumulate before the breakdown of a massive obstacle, followed by the release of a large event. This fact gives a new aspect to at least three of the basic parameters: coefficient of friction, cohesion, and external forces (for instance the pressure of a glacier acting as abutment) may be *"retaining" instead of "driving" causes.*

Besides the just-mentioned force of a glacier, *retaining causes* may be, in the first instance, the incomplete segregation of joint surfaces or a particularly high coefficient of friction due to interlocking between the mass and the ground. Another possibility is the wedging effect of a mass between lateral solid rocks which leave it a free passage narrowing from top to bottom ("non plus ultra" example: rockslide of Aegerti, Bern, Switzerland, released and immediately arrested in 1901, Heim 1932, 158–162). A similar effect also can occur between partial masses tending to move in convergent directions and thus blocking each other mutually. Examples for the collapse of abutments in the true sense of the word are the large events of Cerro Mesón Alto in Chile and Totalp in the Grisons, Switzerland (2 and 0.4 km^3 respectively) described by Abele (1984, 168–169, 172) where barriers of stable crystalline rock had held in place masses of less resistant material (andesite, serpentine).

So a *pragmatic attribution of any cause* (be it driving or retaining) to at least one of the presented parameters (slope, friction, cohesion with stable rock, external force) is obtained. It has to be borne in mind that a given cause must not necessarily act upon one single parameter. Undercutting of the distal end of a mass by a river may, for example, simultaneously increase slope (by taking away material lying on approximately horizontal ground) and reduce cohesion with stable rock (by taking away material abutting against the opposite slope).

To sum up the discussion of causes, a general statement is perhaps appropriate. The search for the geomorphological circumstances favouring the release of a given mass is nothing but an *application of the four basic parameters* to the features of the particular case (materials, arrangement of joints, faults, or other results of disintegration, as well as bedding planes etc.). It is needless to say that the regularly repeated term "sufficient" in the sentences to follow hereafter means as much as "sufficient in combination with the other parameters claimed to be sufficient". There must be a pre-existing or easy-to-create surface sufficiently inclined to allow motion. This surface must be sufficiently plane and/or offer a combination of materials granting sufficiently low frictional resistance. There must be a sufficient separation of the presumptive rockslide or rockfall mass from surrounding stable rock. The mass, if its gradual accumulation otherwise would be surpassed by gradual loss of material, must be kept together for a sufficient time by retaining mechanisms. If the combined action of these preconditions does not suffice, release still may occur by the action of sufficiently strong external forces.

*

A great majority of causes, even if obvious for an expert, are difficult to assess with respect to the *time horizon of danger*. In other words, the certainty that a catastrophic event will occur is as good as useless if the question is left open whether the mass will come down in a couple of days or in thousands of years. This fact calls attention to the signals announcing the approach of danger.

The most important (and easiest to perceive) signal given by an impending event is, as a rule, a slow motion, accelerated before a mass is released. So a powerful means for the prediction of release consists in the observation and *quantitative monitoring* of such displacements. The principle has little changed since the times of Heim (1932, 190); however, the technical means have developed rapidly. Instead of the installation of a simple bar with marks to measure the width of a growing joint, extensometers are installed in more than one single direction, inclinometers show angular motion, piezometers determine the ground water level, and telemetric transmission makes it possible to obtain a synoptic view of the displacements right at the headquarters of operation. So, in a well-monitored case, decision-making (for evacuations etc.) is possible on the basis of rather complete real-time knowledge of the situation.

It was, by the way, a macabre irony of fate that wrong interpretation and not a lack of monitoring caused the catastrophe of Vaiont (Sect. 2.6). Fortunately *successful monitoring*, by no means impossible even in this tragic case, is known from numerous other events. Impressive examples are reported, for instance, about the risk management of two rockfalls in Switzerland which both threatened the unique roads connecting villages Amden (St. Gallen, 1974) and Zermatt (Valais, 1991) to the respective main valleys (in the last-mentioned case also the hamlet Randa and the rack railroad). Jäckli and Kempf (1975) describe the event of Amden which could be managed with relatively simple means. For the giant twin-stroke rockfall of Randa (Noverraz and Bonnard 1991) high-tech equipment was used and extensive work had to be done: evacuation of Randa, outlet to avoid inundation, helicopter service to Zermatt, etc.. Mechanisms and measures are shown in an instructive video record (GEOTEST/BWW 1992). A last example: long-term prophylactic monitoring in a densely populated valley is permanently installed in the region of the Val Pola event where an unstable mass has remained near the scar of the main slide (Sect. 2.5; I.S.M.E.S. 1990; Costa 1991; Crosta 1991). More general remarks on monitoring have recently been published by Bhandari (1990), Kovari (1990), and Krauter (1988).

Besides the mentioned successes, the importance of monitoring is stressed by the fact that Heim's dictum postulating for any rockslide a *previous signal* by slow motion (1932, 188–189) almost unconditionally holds good since more than 60 years. In their investigation of the literature, embracing more than 200 references, the authors encountered only one publication describing a case in a well-observed region for which "*There were no signs signalling the imminent slide*" (Wildberger 1988, 1381). Of course, there were several cases in which no signal was observed, but then observation was excluded by plausible reasons, as, for instance, the absence of observers in an uninhabited region (example: Pandemonium Creek, Sect. 2.2; Evans et al. 1988).

*

So far, the present section, as part of a rigorously focused chapter, could be set up with no more than approximately 20 quotations, a modest number when compared with the comprehensive literature treating various aspects connected with the stability of slopes and the release of masses. To assist readers wishing to dig deeper in this context, an additional *review of references* is presented hereafter. The particular highlights of each publication, as far as connected with the scope of this section, are given in bracketed telegraphic-style comments.

Publications containing *basic information*, be it theoretical or practical: Abele 1974 (chapter on release and its causes); Cruden 1976 (Canadian Rockies; engineering aspects of friction and effects of water); Eden 1975 (main concern: Leda clay, Canada); Hutchinson 1993 (review of triggering and motional mechanisms); Krauter 1994 (anthropogeneous causes; accent on soil and mud); Law and Lumb 1978 (progressive failure as important cause); Scheidegger 1970 (chapter on slopes, including stability); Scheidegger 1984 (state of the art; importance of water); Selby 1987 (fundamentals of slope stability); Ter-Stepanian 1977 (accent on clay; compound, complex slides etc.).

The following authors point out the *importance of water* (as well as already some of the above-mentioned): Costa 1991 (Val Pola, Italy; s. also monitoring); Crosta 1991 (Val Pola, Italy; s. also monitoring); Cruden and Antoine 1984 (Mt. Gronier, France; side glance to friction); Hewitt 1988 (Karakoram, Pakistan; freeze/thaw cycle as trigger); Kojan and Hutchinson 1978 (Mayunamarca, Peru; reflections on friction); Ruby and Hubbert 1959 (influence upon slope angles).

Volcanic activity and *earthquakes* also belong to the most frequently cited causes of rockslides and rockfalls: Bull et al. 1994 (New Zealand and California; slides make possible lichen dating of earthquakes); Crozier 1991 (determination of paleoseismicity from slides); Eisbacher 1977 (Mackenzie Mountains, Canada; incomplete release is normal); Francis et al. 1985 (Sacompa, Chile); Muir 1912 (Yosemite, U.S.A.); Nicoletti et al. 1993 (Scanno, Italy; earthquake report by Titus Livius?); Voight et al. 1983 (Mt. St. Helens, U.S.A.); Watson and Wright 1967 (Saidmarreh, Iran; water considered as important).

Of course, there are various aspects *fitting into none of the groups* mentioned hereabove. So the present section is closed with three extremely different references, one dealing with a very slow, though essentially local process, the second with perhaps the fastest imaginable (and somewhat exotic) one, while the third considers mass movements from the fundamentally different viewpoint of tectonics: Samalikova 1977 (weathering mechanisms preparing failure of crystalline rock); Surenian 1988 (meteorite assumed to have triggered Köfels; Sect. 2.4); Scheidegger 1998 (tectonic predesign).

Still, even in the relatively narrow sector of local-scale release an enormous field is open for research.

3.2 From Cohesion to Motion

Cohesion with surrounding stable rock may refer to the bottom surface of a mass as well as to its (often more or less vertical) lateral and proximal surfaces. In other words: also in the presumptive sliding surface sheer friction is not necessarily the unique force that inhibits the start of motion.

At first sight, the crucial question in the context of cohesion seems to be nothing but trivial: will the *material's strength* suffice to resist the load exerted by the mass as a consequence of gravity and – if any – of other forces? And since centuries qualitative "danger indicators" are known. If, for instance, at the top of a suspicious-looking mass of rock there is a growing crack, and if this crack – after a period of slow, steady growth – starts to open at a distinctly accelerated pace, it is a high time to evacuate persons and goods from the threatened area. It will become manifest in the further course of this section that the present state of the art, except for a better understanding of the phenomena observed and the improvement owed to modern monitoring equipment (Sect. 3.1), does not fundamentally differ from such traditional methods.

*

Quantification of the approach to cohesion, despite the last remark an important task, depends in first instance on the available knowledge of the material's strength. This is by no means an easy task: rock is the product of a long and complex process with countless parameters varying within wide limits. So for apparently one and the same material the scatter band of strength may be remarkably extended. Selby, in his excellent approach to this problem (1987, 476–478), remarks that in situ the modulus of elasticity may be 20 times lower than that obtained from perfectly intact specimens. He also describes a method to determine the quality of rock on the basis of a scoring system that takes into account eight parameters (intact strength, weathering; parameters concerning joints: spacing, orientation, width, continuity, infill, outflow of water), assessed each in a five-class rating (continuity and infill being rated together). This is a valuable engineering tool yielding useful (though certainly not absolutely accurate) results. It suffers, however, from being to a large extent a "black box" system and thus leaving in the dark the mechanisms behind the rating. Scientifically – and in the long run also practically – this situation is unsatisfactory.

*

At least partially – namely in the quantitative consideration of joints or cracks – this gap can be closed by introducing a tool of material sciences, known since a considerable time, but hitherto little used in connection with the initiation of rapid mass displacements. So a short excursion to *fracture mechanics* is justified, although it is impossible to expose the subject in extenso and although the method at the present time cannot substantially enhance practical work of mitigation; as a valuable counterweight it yields some insight into the causal connections between strength and shape.

The *history of fracture mechanics* is typical for the destiny of an early finding. In the endeavour to explain the behaviour of brittle materials like glass, Griffith (1920) found a mathematical expression for the fracture of such a material in presence of a flaw, idealised by a sharp-edged crack of given length. At the time glass was considered as a material of marginal importance as far as strength was at stake. So Griffith's outstanding work did not raise the interest it merited, and the study remained more or less forgotten for almost 40 years. In the meantime, especially under the influence of a rapidly developing aircraft industry, the strength of brittle materials (in first instance cast and/or heat-treated metals) more and more became a serious concern in engineering. And the presence of flaws (for instance shrinkholes in castings or fatigue-generated cracks) never could be excluded. So when Irwin (1957) revitalised Griffith's ideas and gave them a form appropriate for practical application, a world-wide echo was evoked and fracture mechanics became one of the most important methods (Liebowitz 1968), first in mechanical engineering and somewhat later in certain fields of civil engineering and even mining (Gramberg 1989).

The *philosophy of fracture mechanics* is based on the fact that in the vicinity of a flaw, again idealised by a sharp-edged crack, a massive local stress and strain concentration reduces the strength of a body so that conventional calculations can yield dramatically wrong results. And even the introduction of so-called notch factors, in spite of being successful in certain particular cases, fails to solve the problem. This is due to the fact that stress concentration depends on the size of the flaw. In the basic formula of fracture mechanics, the so-called *stress intensity factor*

$$K = s\sqrt{(c\,a)} \tag{3.1}$$

is linked both to the mean stress $s = F / A$ (F being the total force exerted upon the body and A its sectional area in the plane of the crack) and to the crack length $2\,c$ (Fig. 3.1). The dimensionless constant factor a depends on the mode (s. hereafter) and the geometry of the considered body (for mode I and a body large enough to represent quasi infinite size $a = \pi$ is valid). Obviously K has not the dimension of a stress but that of a stress multiplied by the square root of a length (Pa$\sqrt{}$m). With respect to the crack length it must be borne in mind that the equation is correct for a perfectly sharp-edged crack and that a discontinuous widening located near the sharp edge of the crack (Fig. 3.2) sensibly reduces the length valid for the equation. It is, by the way,

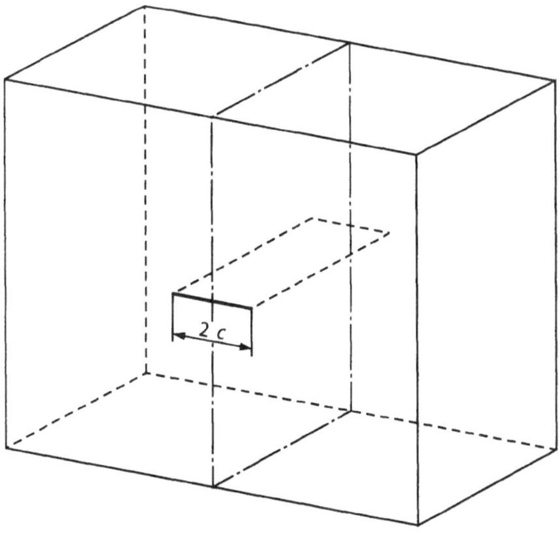

Fig. 3.1. Basic concept of fracture mechanics: plane, sharp-edged crack of length $2c$ in a sufficiently large body of sufficiently brittle, homogeneous, isotropic material. To deal with cracks starting from the surface of a body, the crack length is halved along the plane marked by dash-and-dot lines (sketch by Erismann)

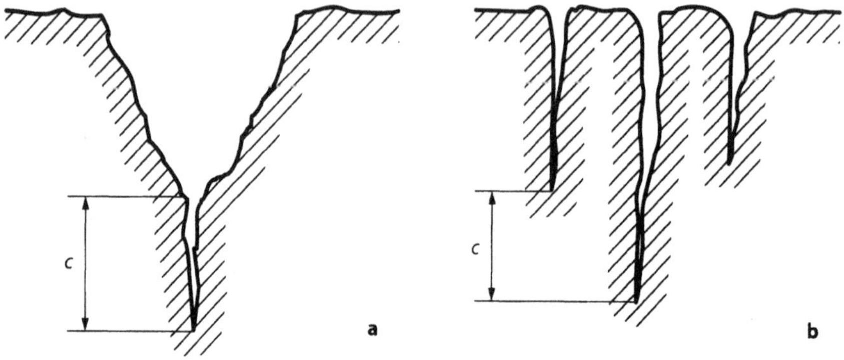

Fig. 3.2. Effective crack length *c*. Irregular crack shape (**a**) and/or nearby cracks (**b**) may distort stress distribution and reduce the crack length valid for fracture mechanical considerations. Resulting errors are on the safe side: strength is higher than estimated (sketch by Erismann)

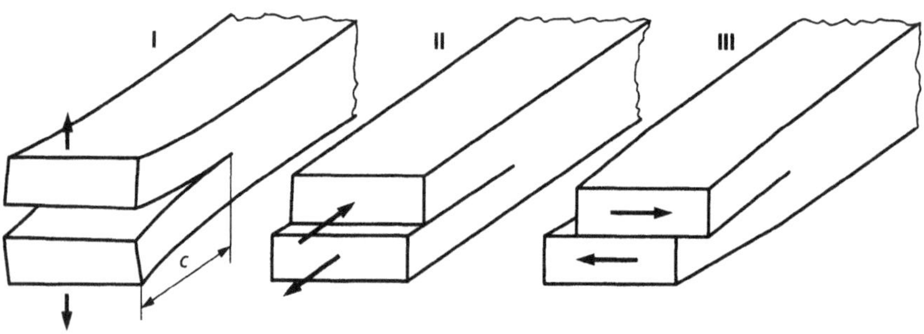

Fig. 3.3. Fracture mechanical loading modes *I, II, III*. Mnemonic keyword marks typical attitude of animal's mouth: *I:* alligator; *II:* bulldog; *III:* cow. *Arrows* indicate loading forces (sketch by Erismann)

a strong point of fracture mechanics that such imperfections, as a rule, are on the safe side: usually strength, when calculated with idealised conditions, is underestimated.

Basically, the wide field of applications of fracture mechanics is divided into *three modes* (Fig. 3.3). The key-word annexed hereafter to each mode has a mnemonic background: as the configurations belonging to the respective modes were easily mixed up by students, one of the authors, in his lectures on the subject, introduced these alphabetically significant keywords linking each mode to a typical attitude of an animal's mouth.

- Mode I (keyword: Alligator): a tensile stress, acting at right angles to the plane of the crack, tends to open the crack.
- Mode II (keyword: Bulldog): a shearing stress, acting in the plane of the crack and at right angles to the edge of the crack, tends to pull one of the jaws of the crack away from its edge and to push the other jaw in the opposite direction.
- Mode III (keyword: Cow): a shearing stress, acting in the plane of the crack and parallel to the edge of the crack, tends to displace the jaws of the crack into opposite directions parallel to the edge of the crack.

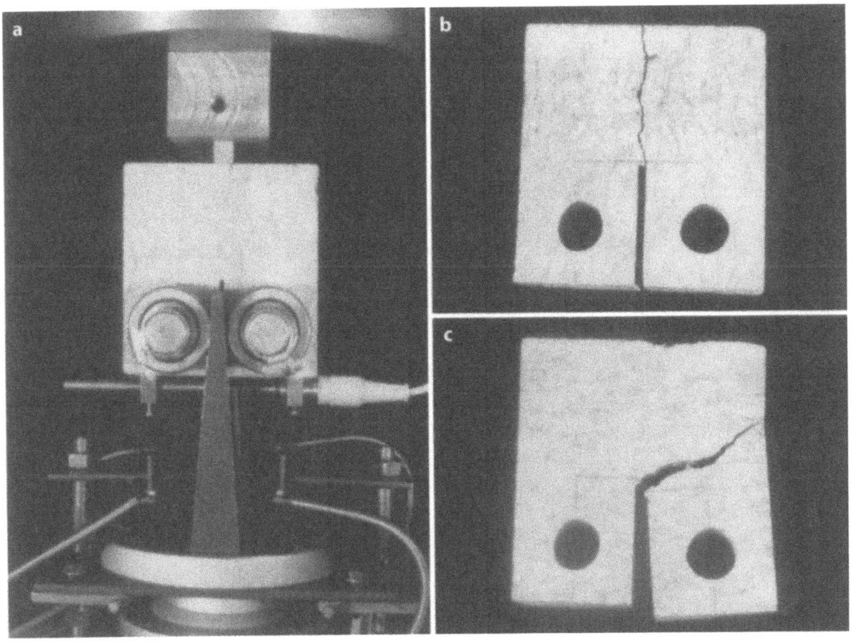

Fig. 3.4. Mode I fracture mechanical test of Köfels gneiss specimen (s. Sect. 2.4) between the circular compression platens of a testing machine (**a**). To minimise the elastic energy stored in the machine, the load is transmitted to the specimen by wedges driving apart roller bearings fixed to the specimen. Anisotropy may favour (**b**) or exclude (**c**) correct testing (by courtesy of H. K. Hilsdorf, Institut für Baustoffkunde, University of Karlsruhe, Germany)

In a somewhat simplified approach the cohesion between particular zones of the mass and the surrounding solid rock can be attributed to the three modes as follows: proximal to mode I, bottom to mode II, lateral to mode III. It is not excluded that mixed modes – a mathematically rather demanding field emerging from neglect since a few years – may play a certain role in days to come. The present state of general knowledge about rock failure makes such sophistications obsolete for the time being.

One of the most important findings of fracture mechanics is the fact that, for a given material, K has a finite maximum value, the critical stress intensity factor or *fracture toughness* K_c (usually written as K_{Ic}, K_{IIc}, or K_{IIIc} to indicate the particular mode). This value is a strength-limiting parameter of the material and may be influenced by other parameters, in first instance temperature and/or humidity for the materials of mechanical engineering; for rock, weathering is, perhaps, the most significant of Selby's parameters in this context. K_c can be determined experimentally in testing machines (Fig. 3.4), using specimens of standardised shape and size (ASTM 1974). It should be added in parentheses that, owing to various refinements, fracture mechanics nowadays can be applied far beyond the rather narrow set of conditions for which it originally had been conceived.

Users not familiar with fracture toughness as an everyday working tool may prefer inversion of Eq. 3.1 so as to obtain a critical stress

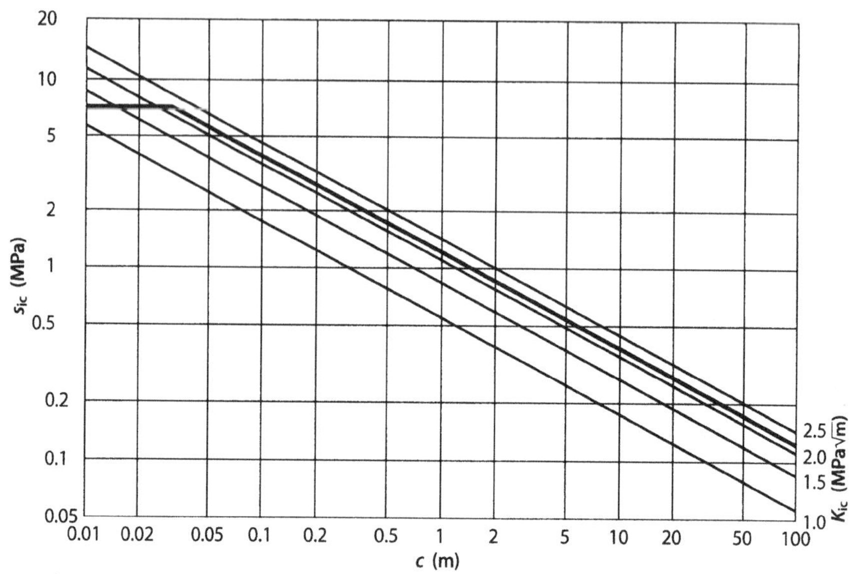

Fig. 3.5. Critical fracture mechanical strength s_{ic}, represented as a function of crack length c and fracture toughness K_{ic}. Bold line: strength of material with $K_{ic} = 2.2$ MPa$\sqrt{\text{m}}$, limited to conventional strength of 7 MPa in extension. Attention: functions are calculated from Eq. 3.2, assuming near-to-ideal geometry ($a = \pi$). As a rule, real geometries are on the safe side: they yield slightly higher strength (sketch by Erismann)

$$s_{ic} = \frac{K_{ic}}{\sqrt{ca}} \tag{3.2}$$

where the index i stands for the respective mode I, II, or III (s, in such instances, can be a tensile or a shearing stress). Besides the trivial advantage of allowing work *as if conventional stress were used*, this simple trick also makes possible a direct comparison of conventional and fracture mechanical approach, an important point in view of a particularity of the Eq. 3.1 and 3.2: as they completely leave aside the aspect of conventional strength, they do not exclude, in case of very short cracks, stresses exceeding those of conventional strength. The risk of erroneous results thus obtained can be excluded by an appropriate graphic representation (Fig. 3.5) containing the necessary limitation of stresses below the conventional value.

Obviously Eq. 3.1 and 3.2 signal a *size effect:* larger bodies with larger crack length c are less resistant than smaller ones. This effect has to be added to the trivial size effect given by the rules of similarity: as in geometrically similar conditions the sectional area ($A = B\,L$ in Fig. 3.6) is proportional to the square and the volume ($V = B\,L\,H$) to the third power of the linear dimensions of a body, the mean stress ($F\,/\,A$) turns out to be size-proportional. So, as both size effects act in the same sense, there is no question that a large mass will undergo fracture-generated release at a far lower slope than a smaller one.

The second size effect merits a special comment. As a matter of fact, this "*area-to-volume*" rule is ubiquitous in nature and human activity. To quote only two among

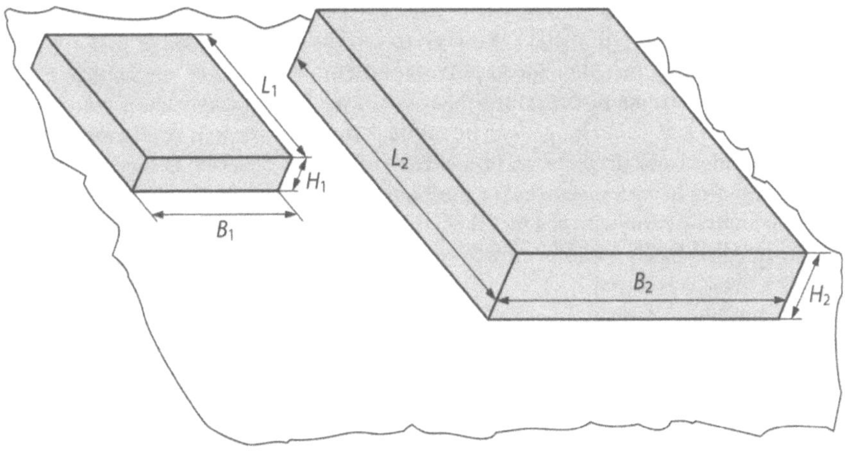

Fig. 3.6. Trivial size effect ("area-to-volume" rule) in conventional fractural release of two geometrically similar bodies. The gravitation-generated driving force is proportional to the volume $L_i B_i H_i$, the resisting strength to the contact area $L_i B_i$ (index i is 1 or 2, according to the respective body). Thus the average stress in this area is proportional to H_i. In a fracture mechanical consideration the probability of longer cracks increases the effect (sketch by Erismann)

innumerable examples: it forces an elephant to move on column-like legs, while a daddy long-legs comfortably rests on almost incredibly thin extremities; and it explains to a certain extent why a champion in a skiff is slower than a mediocre crew of eight. The importance for rockslides will entail repeated mentions in sections to follow (Sect. 5.4, 6.2).

*

Fracture mechanics in its present form was developed in first instance for metals and plastics, two classes of materials available in a highly homogeneous and, if not reinforced, also isotropic state. For the *application to rock* certain precautions are required. Rock is nonhomogeneous, and therefore its strength may differ in different locations of one and the same piece; rock is normally anisotropic, and therefore in certain planes its strength may be unexpectedly high or low (Fig. 3.4c; Hillemeier 1976; Müller and Rummel 1987). In accordance with Selby's (1987) line of thought, the practical consequence is that the physical parameters of rock vary in far larger scatter bands than those of materials having passed through a carefully supervised industrial process. In particular a weak plane may play the part of a "hidden crack" or at least impress its low fracture toughness upon the surrounding zone of a body.

In order to make plausible the fracture mechanical aspect of release it is essential to recall the differences between experiments and reality as encountered in a ready-to-move mass. For the sake of accurate K_c determination it is desirable in experimental fracture mechanics to obtain a stable crack growth not only until K_c in Eq. 3.1 is reached but also beyond this point. As K in such instances necessarily must remain constant ($K = K_c$), crack length only can grow at the expense of a diminishing force F. Hence testing machines for fracture mechanics are conceived and programmed to keep constant the pace of deformation, not of force (in this context, Selby's belief in the possibility of making up for deficient stiffness of a testing machine by servohydraulic control is somewhat optimistic 1987, 478). In a ready-to-move mass of rock, by contrast with a testing machine, the gravity-dependent driving force – at least in a first approximation – remains constant so that, when the critical condition $K = K_c$ is trespassed, Eq. 3.1 and 3.2 no longer can be fulfilled. In other words: if at all a crack starts growing under constant force, an immediate collapse of cohesion is inevitable. Fracture mechanics in such instances is a mechanism of "sudden death". This statement is in open contradiction with the fact that, as a rule, continuous crack growth, in many cases signalled by continuous growth of a crack opening at the surface, is observed before a mass is released.

One possible approach to solve the dilemma consists in introducing *fatigue*, the *property of a material to fail after many repetitions of a deforming stress, which by itself is not high enough to cause failure* (Visser 1980, p. 348, Item 4674). Most of the experimental work done in the context of fatigue deals with metals and reinforced plastics, and a good portion of it is aimed at phenomenological questions. Furthermore, our knowledge about fatigue mechanisms in rock is rather poor so that the definition has to be accepted as another black box. In parentheses, it should be added that fatigue may occur in conjunction with other phenomena like corrosion, abrasion, weathering, etc. and that normally the resulting damage is increased in comparison with that due to fatigue alone.

To obtain a general idea of *fatigue crack propagation* Paris' hypothesis (1961), in spite of a somewhat deficient scientific foundation, is useful as long as it is not employed to describe the process under very low (damage-free) stress amplitudes. In

$$\frac{dc}{dN} = (C\Delta K)^n \tag{3.3}$$

the left side expresses the growth of crack length c per load cycle (cycle number N), C and n are constant values (to be determined experimentally), and ΔK is the variation of stress concentration in the considered load cycle. As C and n are scarcely available for rock owing to lacking experimental results, the practical value of the equation is mainly qualitative. In combination with Eq. 3.1 it makes obvious that, as crack length grows cycle by cycle, also ΔK will grow in a manner to induce step-by-step increase of $dc\,/\,dN$. This means in clear that under more or less constant force cycles the crack length – and consequently also the crack opening displacement (i.e. the crack width) – will grow at an accelerated pace. This statement is in perfect coincidence with the already mentioned field experience that acceleration of crack growth signals danger.

*

Fatigue is not the only mechanism allowing very slow mass displacement, accelerated before release. In fact, *progressive failure of interlocking contacts* between the surfaces of mass and ground is, at least in its phenomenological appearance, analogous to that of a growing crack under fatigue load. In contrast to fatigue, however, the reach of such motion is not limited by the necessity of not losing cohesion with stable rock. The term "interlocking" is used here in a sense not explicitly defined in Visser's "Geological Nomenclature" (1980, 60, Item 0816; 225, Item 3118; 303, Item 4098), but it can be found as an implicit constituent in the definition of the term "apparent coherence" (p. 350, Item 4693) which begins by "*– may be observed along rock separation planes* *due to interlocking of surface roughness...*" The sentence obviously implies the ability of meshing asperities in two surfaces, to transmit a substantial shearing force from one surface to the other. Hereafter, force-transmitting asperities of this kind will be denominated as "*locks*". Theoretically the original geometry would allow only for one or two active locks in a pair of interlocking surfaces. But local crushing as well as small elastic deformations will in any case distribute the total retaining force upon a multitude of locks. Still massive stress concentrations near the locks are inevitable, an analogy to those near the edge of a crack in fracture mechanics and fatigue.

Now assume that, by a slight variation of the acting forces, comparable to that of a fatigue cycle (examples are quoted hereafter), *a highly stressed lock fails*. The consequence is a redistribution of loads in the remaining (and possibly also in newly loaded) locks which, according to the local circumstances, may increase or reduce the stability of the mass. Of course, reduction is more probable than increase because the failure of a lock reduces the number of possible load-bearing links and thus can be considered as a germ of "track-making". The next cycle of lock failure(s) and redistribution of forces is initiated by the next sufficiently strong variation of the acting forces. Owing to the redistribution the conditions of failure may substantially change from cycle to cycle.

It is obvious that each readjustment of stresses and forces, as entirely depending on gravitational energy, occurs at the expense of an increment in altitude and requires a certain time. So a slow step-by-step downhill motion of the mass results, in other words a *rock-on-rock creep* as insinuated under Heading 2.6. It is essential that any mechanism able to alter the distribution of forces is potentially able to run the process and that after each failure a redistribution of forces takes place. So a quasi-continuous motion becomes possible. The principle, though in the represented case far from quasi-continuity, is shown in Fig. 3.7 with a minimum of locks. (It is remarkable that Cotton 1960, 29, in describing soil creep, points out some remarkable analogies: "*Small to-and-fro movements... are always going on as the result of... heating and cooling, freezing and thawing, wetting and drying. Owing to the constant pull of gravity there is a predominance of downhill over uphill movements, and this results in downhill creep...*").

As the number of effective locks decreases at every step, the step length (i.e. the displacement until a new set of sufficiently strong locks is "found") normally will increase and, in accordance therewith, a tendency to *acceleration* of the average velocity has to be expected – a striking analogy to the process described for fatigue crack growth. Eventually, if interlocking is the dominant retaining force, release takes place as soon as a new equilibrium of forces no longer can be established after the last failure of a lock. It has to be taken into account in this context that, owing to the increasing velocity at the end of a step, possibly even the kinetic energy no longer can be neglected: it may help to cut away locks.

To make things clear, two *comments* should be made before closing the discussion about progressive failure. (1) Both fracture mechanics and fatigue may occur in the process of progressive failure: a local fracture initiating a step of progressive failure very probably is brittle, and it may be the result of several fatigue cycles as well as that of one single load peak. (2) The presented form of progressive failure is not identical with that described by Law and Lumb (1978); yet both mechanisms are based on redistribution of stresses after local failure.

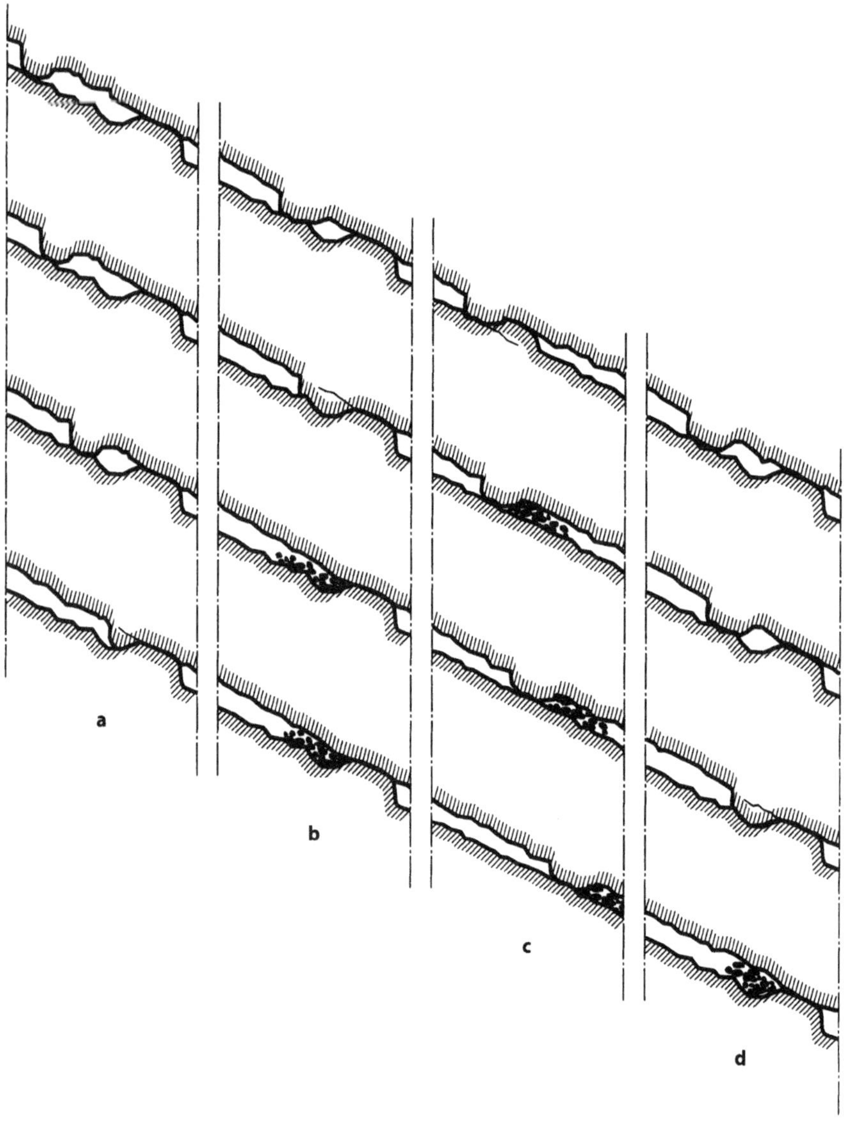

Fig. 3.7. Progressive failure, working principle. **a, b, c, d**: Sections through contact planes between ground and creeping mass, showing from top to bottom four positions of a pair of locks. Each position is synchronous with the respective positions of the other sections. Changes are due to slight variations of stress, peak loads exceeding the strength of a lock and thus provoking a minute displacement and a redistribution of loads. In reality the mass is kept in place by many locks, and not every change involves loading of new locks. Note cracks before and debris after failure (sketch by Erismann)

Anyhow, both fatigue and progressive failure essentially depend on mechanisms of *repeated load variations* in critical zones of a presumable rockslide mass, be it by continued attack at one single location (as in fatigue), be it by currently changing the location of attack (as in progressive failure). So it is worthwhile to sort out the mechanisms capable of substantial effects by repetition. The following comments begin with a low-stress mechanism unable to generate a considerable mass displacement in one single cycle; later more powerful mechanisms will follow, and single stroke action no longer will be excluded a priori.

The generation of cracks by a stress due to *thermal elongation* is possible when a cold surface is forced into tensile elongation by a warmer core of a coherent mass. A variation of approximately ±5 °C against the average is obtained at a depth of 0.05 m in a 24-hours cycle and at 2 m in a seasonal cycle (Schindler and Nievergelt 1990, 241; the figures, obtained in dry sand under moderate climatic conditions, give an order of magnitude). Now a variation of 10 °C from minimum to maximum, with a modulus of elasticity around 15 GPa and a specific thermal elongation of 0.002 mm per m and °C, means an elongation of 20 mm per km length and a stress variation of about 0.3 MPa, a rather poor basis both for fracture mechanics and progressive failure. In fact, according to Eq. 3.1 a perfect crack $2c = 10$ m long would be required to produce single stroke fracture with a K_c value of 1.2 MPa$\sqrt{\text{m}}$. This is substantially more than the depth at which significant temperature variations normally occur. In other words: if circumstances are not particularly favourable, thermal elongation may be good enough for seasonal-cycle fatigue crack growth near the surface of a mass (immediately below the surface also 24-hours cycles may be effective).

The just-mentioned particularly favourable circumstances merit a short comment. With a massively jointed geometry as represented (for other purposes) in Fig. 3.2b the variations of temperature, owing to reduced distance for heat transfer and greatly enlarged surface, may get nearer to the edge of a crack. Another powerful mechanism is heat transport by flowing interstitial water. And a third – perhaps the most promising one – consists in the pre-existence of strata-bound weak planes (with extremely low K_c values). In spite of the obvious increase of its effectiveness by such mechanisms it is not probable that thermal elongation can dramatically reach beyond the limits postulated here-above.

Faster progress can be expected from phenomena exerting higher stresses and, in addition, occurring more frequently. Hydrostatic *pressure of water* penetrating into a crack is one candidate. Its effect largely depends on the water depth (yielding a pressure of about 0.1 MPa per 10 m). It is well known that interstitial water pressures may reach values several times higher than 0.1 MPa. Peak depths normally occur after heavy precipitation, so that, according to the climatic conditions, several peaks per year may be the rule. In addition it has to be pointed out that the generated stress is not equal to the hydraulic pressure. To quote an example: in a water-filled crack (Fig. 3.8) it becomes evident that the tensile force exerted upon the mass is, roughly speaking, proportional to the square of the crack length, while the sectional area undergoing this force is proportional to the difference between total thickness of the mass and crack length. So, if cracks reach deep, the average tensile stress may be substantially higher than hydrostatic pressure, and, in contrast to thermal elongation, the mechanism turns out to be particularly effective in the late phases which precede (or include) triggering. Numerous disasters in the course of or shortly after intense precipitation are noteworthy evidence in this context.

On the other hand, it has to be taken into account that increasing crack opening normally results in widening escape sections for the water so that the direct driving effect is reduced or lost owing to deficient depth. Still an indirect drive can occur from

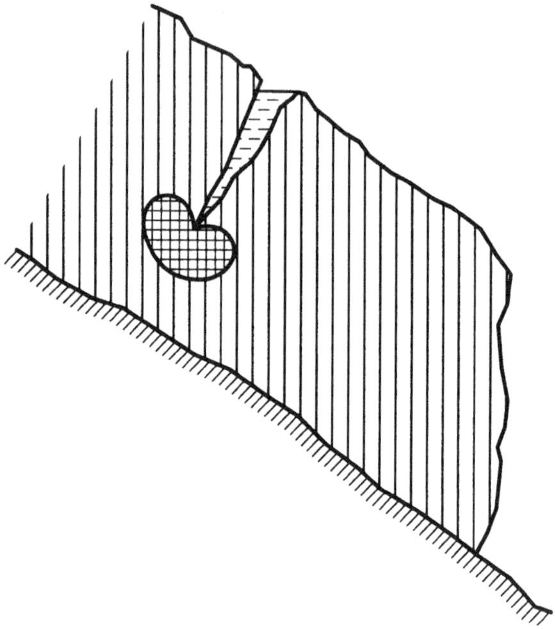

Fig. 3.8. Stress concentration (*chequered area*) near the edge of a water-filled crack. Attention: stress intensity (Eq. 3.1) is proportional to water depth and square root of crack length (sketch by Erismann)

the sliding surface by a reduction of friction due to the lifting effect of water acting as a pressurised lubricant. This mechanism is assumed for Vaiont (Sect. 2.6) and it might be the source of repetitive drive maxima in cases similar to that of Fig. 3.7.

In regions with sufficiently cold winter climate a further mechanism connected with water is *freezing*. And usually there is a tendency of cumulation of the stress thus generated with that coming from thermal contraction by low temperature: both drive the jaws of the crack apart, thus increasing the local stress concentration. It must, however, be taken into account that a sharp-edged crack necessarily is extremely narrow so that a gain in volume of about 8% (as occurring in the process of freezing) cannot produce a substantial displacement. In this respect the mechanism somewhat resembles that in a very stiff testing machine under load: only little energy is stored in the device driving apart the jaws of the crack. Thus a single stroke release in a majority of cases is not probable, and freezing normally is reduced to play its part as a motor in a mechanism of fatigue or progressive failure. Also this fact is in perfect coincidence with field evidence: rockslides seldom occur in the cold season. Theoretically the resulting stress might be in the range of 20 MPa if crack growth were excluded; yet, as soon as the crack starts growing, the stress immediately decreases. An example of this mechanism acting by mere dilatation is given by Howe (1909, 52) for the famous Frank slide (Alberta, Canada): "*The heavy frost on the morning of the slide, which followed hot, summer-like days, appears to have been the force which severed the last thread and precipitated the unbalanced mass.*"

A completely different mechanism connected with freezing was described by Hewitt (1988, 242) for rockslides he had observed in the Karakoram. The respective sentences run as follows: "*The landslides occurred during the melting of a larger than normal snowpack and during a period of heavy rainstorms. The regulation of meltwater sup-*

ply by diurnal freeze-thaw cycles seems to have been the trigger for the sequence of failures on successive days and late in each day."

*

One of the most referred-to release mechanisms is due to the *dwindling of glaciers.* This phenomenon has been known for a long time: Heim (1932, 185) describes it as follows. "*By the melting process of the glaciers the mountainsides were freed from the counterpressure of the ice, and rock masses, loosened over the course of hundreds or thousands of years, broke off in particularly large portions.*" (translation by Erismann). Obviously the stress thus generated in the rock can be much higher than in case of the above-discussed mechanisms and, under adequate geometrical conditions, it can equal the hydraulic pressure of the ice (Fig. 3.9). In other words, the stress at the bottom of a mass may be somewhat less than 1 MPa per 100 m of ice thickness. So in a large glacier hundreds of meters thick it can reach remarkable values. The primary influence of this stress upon the rock is compression, and only few cracks are generated or enlarged within the presumptive rockslide mass. Nevertheless, besides the never-excluded underground activity of water and, above all, the (depth-dependent) undercutting of slopes by the action of glaciers, at least three mechanisms of damage are possible, namely fatigue and progressive failure due to repeated variations of ice thickness and, if the pressure acts during a sufficiently long time, a redistribution of strain and stress within the rock owing to relaxation. This last-mentioned process shifts the non-stress conditions so as to induce extension as soon as the pressure decreases substantially. Hence cracks may develop, and both single stroke and fatigue scenarios may result.

Anyhow, it is nothing but natural that glacial periods are, as a rule, followed by frequent rockslides. The *time scale of this process*, however, is undergoing some corrections at the present time. For many decades it was an established school of thought that a majority of the large Alpine rockslides came down more or less soon after the recession of the ice. In fact, a look at Fig. 3.9b gives the impression of an immediately impending rockslide, and it appears plausible to consider glacier recession as a typi-

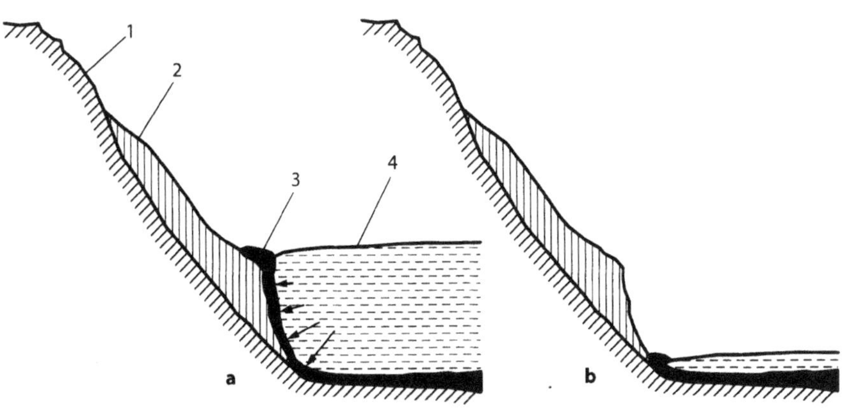

Fig. 3.9. Quasi-hydrostatic pressure of glacier upon rockslide-endangered mass. Cross-sections through valley before (**a**) and after (**b**) intense melting. *1:* bedrock; *2:* mass; *3:* moraine; *4:* glacier. *Arrows* indicate approximate direction and amount of pressure (sketch by Erismann)

cal single-stroke mechanism. But since it has become current practice to determine the age of the events by physical means (in first instance by radiocarbon dating), various assumptions had to be revised (Abele 1997b). Poschinger (1997) gives the ages of four Tyrolean and Bavarian events originally assumed to have occurred in late glacial or, at the best, in immediate postglacial time (for Köfels the most recent dating is quoted hereafter).

- ~8 700 B.P.: Köfels, Tyrol (Sect. 2.1, 2.4; Ivy-Ochs et al. 1998);
- ~3 700 B.P.: Eibsee, Bavaria (Jerz and Poschinger 1995, 389–391);
- ~3 500 B.P.: Hintersee, Bavaria (Poschinger and Thom 1995, 408);
- ~2 900 B.P.: Tschirgant (Sect. 2.1, 2.2; Patzelt and Poscher 1993, 209).

On this basis Poschinger raises the question of the *future development* of rockslide and rockfall catastrophes. He comes to the speculative conclusion that certainly a number of events followed the recession of the glaciers rather soon, but that in the meantime other causes have had the time to create new dangerous situations so that an increase of risks may be expected in future (all the more in case of further recession of the glaciers). It is not easy to judge this idea at the time being. As a matter of fact, only an evaluation of a far larger number of events might yield a statistically relevant result. In the hope that such an investigation will be carried out one day, it remains an open question, how far recent events are the rearguard of glacier recession (as insinuated by Noverraz and Bonnard 1991, 169, for the giant rockfall of Randa) or forerunners of phenomena developing clandestinely. When taking into account the possibility that, for instance, a series of fatigue loadings at the beginning of glacier recession may have been interrupted at any degree of damage accumulation, the rearguard hypothesis cannot be rejected without further ado.

*

In this context, besides the recession of glaciers, the problem of *permafrost* can be especially important. It is not intended here to go into the details of Büdel's (1977, 19–87) reflections on "Eisrinde". Yet, one has to be aware of the fact that ice in permafrost conditions is a resistant rock and that its thermal destruction can reduce in a catastrophic manner the stability of certain mountain-ranges. Gravity-driven events thus resulting often tend to come down in relatively small portions (Stötter et al. 1996). The endeavour, in some recent studies, to consider expected events from the standpoint of the material's stress resistance is most promising in the context of release prediction (Wegmann 1998; Wegmann et al. 1999; Funk, in press).

*

One of the most important force-generating mechanisms is due to *earthquakes*. Here the input of energy into a mass is influenced not only by the (longitudinal or transverse) earthquake waves but also by the relation between their frequency and the natural frequency of the mass. Obviously, if the natural frequency is definitely higher than the wave frequency, the forces exerted upon a mass essentially depend on the accelerations of the waves; in contrast thereto they depend on the spatial wave amplitudes if the relation of the frequencies is inverse; in between is the field of resonance (or, in view of the irregular sequence of earthquake oscillations: quasi-resonance) where the resulting oscillations of the mass reach their maximum. It is, therefore, worthwhile to consider the orders of magnitude of frequencies. The conditions of excitation (in particular the direction of the acceleration with respect to the mass) are assumed to be so as to produce the largest possible effect.

The highest amount of superficially tangible earthquake energy is observed in transverse waves with amplitudes decreasing in function of depth (Rayleigh and in particular Love waves). In the vicinity of the epicentre several shocks separated by seconds or minutes may be observed. Therefore quasi-resonance is due rather to the variety of wave paths between metacentre and considered mass than to the sequential order of shocks. Frequencies in the range above 5 Hz are the rule, and if such is also the natural frequency of a mass, resonance magnifications up to 2.5 can be observed. To quote an easy-to-follow example: a large mass as shown in Fig. 3.6, coherent in itself but having no lateral or proximal coherence with solid rock, obviously will assume a *shearing oscillation* and its natural frequency will be

$$f = \frac{\sqrt{G/\partial}}{4H} \tag{3.4}$$

where G is the shear modulus of the material, ∂ its density, and H the thickness of the mass. It will be observed that neither length nor breadth of the mass are relevant as long as shearing is the dominant mode of deformation. With $G = 5$ GPa and $\partial = 2500$ kg per m^3 a frequency of 5 Hz is approximately correlated with a thickness $H = 70$ m. This means in clear English, that quasi-resonance is not excluded and, under extreme circumstances, the maximum acceleration of a very heavy earthquake (acceleration 0.5 g) has to be multiplied by 2.5, thus yielding 1.25 g. There is, of course, no need to go into further details to demonstrate the dramatic possibilities of great earthquakes as single stroke release mechanisms. For the sake of completeness, however, it should be recalled that analogous results are easy to obtain with resonance in *bending oscillations* and that even in case of far smaller magnitudes still an efficient fatigue mechanism can be at work, especially in regions with frequent earthquakes.

Again the coincidence with field experience is obvious: gravitational mass displacements are an almost regular consequence of heavy earthquakes. If there is any need for illustration, a most vivid one is given in Muir's description of the great earthquake of March 26, 1872 that he had observed in Yosemite Valley, California, U.S.A. (1912, 59–65); once more, the catastrophe of Huascarán/Yungay is tragic evidence (Sect. 2.7); and one of the largest slides known on Earth, Saidmarreh (Iran), probably was triggered by an earthquake (Watson and Wright 1967, 126).

Eventually, it is important to note that also in the context of earthquakes a *size effect* is obvious, the stress for a given acceleration being proportional to the thickness of the material. This size effect, however, is superposed by resonance (if any) so that particular geometries may result in partial reversal of the general tendency.

*

Of course, volcanic activity connected with earthquakes can *change the slope*, a parameter up to now tacitly assumed as constant. In certain (and by no means rare) cases, however, much less is needed to steepen a slope from harmless to dangerous in spite of unchanged values for the impeding parameters (cohesion and/or coefficient of friction). If, for example, the slope diminishes from the head to the toe of a mass (Fig. 3.10), small events (e.g. erosion, rockfalls, etc.) may increase the effective slope angle: both the load of additional material at the head and the loss of material at the

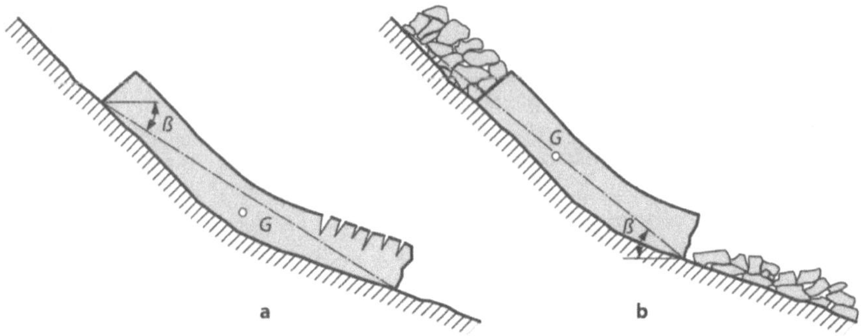

Fig. 3.10. Mass on slope decreasing from head to toe, situation (**a**) before and (**b**) after loading at the head and unloading at the toe, both resulting in a displacement of the centre of gravity G and an increase of the average slope angle β. Note that boosting is possible by dynamic effects at the head and rapid removal of debris at the toe (sketch by Erismann)

toe tend to displace the centre of gravity of the total mass uphill to the region of high slope (Hutchinson and Bhandary 1971). Obviously the static effect of the mechanism (slow, quasi-continuous loading/unloading) may transiently be boosted by dynamic conditions (sudden loading by material moving fast) or by conditions favouring rapid removal of the debris at the toe. And obviously an inverse (stabilising) effect occurs on a slope increasing from head to toe: a loss of mass at the toe, as far as it occurs in portions small enough to avoid catastrophic consequences, reduces the average slope and thus helps the main mass to remain on top.

In mentioning rapid removal of debris at the toe perhaps the most important mechanism of slope steepening is mooted. In fact, *flowing water* is responsible for so many displacements of loose rock material that its high transporting capacity, for the moment, needs no further explanations (it will be treated in more detail in another context under Heading 3.3). To state the essential point in other terms: a moderate volume of material removed by water may have played the part of an abutment for a far larger volume that, after having lost this last hold, will come down.

*

Perhaps it is useful to give a short summary of the *conclusions* obtained in this somewhat extended section.

11. As rocks usually are brittle and often rich in pre-existing cracks, the quantitative approach to cohesion should not be undertaken without at least a side glance to fracture mechanics and taking into account the special features of rocky material (in first instance nonhomogeneity and anisotropy).
12. However, in view of the large scatter bands of parameters in rock, the fracture mechanical approach means rather improved understanding of mechanisms than accurate calculation of particular cases (this situation may improve with better knowledge of materials).
13. Under the constant drive of gravity fracture mechanics would display a single-stroke character, thus contradicting the slow opening of cracks as normally preceding release. This dilemma can be solved by introducing fatigue, or, in cases of

creep over more extended distances, by progressive failure of interlocking surfaces. The last-mentioned mechanism is not necessarily based on fracture mechanics.

14. Acceleration of fatigue crack opening or creep by progressive failure is a signal of increasing danger.
15. The following mechanisms are able to generate repetitive loading as required for fatigue or progressive failure: thermal elongation; hydrostatic water pressure; freezing (sometimes combined with thawing); recession and re-growth of glaciers; earthquakes. At least the two last-mentioned mechanisms also may be strong enough to act by one single stroke. Slope-steepening mechanisms are essentially acting in a single stroke. Other mechanisms are rare in comparison with those mentioned.
16. In conventional, all the more in fracture mechanical cohesion as well as in several driving mechanisms (water pressure, melting of glaciers, earthquakes) size effects, all acting in the same sense, are present: if geometrically similar cases are considered, large events have a better chance of release than smaller ones.

More exotic possibilities of release are described under Heading 3.3.

3.3
Particular Mechanisms

The present section mainly deals with release mechanisms not very often encountered in connection with rockslides and rockfalls. Several particular cases even must be considered as definitely exotic and are presented without any pretension to practical use. Their interest lies rather in the fact that the immanent logic of an outstanding, perhaps impossible process may be helpful in grasping the essence of more normal cases. Other mechanisms treated hereafter have demonstrated their practical importance at rare but spectacular occasions. And two mechanisms connected with water (in one case flowing, in the other statically pressurised) even can play an important role in the frequent processes of slope creep and slope steepening (Sect. 2.6, 3.2).

*

To begin with one of the occasions mentioned as "rare but spectacular": the material of the Huascarán event (Sect. 2.7; Fig. 3.11) – as far as ice and moraine from the summit region are concerned – had been transported by the *external locomotive power of a glacier* to the top of the almost vertical wall where, shaken by the terrible earthquake of May 31, 1970, it lost balance together with further rock masses. The previous analogous events (January 10, 1962, and prehistoric), as well as the configuration having caused the tragic outcome, have already been discussed (Sect. 2.7).

*

While in the mechanism of undercutting a slope the effect of rapidly moving water is obvious (Sect. 3.2), a direct mobilisation of rock by pushing it over the edge on top of a steep slope hardly can reach a substantial scale: it normally cannot release

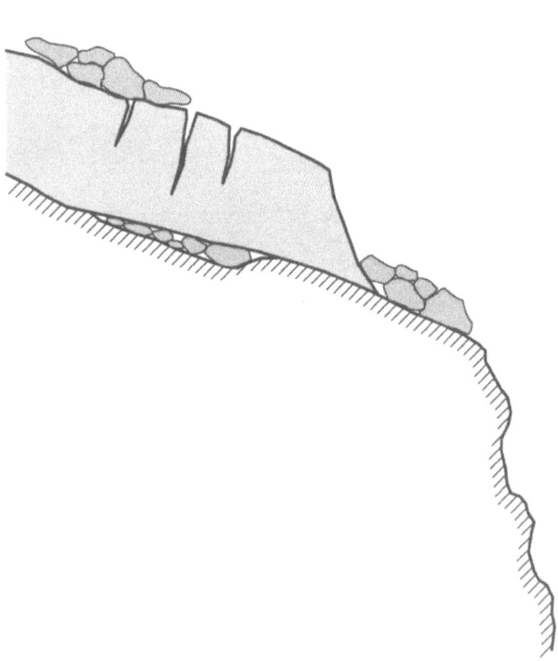

Fig. 3.11. Glacier transporting moraine and other rock material near the edge of a steep slope. Fractures due to cracks in the ice may break through and thus initiate a catastrophic event. In the case of Huascarán (1970), the triggering earthquake was strong enough to detach additional rock below the ice (sketch by Erismann)

more than single blocks or – at most – a rockfall of moderate size. Nevertheless the *transporting capacity of water* should not be underestimated as can be demonstrated by the following calculation.

The force F_v exerted by a turbulent flow of velocity v upon a completely immersed block with a sectional area A at right angles to the vector of v is

$$F_v = 0.5 A c \partial_w v^2 \tag{3.5}$$

where c is the dimensionless, shape-dependent drag coefficient and ∂_w the density of water. To move the block on approximately horizontal ground the friction

$$F\mu = Vg\mu(\partial_r - \partial_w) \tag{3.6}$$

has to be overcome. V is the volume of the block, ∂_r the density of its rock, g the gravitational acceleration, and μ the coefficient of friction. By equating the forces F_v and F_μ the critical velocity

$$v_c = \sqrt{\frac{V\mu g(\partial_r - \partial_w)}{2A\partial_w c}} \tag{3.7}$$

can be determined. It marks the transition from standstill to motion.

The difficulty in applying this formula consists in finding *realistic values for the parameters* involved. Only g is practically constant; ∂_r and ∂_w vary in relatively narrow bands (∂_w according to the amount of suspended fine-grained particles in the water); V and A can be measured or, at least, estimated for a particular case. The most problematic parameters are c and μ, the dimensionless coefficients. c depends on the more or less streamlined shape of a block (taking into account its angular position with respect to the vector of v) and can assume values between 0.2 and unity. μ is a function of the local conditions in the water-lubricated contact surfaces so that a range from 0.15 to 0.5 might be realistic.

In spite of such uncertainties, Eq. 3.7 may be useful to obtain the order of magnitude of v_c (and more if applied with due care). In Fig. 3.12 volume V is shown as a function of critical velocity v_c for a completely immersed, approximately spherical body, but reposing on a sufficient surface to exclude rolling. The chosen parameters are: $\partial_r / \partial_w = 2.5$, $c = 0.5$, $\mu = 0.25$. When interpreting the figure it must be borne in mind

1. that parameters in the middle of the considered ranges are used so that, if worst comes to worst, v_c may assume half the value suggested by the figure (as might be expected for a sharp-edged cube on slippery ground); and that in the opposite case a well-streamlined block may remain in place even at somewhat more than twice the velocity obtained from the diagram;
2. that dramatic changes of the critical velocity are possible if the block initially rests on inclined ground with a – positive or negative – slope angle β (for soft slopes μ may be replaced by $\mu - \tan\beta$);
3. that in case of partial immersion corrections have to be made – a somewhat tricky business as c may suffer unexpected changes, for instance owing to formation of waves;

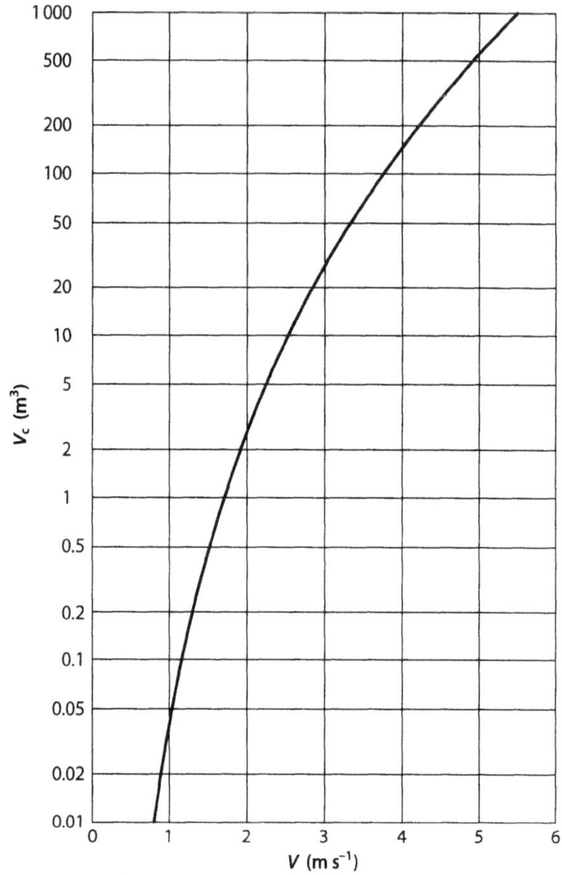

Fig. 3.12. Volume V of immersed, approximately spheric, but non-rolling block against critical velocity v_c of horizontally flowing water just able to overcome the block's resistance against sliding (Eq. 3.7). Main parameters: rock-to-water density quotient = 2.5; hydrodynamic drag coefficient $c = 0.5$; coefficient of friction $\mu = 0.25$. Scatter band of velocity for given volume is approximately between 0.5 and 2.0 of the values indicated by the curve (diagram by Erismann)

4. that appropriately shaped blocks (especially if completely immersed) may be displaced by rolling instead of sliding, thus making Eq. 3.7 obsolete and possibly reducing the critical velocity in a dramatic manner (for the sake of quantification it should be remembered that rolling on horizontal ground requires $\mu > \tan 45° = 1.0$ for a square section and $\mu > \tan 30° = 0.577$ for a regularly hexagonal one; s. also Sect. 5.3).

Anyhow, the figure strikingly demonstrates the remarkable *capacity of rapidly flowing water* to move even rather large boulders.

*

Not necessarily the first displacement after release of a block must occur by sliding or falling. As illustrated in Fig. 3.13, the step-by-step release of a mass affected with overhanging transversal cracks results in a *toppling* motion of slab-shaped elements. A further possibility of toppling release of an initially well-balanced tower-like mass may be due to yielding of the material at the tower's bottom, for instance owing to local erosion (Fig. 3.14a). And even after having started a sliding motion, a sufficiently tall mass may lose balance owing to variations of slope and/or friction. On the other hand, tower-shaped peaks are not necessarily released by toppling, especially if the

Fig. 3.13. Step-by-step release by toppling of slab-shaped blocks separated by cracks. Substantially larger distances between the fracture-generating cracks would result in sliding (sketch by Erismann)

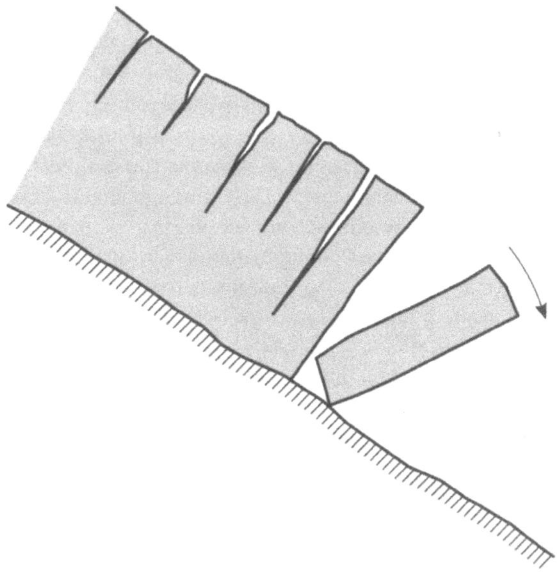

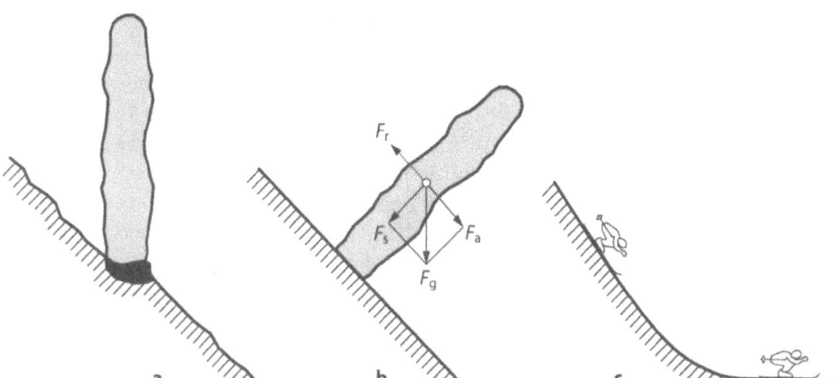

Fig. 3.14. Preconditions of toppling. **a** Tower-like mass with eroded zone (*black*) at its bottom; **b** mental experiment with slender tower sliding "upright" on steep, plane, perfectly frictionless slope. F_g = gravitational force; F_s = component perpendicular to slope; F_a = component accelerating mass parallel to slope; F_r = inertial reaction of mass balancing F_a; **c** skier in slope-adapted (not gravity-adapted) positions as illustrative approximation to **b** (sketch by Erismann)

supporting material has a tendency to sink and numerous pre-existing joints make a more internal collapse possible (Lehmann 1926, 280–281; 291–294).

By no means is forward toppling the only possible mode: every person unsuspectingly having set foot on an icy slope knows from experience that *backward toppling* may be an important alternative. And Howe, in his excellent description of slides in Colorado (1909, 53), remarks: "… *when large blocks slipped down their upper portions have fallen backward…*". In spite of being remarkably fascinating, the discussion of a general theory of combined sliding and toppling would go beyond the frame of this book. One example, however, incongruous at first sight but trivial after due consideration, may serve as a spotlight. In case of negligible friction and sufficiently smooth transitions between different slopes, a well-shaped, tall prismatic body never will lose its equilibrium, whatever the slope angle will be (Fig. 3.14b). An impressing illustration is given by the fact that downhill skiers even on the steepest slopes stand "upright" with respect to their skis (and not to gravitation, Fig. 3.14c). A marginal note should be made in this context: as a consequence of the almost complete absence of friction the moving body, on a sufficiently long slope, accelerates until gravitation is balanced by air drag; but even then stability is not lost as long as the drag centre is located not too far from the centre of gravity…

In itself, toppling is, as a rule, not of fundamental importance for the further displacement of the mass. Of course, a certain amount of kinetic energy results from the rapid loss of height. But a substantial part of this energy is dissipated in the collision with the ground. Significant effects are rather due to the mode of *disintegration after toppling*. Anticipating the considerations of Chap. 4, three possibilities are presented.

11. A coherent mass may remain intact after toppling and continue its travel by sliding (Fig. 3.15b). Then, owing to the contribution of the energy acquired in toppling, its kinetic energy (in other words: its velocity) will slightly exceed that of an analogous (though not very probable) sliding process in an upright position (Fig. 3.15a).
12. The mass may be broken to pieces when hitting the ground, thus being transformed into smaller blocks which will roll, slide, or fall according to the local circumstances (Fig. 3.15c). This means an extremely early transition from coherent to disintegrated motion with all consequences resulting therefrom (Chap. 4; Sect. 5.1).
13. If coherence at the top of the mass is bad right from the start of toppling, the interesting phenomenon of "catapulting" is not excluded: a block on top of a tower-shaped mass, interlocked though not coherent with the rest of this mass, may take part in its toppling motion until the centrifugal force pulls it off. Such a block can reach a sensibly higher horizontal velocity than the bulk of the mass (Fig. 3.16).

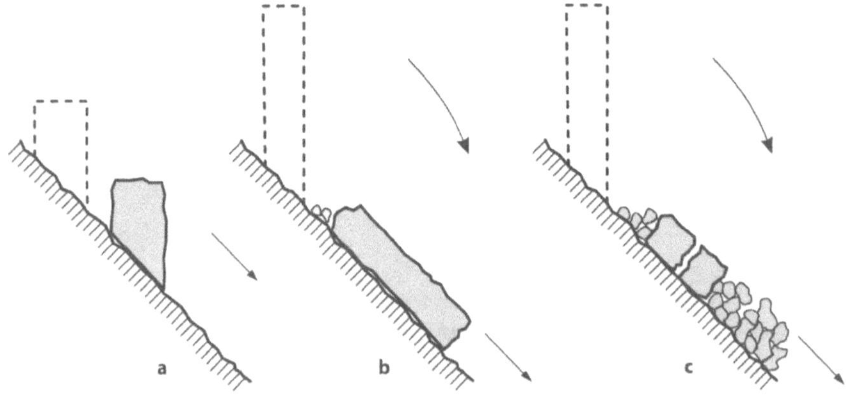

Fig. 3.15. Possible case histories of tower-like masses. **a** Sliding (only in particular cases, if friction and slenderness low); **b** toppling and sliding in coherent state; **c** disintegration in toppling (most probable case) and displacement as debris (sketch by Erismann)

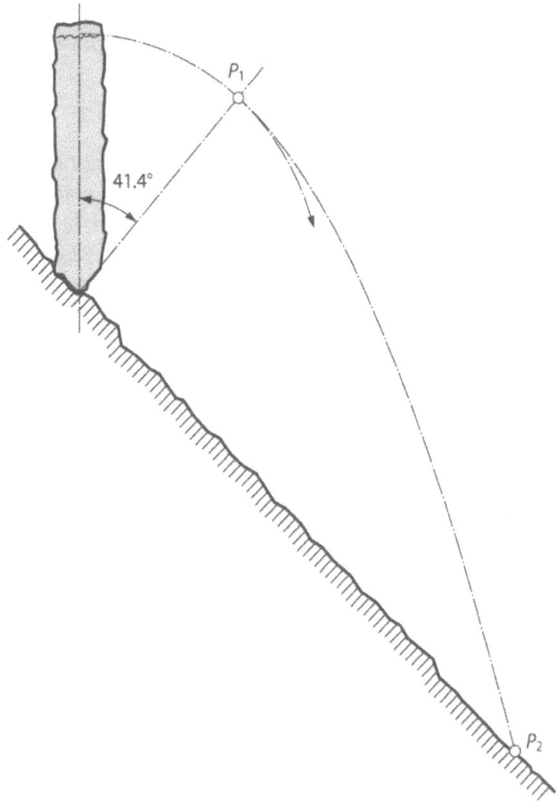

Fig. 3.16. Catapulting of block, interlocking though not cohering on top of toppling prismatic tower. P_1 = taking off point; P_2 = landing point where the velocity of the block may be approximately twice that of the main mass, in case of conical towers even more. Note: for ideal (infinitely slender, frictionlessly rotating) prismatic towers take-off angle is independent from height (sketch by Erismann)

The key-word "*catapulting*" signals a threat: it is not excluded that after toppling relatively small – though still dangerous – parts of a mass will move faster than the rest and possibly also reach farther. Thus it is justified to consider the mechanism a bit more in detail. As a typical (and easy-to-treat) example a prismatic rock column with a comparatively light block (of dynamically negligible mass) on its top is chosen. And for the sake of simplicity only the most probable configuration is analysed in which a backward displacement of the column's bottom is excluded. (As a curiosity it might be remarked that in case of a frictionless contact with a horizontal ground toppling would consist of a rotation of the column around its centre of gravity which would fall vertically.)

Assume in the configuration of Fig. 3.16 the prism to be "infinitely" slender and to rotate without friction around its bottom. Then, by equating its gravitational to its kinetic energy the velocity at its top

$$v = \sqrt{3gL(1 - \cos\beta)} \tag{3.8}$$

can be determined. g is the gravitational acceleration, L the length of the column, and β the angle of its momentary inclination. In the moment of taking off the centripetal acceleration of the block on top of the column, $3g(1 - \cos\beta)$, must equal the centripetal component of gravitational acceleration, $g \cos\beta$. The resulting equation

$$3(1 - \cos\beta) = \cos\beta \tag{3.9}$$

yields $\cos\beta_c = 0.75$, $\beta_c = 41.4°$, the critical velocity $v_c = 0.866\sqrt{(gL)}$, and the horizontal component thereof, $v_{cx} = 0.650\sqrt{(gL)}$. In order to obtain a reasonable comparison with the movement of the column, the maximal horizontal component of velocity of its centre of gravity $v_{xm} = 0.333\sqrt{(gL)}$ is calculated by appropriate differentiation of Eq. 3.8. The respective angular values are $\cos\beta_m = 0.667$ and $\beta_m = 48.2°$. These figures mean in clear English: under the assumed conditions the block on top of the column attains an initial horizontal component of velocity 1.95 times higher than the column itself.

This example, in itself neither typical nor even particularly probable, reveals the fact that single blocks, initially perching on top of tall rock bodies, may after toppling precede the rest of the mass at approximately the double starting velocity. So, if the following travel of the mass is not too long and not too uneven, it is by no means excluded that such blocks never will be caught up by the bulk of the mass and thus may produce damages at *unexpectedly distant* locations.

*

Besides other effects of *hydrostatic water pressure* as mentioned in Sect. 2.2, 3.1, and 3.2, particular importance was attributed to the circumstances preceding the descent of the Vaiont slide (Sect. 2.6) where a large mass of rock lost its stability under conditions resembling in more than one respect those of floating. So it is perhaps worthwhile to look into the details of this phenomenon that certainly occurs more frequently on a far smaller scale.

Consider the conditions required to start the downhill *displacement of a slab-shaped block*, partly supported by the pressure of a coherent water cushion and thus enabled to get over an arresting obstacle (Fig. 3.17a). Assume further that somehow the water is enclosed below the block and independent from external pressure sources. Eventu-

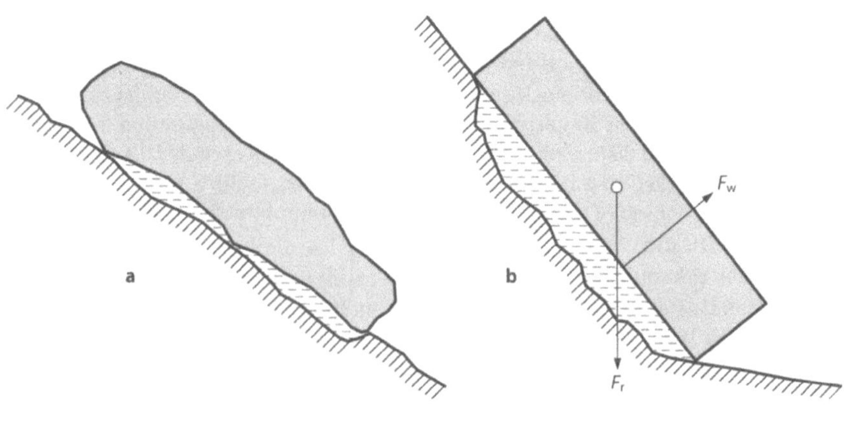

Fig. 3.17. a Water-filled void under flat block as mechanism favouring release by pushing the block off the slope; **b** rectangular slab as easy-to-calculate model. F_r = gravitational force of rock; F_w = force generated by hydrostatic pressure of water (sketch by Erismann)

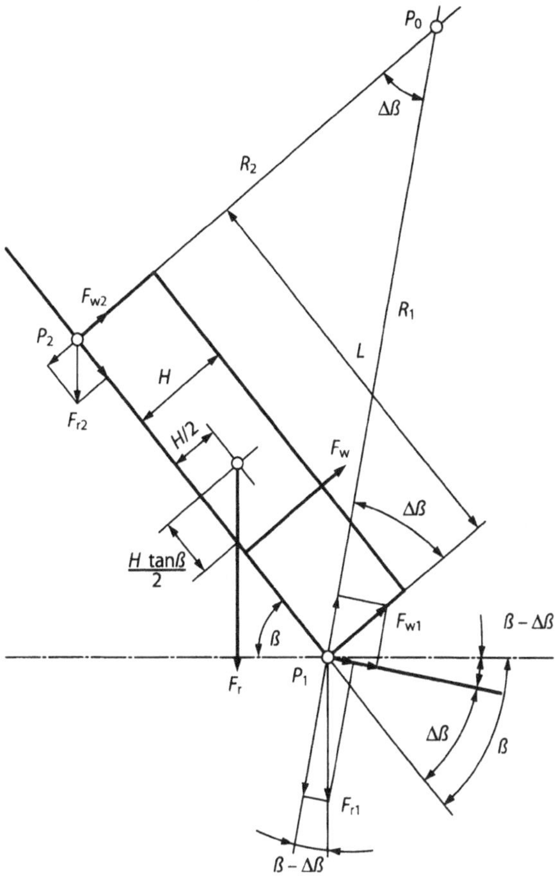

Fig. 3.18. Forces F_r and F_w (s. Fig. 3.17) and their components F_{r1}, F_{r2}, F_{w1}, F_{w2} in the supporting points P_1 and P_2. H and L = thickness and length of slab, respectively; P_0 = centre of incremental rotation in case of sliding; R_1 and R_2 = radii $P_0 - P_1$ and $P_0 - P_2$ respectively; β = slope angle; $\Delta\beta$ = additional angle representing obstacle (sketch by Erismann)

ally, to simplify things without serious violation of the basic conditions, replace the block by an appropriate rectangular slab of length L and thickness H (Fig. 3.17b). By exception of the reactions from the ground and friction, only two forces are active, namely weight F_r and buoyancy F_w. As pressure is proportional to the local water depth, buoyancy is obviously excentric. The situation (similar, though extremely simplified in comparison to that observed in Vaiont valley), when considered in Fig. 3.17b, gives the impression of being quite simple.

First doubts with respect to such seeming simplicity will turn up when looking at Fig. 3.18 where the basic information required to calculate the conditions of motion is presented. It is not intended to bother the reader with a detailed deduction of the somewhat complicated Eq. 3.10. For the points of contact P_1 and P_2 this formula yields the *critical coefficient of friction* (assumed as being equal for both points),

$$\mu_c = \frac{\left[6(h+h^2\tan\beta)\sin(\beta-\Delta\beta)+4q\sin\beta\sin\Delta\beta\right]_1+\left[6(h-h^2\tan\beta)\sin\beta\cos\Delta\beta\right]_2}{\left[6(h+h^2\tan\beta)\cos(\beta-\Delta\beta)-4q\sin\beta\cos\Delta\beta\right]_1+\left[6(h-h^2\tan\beta)\cos\beta\cos\Delta\beta-2q\sin\beta\cos\Delta\beta\right]_2} \tag{3.10}$$

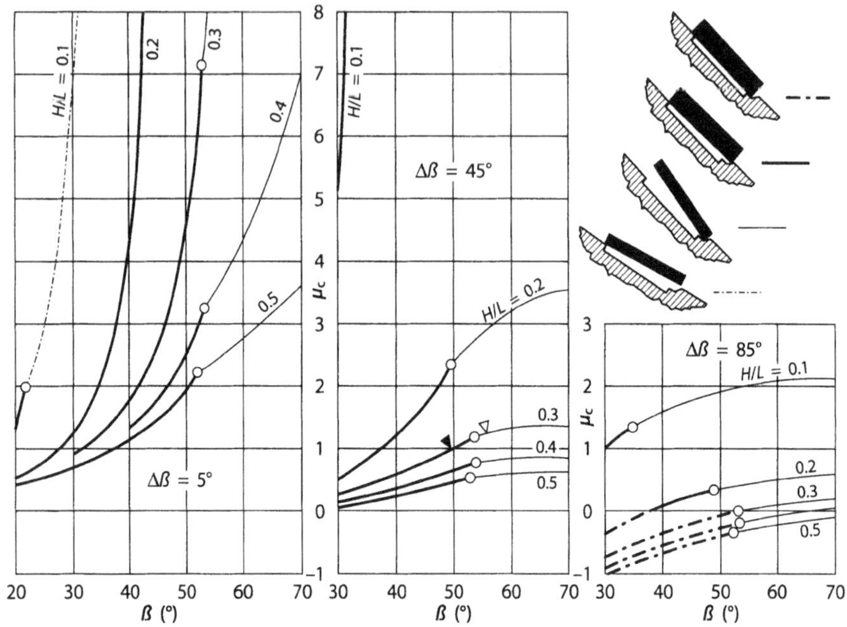

Fig. 3.19. Quantitative effects of water-filled void under flat block (s. Fig. 3.17 and 3.18). Critical coefficient of friction μ_c (above which no sliding is possible) as a function of angles β and $\Delta\beta$, and relative thickness $h = H / L$. Curves: *bold dots and dashes*: immobility; *bold*: sliding; *fine*: lifted at proximal end; *fine dots and dashes*: lifted at distal end (according to qualitative character shown on top right). *Triangles* in central graph refer to example presented in text (diagram by Erismann)

below which the slab will move. Partial results (s. brackets with indices 1, 2) signal possible lifting of P_1 or P_2. To use the results of the equation it is not necessary to go into further mathematical details. For a water-to-rock density ratio of $q = 1:2.5$, Fig. 3.19 gives an idea of what can be expected if the approximate relative thickness $h = H / L$ of a slab-like block, the main slope angle β, and the differential angle $\Delta\beta$ are known. In the respective graph (given by $\Delta\beta$), the critical coefficient of friction μ_c is simply the value of the ordinate for the given configuration of parameters. An example: assume $H / L = 0.3$, $\beta = 49°$, and $\Delta\beta = 45°$; then μ_c is found in the central graph at left of the crossing of curve 0.3 with the vertical line for $\beta = 50$, i.e. at a value of 1.0 (*black triangular mark* in Fig. 3.19). As usual coefficients of friction definitely lie below this figure, there is a good reason to expect that the block, in this particular case, will move.

Additional information is given by the character of the curve at the considered point. In the case assumed here-above the curve is bold. The meaning thereof is shown on top right in the figure: the block accomplishes pure sliding. Now let the slope angle increase from 49° to 55° (*contoured triangle*). The critical coefficient of friction is increased by about 0.3, but the line is changed from bold to fine. This signals that the water pressure is sufficient to *lift the block* at its proximal end, with the possible consequences described under Item 5 hereafter.

For readers wanting to know more about the *line of thought behind Eq. 3.10* and Fig. 3.19, the following remarks, as well as some details of the equation, may be helpful.

21. $q = \partial_w / \partial_r$ is the relative density of water referred to that of rock; $h = H / L$ is the relative thickness of the slab; $\beta - \Delta\beta$ and β are the respective angles of the slopes on which points P_1 and P_2 repose. It will be observed that all parameters used in the equation are dimensionless. In addition, neither the absolute size nor the width of the slab appears in Eq. 3.10. These conditions hold good only if the coefficient of friction is not affected by size and the relation between supply and escape of water is sufficient to keep the space below the slab well-filled.
22. An incremental displacement of the slab is considered as a small rotation around point P_0 (Fig. 3.18). So equilibrium is given if the total momentum around P_0 is zero. This momentum is the sum of two partial momenta, one generated in P_1, the other in P_2. Each of these momenta is the product of the respective radius R_1 or R_2 and the total force acting in P_1 or P_2 parallel to the local slope. And finally, each of the last-mentioned forces consists of the locally acting components of F_r and F_w parallel to the slope and of the friction generated, mediating μ_c, by the components of F_r and F_w perpendicular to the slope.
23. To visualise the contributions of F_w and F_r at P_1 and P_2, Eq. 3.10, beyond its mathematical significance, bears arithmetically insignificant information: the square brackets and their indices 1 and 2 signal points P_1 and P_2 respectively. The brackets in the numerator represent the components of F_w and F_r acting parallel to the respective slope; the brackets in the denominator represent the friction-generating forces perpendicular to the slope.
24. All terms containing q are consequences of F_w; all terms containing h are consequences of F_r.
25. Friction only can occur if the slab is pressed against the slope, i.e. if the expression in the respective square bracket of the denominator is positive. Now negative values are not excluded, and they have an important physical meaning: as soon as the sign changes to negative, the slab is lifted off the slope at the point signalled by the index 1 or 2. So the four possibilities represented in Fig. 3.19 on top right are obtained (from top to bottom: remaining in place; sliding; sliding at P_1 alone; sliding at P_2 alone).
26. It has to be taken into account that lifting at one of the mentioned points is most probably an unstable process: it normally opens so large a passage for escaping water that, at least for a short time, F_w breaks down and the slab falls back upon the slope. A repeated ("rattling") cycle of being lifted and falling back is not excluded. And, of course, this phenomenon may be accompanied by step-by-step creeping.

As local small-scale mechanisms, the described effects of buoyancy may perhaps be harmless. But they also can effectively contribute to the step-by-step (creeping) descent or even to the collapse of a talus, especially if, in presence of abundant water, many slab-like blocks are involved.

*

Perhaps it is not entirely correct to consider events caused by *human activity* as rare. Certainly they are rare if compared to some other classes, but their percentage of tragic consequences is definitely higher than proportional. As a matter of fact, some of the heaviest catastrophes in history would not have taken place if mankind not had oversteepened slopes which might have remained stable for centuries or millenia without human intervention. The reason of the particular heaviness of such events is easy to understand: to impair seriously the stability of a mass weighing many millions of tons is a rather substantial piece of work unlikely to be done (and the results to be sold) by people dwelling far away. So, as a rule, man-caused tragedies happen in areas with a comparatively dense population.

Let – besides Vaiont, the heaviest European rockslide disaster of the 20th century (Sect. 2.6) – three other famous *historical examples* stand for a number of less spectacular cases.

In the 17th and early 18th century the municipality of *Piuro* (in Latin and German publications: Plursium and Plurs, respectively) in Val Bregaglia (German: Bergell), northern Italy, flourished owing to a successful steatite quarry and the trade connected therewith (Fig. 3.20; Bertrand 1757, 54; Heim 1932, 180–181, 186–187; Martin 1963/65, 9; Abele 1974, 193). From the viewpoint of present knowledge it was nothing but natural that the location of the quarry, the northern slope of Monte Conto, facing the town, more and more became oversteepened. Nevertheless, from the viewpoint of the inhabitants (who had seen several rather harmless rockfalls) it also was natural not to think of escape when, on August 25 (some authors speak of September 4), 1618, more stones than usual fell. And even the warnings of terrified dairymen from the mountain pastures were received with derision. The result, a couple of hours later, was an almost completely destroyed town and a large number of killed persons (the reports ranging between 900 and 2500).

A rather similar situation is reported from *Elm* (Glarus, Switzerland), owing to Heim's classical descriptions (1882a, 1882b, 1932, 109–111, 199–207) perhaps the most-cited Alpine rockslide. In this case the object of the quarry was an excellent slate-schist in the – more and more undercut – northern slope of the Tschingelberg. And also the reactions of the inhabitants were similar: on November 11, 1881 (a Sunday), in spite of frightening signals from the region of the quarry (repeated rockfalls in the course of several hours) many curious persons came from neighbouring villages directly into the threatened zone "*to have a look at the mountain...*". Luckily only part of the village was engulfed by the mass so that the death toll did not exceed 119 persons.

In the case of the *Frank* rockslide of April 29, 1903 (Alberta, Canada) not all the circumstances were similar to those reported for Piuro and Elm (McConnell and Brock 1904; Cruden and Krahn 1978). In particular different opinions have been expressed with respect to the proportional contribution of human activity in starting motion. Kent (1966, 81), for example, considered it as "*... possible that a mine subsidence triggered the landslide...*", while Howe (1909, 52) presented the more cautious opinion that "*... opening of large chambers in the coal mine... may have been a contributory cause...*", and, several lines later, the very trigger was identified as the heavy frost on the morning of the slide (citation in Sect. 3.2). Anyhow, a certain role of the mining activity near the sole of the steep eastward-dipping slope of Turtle Mountain seems to be a fact. And, in parallel with Piuro and Elm, mining had been an important reason for the development of the town of Frank, especially in view of the Canadian Pacific Railway crossing the Crownest Pass nearby. A lucky similarity with Elm: Frank was only partly destroyed, and no more than about 70 lives were taken.

It is an open question how far, despite the world-wide growth of population and its increasing density in zones potentially exposed to risks, improved knowledge will sooner or later counterbalance the notorious *human indifference* to hidden (and sometimes even to clearly perceptible) danger. The examples of Piuro, Elm, Frank, and Vaiont

(Sect. 2.6) cannot be regarded as favourable auguries. And developments like the uncontrolled destruction of woods in the Himalaya are not a good omen, at least as far as relatively small events are considered. On the other hand, Val Pola (Sect. 2.5) shows that determined, well-founded, and well-organised interventions can save many hundreds of human lives.

*

Like volcanoes and earthquakes, *extraterrestrial causes* lie at the other end of the scale of human (in)ability to interfere. Obviously, vibrations coming from the impact of a meteorite are able to trigger a rockslide as well as those provoked by an earthquake. The problem is not to show that the mechanism is possible, but rather to find out the – comparatively rare – cases in which a celestial body can be inserted into the causal chain preceding a particular mass displacement. This question was answered in connection with Köfels (Sect. 2.4) after a critical analysis of the site as a potential astroblem and the simultaneous setting up of a hypothesis granting better compatibility with the geological and geomorphological evidence. Things would be far more difficult if an essentially normal event were triggered either by a small nearby incidence or by a more remote larger one. The trouble, above all in the last-mentioned case, consists in the fact that a clear causal link between a meteorite (even if its occurrence is unquestionable) and a rockslide or rockfall is difficult to establish. In such instances, a higher probability often speaks in favour of an easy-to-detect earthquake.

It seems that Köfels exerts a special attraction upon researchers interested in extraterrestrial phenomena. Not only Tollmann and Tollmann (1994) consider Köfels as one of the locations hit by a world-wide series of *comet debris*: also Surenian (1988, 1989) expresses the opinion that certain features of the rocks, in particular traces of shock metamorphism, point to a celestial body as a trigger. These hypotheses lie too far from the essential themes of this book to be discussed here in detail. It may, however, be of some use to comment in short on the main criticisms raised against them. Lyons (1993) and Leroux and Doukhan (1993) showed that the phenomenon regarded as an impact-generated shock metamorphism had been a consequence of mere heating (as could be expected in view of the abundant frictional heat in the event); Deutsch et al. (a group of thirteen distinguished scientists from seven countries, 1994) analysed Tollmann and Tollmann's hypothesis point by point and found a large number of methodical and logical flaws; Heuberger (1996), one of the best-informed experts as concerning Köfels, gave (in German) a short, though complete overview of the problems and added several remarks of his own. To quote an example: the radiocarbon ages of a number of slides claimed to have been released by the impact of Köfels vary, roughly speaking, between 3 000 and 9 000 years B.P.

All this is by no means intended to exclude a priori the possibility (and in some cases the necessity) of rockslides being released by the impact of celestial bodies. It only means that for Köfels a definite proof of such a cause does not exist.

*

Once the discussion being focused on exotic mechanisms, another problem connected with release should not be left aside. From time to time, orally as a rule, the question is raised what happens, in the moment of release, with the *elastic energy* stored in a coherent mass. Chowdhuri (1980, 168) considers the possibility of this energy acting as an additional accelerating booster at the start of a mass. In applying the idea to Elm, he comes to the surprising result: "... *that the portion of released* (elas-

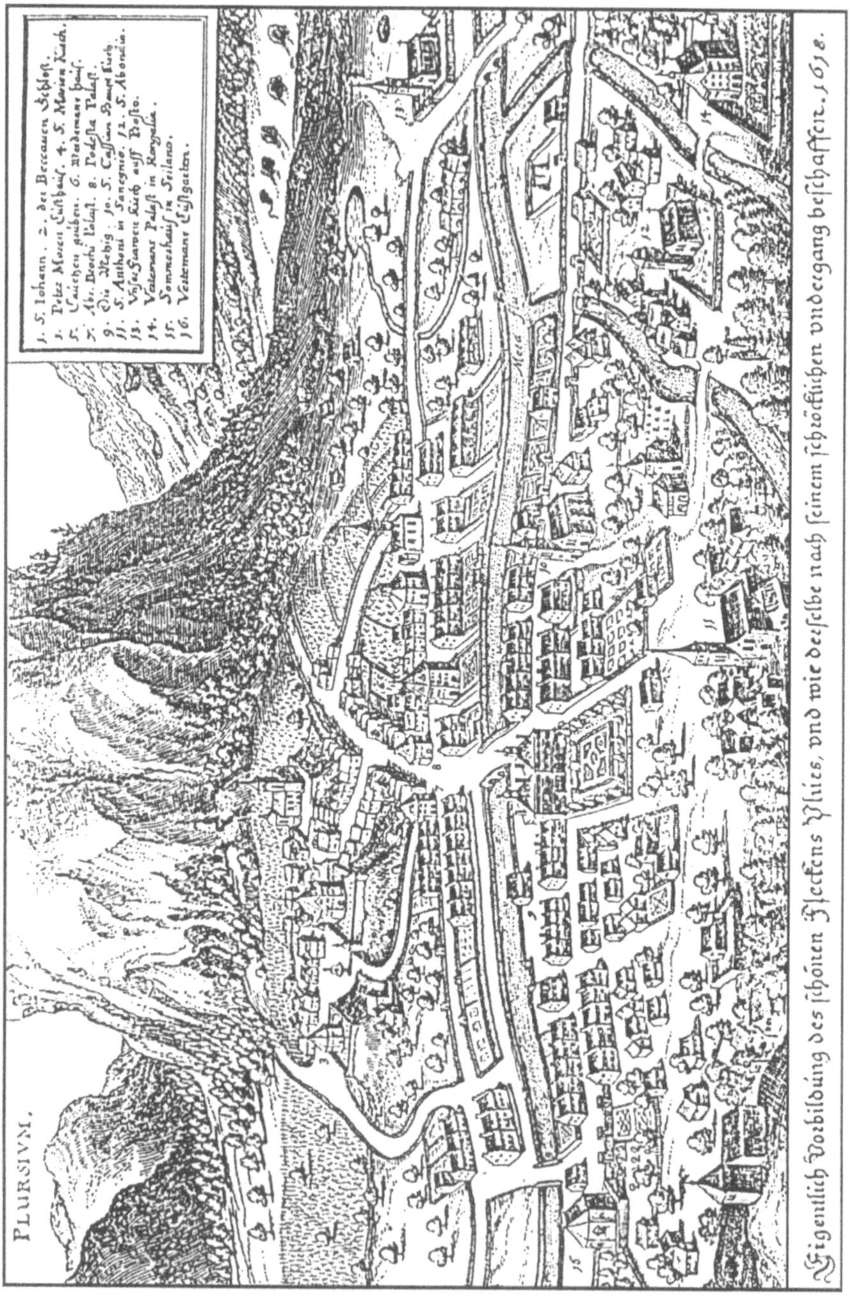
PLURSIVM.
Eigentlich Vorbildung des schönen Fleckens Plüers, und wie derselbe nach seinem schröcklichen Undergang beschaffen. 1618.

a

b

Fig. 3.20. Piuro (Plursium, Plurs) rockfall, general view from north before (**a**) and after (**b**) the event of August 25 (or September 4), 1618. Note (in **a**) the oversteepened, partly overhanging excavations of the quarry in the slope of Monte Conto and (in **b**) the scar, the dimensions of the blocks, and the lake formed by the Mera having been dammed (reproduced from copperplate engraving, Merian 1642/54) tic) *energy would have been sufficient to impart an initial velocity of 53.13 m/sec to the mass.*" In other words: it is assumed that the mass, immediately after release, could have been catapulted to a velocity ranging between 30 and 50 m s^{-1} like a paper projectile by a boy's rubber sling. This statement cannot pass uncommented.

The following criticism is not aimed at minor problems like the adoption of errors committed by Heim (1932, 143–150) and his physical councelor Müller and Bernet in forgetting to use the centre of gravity for calculation of velocity (Sect. 2.8, Item 14; Sect. 6.3). Instead this criticism is aimed directly at the *energetic basis* of the presumed mechanism, and it can be condensed into one single equation. Energetic equivalence between the elastic energy stored in a mass (*E:* modulus of elasticity; *s:* average stress) and the kinetic energy of the same mass in motion (*v:* velocity; *∂:* density) is given if

$$\frac{s}{\sqrt{E}} = v\sqrt{\partial} \tag{3.11}$$

is true. In case of correct application it does not matter what kind of stress (extension, compression, shear, etc.) is considered. As E and ∂ are given by the material, the connection between v and s is utterly simple. The energy is limited by s which has to be below the stress the material can withstand. Now assume, as a first working hypothesis, that a high-strength material is stressed in the direction of its highest resistance and that its parameters are $E = 15$ GPa; $s_{max} = 300$ MPa, $\partial = 2500$ kg m^{-3}. Then 50 m s^{-1} is the absolute maximum of velocity. The problem is, however, how a mass, located in labile equilibrium on an inclined surface and almost having lost its coherence with this surface, should acquire an evenly distributed longitudinal compressive stress of 300 MPa, corresponding to that at the sole of a column, 12 232 m high and made of the proposed material!

After this simple argument further considerations about the physical nature of such excessive stress, about its distribution in the mass, about the efficiency of conversion from elastic to kinetic energy, etc. are obsolete: purely and simply, the elastic energy available in reality, even if it would act in the right direction and could be converted into kinetic form without any loss, still would remain too small by at least two powers of ten.

Perhaps a day will come when it will be recommendable to rack one's brains over extraterrestrial triggering of rockslides. But elastic energy as a booster for launching a cubic kilometre of rock may – at least on the Earth – be forgotten without suffering from a guilty conscience…

Chapter 4

Mechanisms of Disintegration

It is convenient that we consider, before doing anything else, what kind of mechanism takes place in the fracture of a piece of wood or other solid...

Galileo Galilei (original in Italian)

4.1 Static Disintegration

Before being released, a mass may have assumed any possible state of coherence ranging from one single giant block to a heap of comparatively small particles. By contrast, in coming to a rest after displacement, the same mass is, as a rule, substantially more disintegrated. So disintegration obviously takes place in the course of the downhill ride. As, on the one hand, the discussion of pre-event disintegration does not fit into the frame of the present book, while, on the other hand, the degree of coherence (or of disintegration, if looking at things from the other end) can substantially influence the mechanisms of displacement, disintegration might with good reason be attributed to Chap. 5 where these mechanisms are treated. However, there is no doubt that the conversion of a more or less coherent large block into debris marks an important caesura in the descent of a rocky mass so that it cannot be insignificant where the process of disintegration is initiated, how it develops in the course of motion, and what are its consequences. Thus it is equally justified to reserve a short *chapter of its own right* to the questions of disintegration. In deciding to act according to this line of thought, two further facts were borne in mind: even so Chap. 5 remains one of the most extended, and – a more substantial argument – in the sections to follow it will be demonstrated that the fundamental differences in reach between coherent and disintegrated motion are not as dramatic as could be assumed at first sight. Continuity between the questions of disintegration and displacement will be established in the first section of Chap. 5 which will deal with the influence of disintegration upon displacement.

*

Even in a perfunctory review of longitudinal sections at least two important modes of disintegration easily can be made evident: collision-free fracturing of large bodies (for instance by bending) at not too sharp bends of slope, and crashing by collisions (for instance after a bounce or a fall, or in banging against a wall). In all, however, there are *four basic modes of disintegration*: the third and the fourth, in spite of being almost ubiquitous, at least near the bottom of a moving mass, are less important with respect to the disintegrated volume. Still they are able to influence motion to a non-negligible degree by producing large amounts of small-sized particles. One of these modes consists in coherent or already disintegrated rock being crushed under the load of a large overburden (for details s. hereafter), the other in the "ploughing" and "grinding" process between uneven surfaces under friction.

In spite of not having been selected for the particular purpose, five of the key events of Chap. 2 (Blackhawk being exempt owing to lacking information about the details of its initial phases) display excellent possibilities for finding the *locations of disinte-*

gration. In the case of Huascarán, for instance, the bang after the initial fall over an extremely steep wall many hundreds of metres high (Fig. 2.48) almost entirely excluded, right from the start, the continuance of a coherent state. At the other end of the scale is Vaiont where the coherence of a large part of the mass (mainly the "seat" of the initially chair-shaped longitudinal section F–F' in Fig. 2.36; s. also Fig. 2.40) was demonstrated by bridging the gorge of river Vaiont immediately before being arrested. Pandemonium Creek (Fig. 2.4), besides a less dramatic fall than that of Huascarán, suggests a possible bounce at km 2.1 and does not exclude a collision with the opposite slope at km 3.1. Val Pola is similar thereto in having almost angular bends at km 0.68 and 0.91 (Fig. 2.30), and the opposite slope at km 1.97, not to speak of the "bulwark" of Plaz (km 2.4; Fig. 2.31). The mass of Köfels probably came down with a minimum of disintegration until it was cut into an upper and a lower part, the last-mentioned being brutally arrested by the wall of the opposite slope; so the deposits on top contain immense blocks, while the bottom on the floor of Oetztal consists of effectively disintegrated material (Fig. 2.18, 2.20, 2.21a).

*

At first sight it may seem that collisions must play the role of a strictly dominant mechanism. Despite their obvious importance, this opinion needs a certain relativation. In fact, a mass, if released as a giant block of more or less coherent material and accomplishing its travel without encountering a sufficiently massive obstacle, never would be split to pieces if there were no other opportunities of fracture. The topographic background of the mechanism thus postulated is trivial: as the track of a moving mass cannot be perfectly congruent with its bottom surface during the whole travel, stresses in the bulk of an initially coherent mass necessarily will be generated by its own weight (Fig. 4.1). If any, also dynamic (i.e. mainly centrifugal) forces have to be taken into account. And such stresses, especially in the presence of cracks or other weak zones, can exceed the strength of the material. Hence there is, in rockslides more than in rockfalls, a considerable probability that collision-free, *static fracture* occurs before collisions substantially have contributed to disintegrate a mass.

Accordingly, in a publication on the subject (1988), Erismann speaks of "first fracture". This study was not aimed at a general description of disintegration processes. It was, in fact, undertaken in the intention to explain the hitherto enigmatic *variety of*

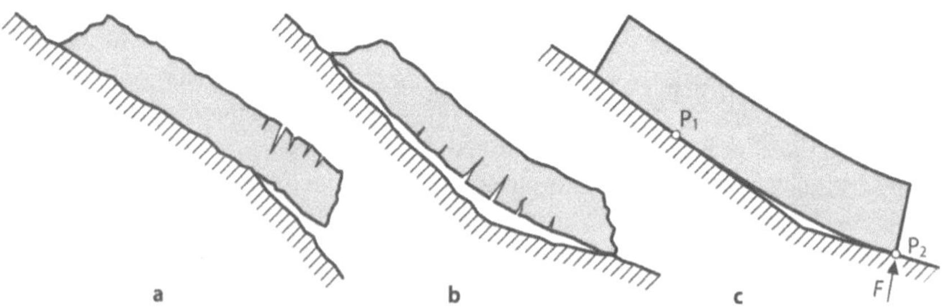

Fig. 4.1. Disintegration of coherent mass by static (collision-free) fracture in bends of increasing (**a**) and decreasing (**b**) slope. Possible locations of fracture are indicated by cracks. **c** simplified model of **b** for quantitative analysis. P_1, P_2: limits of span bridged by mass; F: force acting in P_2. Deformations are drastically exaggerated (sketch by Erismann)

particle sizes in the deposits of large rockslides: giant blocks may be observed in the immediate vicinity of small-sized debris, and the appearance of the material in the respective sites may vary between perfectly solid and sandy. Yet the results turned out to be of a certain interest in a more general sense so that it is considered worthwhile to summarise in short the fundamentals without going into mathematical details.

But before doing so, the *term "static fracture"* requires a comment for readers not familiar with materials testing. It means, roughly speaking, fracture under single-load conditions (no repetitions as in fatigue) in the velocity band above elastoplastic creep, but below high-speed phenomena (e.g. shock waves) affecting strength (e.g. reducing toughness). If, for instance, the maximal strength of a body just suffices to bridge the distance between two points P_1 and P_2 (Fig. 4.1c), the respective critical stress is built up during the motion of the body's front from the first to the second point. The required time, according to the circumstances, may be something between substantial fractions of a second and several seconds. By contrast, in a collision stresses are built up in milliseconds or even faster. Eventually it should be remarked, in parentheses, that a fracture, according to the above definition, is determined as static exclusively by the velocity of the process, irrespective of the nature of the stress-generating force; in particular, fractures of this kind remain static even if provoked by dynamic (e.g. centrifugal) forces.

The following *simplified models* were used in the study.

1. The most probable cases of static fracture are generated by bends of slope. As the normal sequence in the case history of a rockslide is release on a steep slope, followed by reduced slope angles, static fracture is more frequent in upward (decreasing slope, Fig. 4.1b) than in downward direction (increasing slope, Fig. 4.1a). Therefore in the study only the first-mentioned case was considered in detail. By the way, upward bending angles should not be excessively large: otherwise friction might act as a more or less self-arresting mechanism (as shown in Fig. 2.20) and thus induce crashing.
2. The local ground surface was modelled as consisting of two plane sections with slightly different moderate slopes, and the mass was assumed to be of constant rectangular section and consisting of a homogeneous, though anisotropic, brittle material, affected by weak zones (idealised by sharp-edged cracks), in each case parallel to one of the body's surfaces (Fig. 4.1a,b).
3. Although it was obvious that centrifugal accelerations in certain cases cannot be neglected, gravitational acceleration alone was considered as a loading agent (Sect. 5.7). It should be recalled in this context that static fracture is possible even at almost zero velocity.
4. While the neglect of centrifugal accelerations resulted in reducing the calculated forces in case of rapid motion, an opposite effect also was neglected: slope angles were assumed to be small enough to replace their cosine by unity.
5. By no means was bending (Fig. 4.2a) considered as the only possible mode of static fracture: also shearing, both in transversal (Fig. 4.2b) and longitudinal (Fig. 4.2c) direction, was not excluded. For the last-mentioned mode the term "delamination" was used.

As, in spite of using quantitative methods, *qualitative results* were envisaged in first instance, the inaccuracies due to the mentioned simplifications and neglects were amply justified by easy-to-apply algorithms and improved transparency.

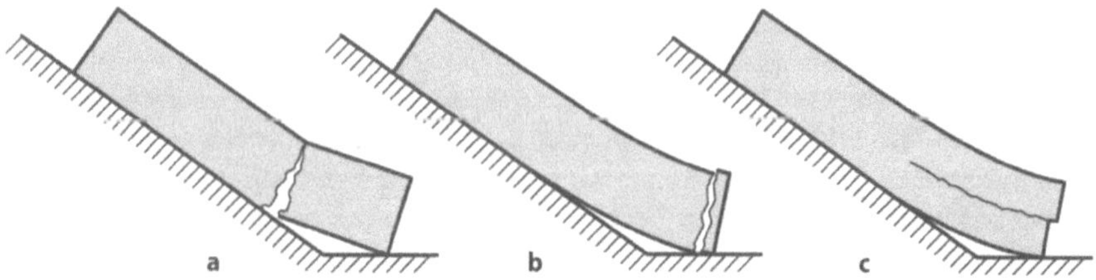

Fig. 4.2. Static fracture according to Fig. 4.1c, basic modes. **a** Bending; **b** transversal shearing; **c** longitudinal shearing (delamination). Each mode may be fracture mechanical or (in extremely crack-free material) conventional. As shown in Sect. 4.2, blocks (**a**) have a better chance to subsist in further motion than slabs (**b**, **c**). Slabs generated by delamination (**c**) may be unstable in bending. Deformations are drastically exaggerated (sketch by Erismann)

The *strength of the loaded body* was calculated both by conventional and fracture mechanical (Sect. 3.2) methods under the assumption of various configurations of pre-existing cracks, and fracture was assigned to the mode that was first to reach the critical stress or stress concentration of the material. So the *best-suited of six modes*, three macroscopic (by bending, transversal shearing, and delamination, Fig. 4.2) and two microscopic (conventional or fracture mechanical) was determined. Then the chances of first-fracture debris to subsist in the further course of motion were assessed. Finally a series of mental experiments was carried out by calculating several particular case histories with parameters chosen so as to yield a maximum of information.

The *main results* thus obtained, as far as relevant in the present context, can be summarised as follows.

11. According to the circumstances of the particular case, any of the six possible modes may, at least theoretically, occur in first fracture.
12. Even small pre-existing cracks (or other flaws) easily become the determining factor of the mode of first fracture. In other words: first fracture is, as a rule, dominated by fracture mechanics if the considered bodies are not almost perfectly crack-free.
13. Bending fracture produces large-size blocks, transversal shearing relatively thin vertical slabs, and delamination horizontal slabs or sheets (Fig. 4.2).
14. Sometimes – in particular if fracture by bending and by delamination have approximately equal chances to occur first – very different final results may depend on small variations of parameters, a statement perfectly accounting for the immediate neighbourhood of large blocks and remarkably small particles.
15. In view of the difficulty to determine the relevant parameters accurately (in particular size and shape of pre-existing flaws) a reliable prediction of granulometric data of debris must be considered as hopeless, at least with the technical means available at the present time.

Readers wanting to know more about the *physical background of Item 13* – perhaps the most important of all – might be interested in the following details. The shearing stress is directly proportional to the bridged length (distance between P_1 and P_2 in Fig. 4.1c), the bending stress is proportional to its square, as obvious from Eq. 4 and 5 of Erismann's study (1988, 260). A trivial consequence of this relation is the following de-

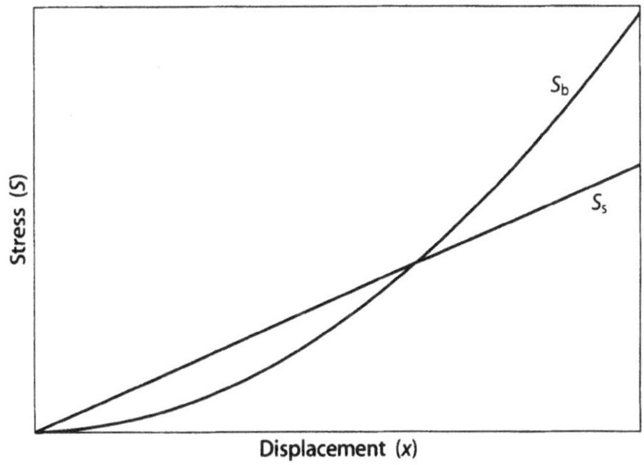

Fig. 4.3. Bending stress s_b and shearing stress s_s in given mass according to Fig. 4.1c, plotted against displacement x. Bending stress is proportional to force F in P_2 and to deformation Δe; shearing stress is proportional to Δe alone. As x, F, and Δe are proportional to each other, the graph of bending stress is parabolic, that of shearing stress is linear. So the domain of shearing failure is weak, that of bending strong materials (sketch by Erismann)

velopment of stresses (Fig. 4.3): immediately after the distal edge has passed the bend, the stress grows faster in shearing than in bending; after a certain displacement, however, bending stress, owing to its parabolic character, necessarily overtakes shearing stress. This means in the clear that weak materials will be fractured by shearing, more resistant ones by bending. And this statement, in its turn, makes Item 13 plausible: fragments obtained by bending necessarily are more bulky than the slabs obtained by shearing.

As, besides their shape (according to Item 13) the *further fate of the fragments* resulting from static first fracture depends to a large extent on the collisions they undergo in the course of displacement, most of this problem will be treated in the following Sect. 4.2. Still there is one particular point which entirely belongs to the field of static fracture.

This case refers to *delamination*. As a matter of fact, it can be shown that this mode of fracture yields laminae which, in their turn, are unstable in shearing (Erismann 1988, 267–268). According to the parameters of the material and the geometry, being "overtaken" by bending (as described here-above and illustrated in Fig. 4.3) occurs sooner or later so that a more or less delaminated mass is the raw material for further – static or dynamic – disintegration.

*

This section should not be closed without a comment concerning *crushing*, a mechanism able to work without substantial relative velocity of the involved particles and thus deserving to be considered as essentially static.

Trespassing the *compressive strength* of a material necessarily leads to a radical deformation. A ductile material may undergo such maltreatment without losing its cohesion (a cylinder of mild steel thus will assume a barrel-like shape); a brittle material will be crushed, i.e. it will be massively disintegrated. Here it is not intended to go into the details of the respective mechanisms. It will suffice to recall that much of the answer of a material to compressive loading depends on the possibility to grow freely at right angles to the direction of the acting force. Two examples: steel rings around a cylindrical concrete column can improve its compressive strength approximately as well as longitudinal reinforcements; and the possibility of making rock ductile by subjecting a cylindrical column to an extremely high lateral hydraulic pressure was elegantly demonstrated by Roš and Eichinger as early as 1949.

Now the compressive strength of the materials involved in rockslides is amply sufficient to withstand the possible averaged static stresses observed in the largest known events. In other words, crushing only can occur in case of massive *local stress concentrations*. So the question arises where the circumstances are best-suited to provoke crushing. The spontaneous answer: "right at the bottom, at the location of the highest average stress", after due consideration, will be revealed as correct, but only for the begin of motion. As a matter of fact, asperities above and below the sliding surface, certainly the main candidates for being crushed, are quickly sheared off so that such surfaces display an almost perfectly planished aspect (Fig. 2.25, 2.26b, 4.6) and the local stress concentrations are effectively reduced. Still there is a possibility of crushing, characteristic just for sliding surfaces: the sheared off asperities, unable to escape from their – anthropomorphously speaking – uncomfortable situation and thus forced to contribute their share in supporting the overthrust load, will again and again locally be overloaded.

Quantification of the mechanism is easy if simplified in a seemingly exaggerated way, turning out as reasonable after due interpretation. Imagine the gap between two perfectly plane, parallel surfaces filled with perfect spheres of equal diameters D, the packing density of these spheres being maximal. In such instances each sphere has to support the hexagonal prismatic overburden that can be circumscribed to the sphere. The Hertzian compressive stress in the points of contact then will be

$$s = 0.6159 \sqrt[3]{F \frac{E^2}{D^2}} \tag{4.1}$$

where F is the force acting upon a single sphere and E the modulus of elasticity of the material. As the sectional area of the mentioned prism obviously is $A = 0.8660\,D^2$ and the load per sphere is $F = A \partial g H$, the product of A with density ∂, gravitational acceleration g, and thickness H, the critical crush-free thickness H_c can be obtained from Eq. 4.1 in the form

$$H_c = \frac{s_s^{\;3}}{0.2023 \partial g E^2} \tag{4.2}$$

where s_c is the maximal admissible stress of the material. In view of the just-mentioned comments concerning the load F, it is nothing but natural that, in the presented model, both equations can be written without using the diameter D.

A remarkable fact should, perhaps, be stressed in this context. The absence of D in Eq. 4.2 postulates that the critical thickness supported by the spheres is independent from their diameters. The practical value of this mental experiment is more important than could be expected from its definitely impossible geometry: in fact, it is easy to show that *independence from the involved radii* is valid not only for spheres, but for all geometrically similar configurations. In other words: if a layer consisting of arbitrarily shaped fragments and transmitting a force by compression of contact radii is converted into a layer of similarly shaped smaller fragments loaded in a similar manner, both layers will begin crushing under equal overburdens. Crushing under such conditions would encounter no fractally given barriers and, if similarity could be sustained, it would continue ad infinitum in a true continuum.

Real crushing is somewhat different, and it is different even for the impossible, but easy-to-understand case of perfect spheres. Assume that, in a primitive approximation to reality, the original set of spheres is converted into an – again perfect and ideally dense – set of somewhat smaller spheres plus a large number of far smaller spheres filling the voids between the larger spheres in an ideal manner. It is clear that in such an experiment the larger spheres are able to support an overburden equal to the initial one; in addition, there will be a large number of further force-transmitting paths through the smaller spheres so that the total crush-free bearing capacity would be increased. And, of course, this *fractal game* can be repeated ad libitum, each time yielding a gain in supporting capacity.

This is, certainly in a far more complicated manner, what happens in the *real process of crushing*: as any crushing process increases the number and, above all, the variety in size of the fragments, it not only multiplies on reduced scale the supporting mechanism active beforehand (without improving supporting capacity); it also creates completely new paths of force transmission by synergetic action of larger and smaller fragments and thus increases the supporting capacity. Once having reached sufficiently small-sized fragments, crushing finds an end.

As the above-discussed considerations show, the model of Eq. 4.1 and 4.2, in spite of its manifest rigidity and other imperfections, is revealed as a useful tool for understanding the mechanisms behind the phenomenon of crushing. Thus it exactly meets one of the requirements claimed in Sect. 1.1: the power of a quantitative approach to improve qualitative knowledge. Nevertheless a *quantitative result* of tentative calculations should not be left aside: for solid rocks a critical thickness H_C of approximately 20 m is obtained from Eq. 4.2. There is no doubt that in reality first crushing at particularly exposed points occurs at lower values.

A trivial concomitant of crushing is its contribution to *gradation*. The intensity of crushing is increased with the compressive stress and hence also with the depth. And as crushing essentially results in a conversion of coarse into fine-grained material, it is clear that crushing acts in the sense of a gradation coarse on top ("very fine on bottom" would describe the process even better).

It should not be concealed that the process of crushing is further complicated by the fact that a part of the fragments, if favourably shaped, may show a tendency to assume a *rolling motion*. It is considered useful to discuss this phenomenon in Sect. 5.3, specialised for the purpose.

*

To close this section, the remark should be recalled that crushing at the very bottom of the moving mass is considered as a phenomenon occurring in the early phases of motion. There are *other locations within a mass* where more time is available during its descent. As will be demonstrated later (Sect. 5.1, 5.4), the comparison of the particles in a disintegrated moving mass with molecules in a flowing liquid is physically questionable, and the relative displacements within such a mass far less dramatic than might be assumed at first sight. As a consequence, there are groups of particles remaining together in more or less similar, though not congruent patterns. In groups of this kind shearing is small in comparison with the zone immediately above the immobile ground. Still, if the considered location is not too far from the sliding surface, the compressive stress may be near to the maximum. In addition the relative displacements may lead to new points of contact so that crushing may be started anew with particles already crushed beforehand.

4.2 Dynamic Disintegration

In the preceding Sect. 4.1 the term "static disintegration" was used to characterise phenomena essentially driven directly by gravitation. Hereafter, by contrast, the complementary expression "dynamic disintegration", again defined in a rather extensive manner, will be applied to mechanisms in which *motion is the immediate source* of the energy that splits larger particles into smaller ones. This definition implies the fact that such mechanisms cannot work before a certain minimum of velocity is attained and thus, at least in the normal course of events, occur somewhat later than the first static fractures.

It is needless to say that in this context *collisions* – highly dynamic processes in the true sense of the word – represent the most important mechanism, able to occur, after due acceleration, in any phase of motion. The following examples are selected to illustrate the width of the span embraced: on the one hand, both falling over an almost vertical slope (Huascarán, Sect. 2.7) and toppling (Sect. 3.3) are normally followed by a vigorous collision with the ground so that disintegration takes place immediately after the start of an event; on the other hand a mass, exempted from severe disintegration during the main part of its travel, may suddenly collide with a massive obstacle. The impact often puts an end to the motion of a rockfall; for the larger mass of a rockslide, however, various modes of secondary displacement, as a rule combined with a run-up, may occur instead of a full stop: for instance a right-angle turn (Pandemonium Creek, Sect. 2.2); or the instantaneous conversion of a substantial fraction of a cubic kilometre of rock into centimetre- or millimetre-sized debris, followed by the separation into a lower (heavily disintegrated) and an upper (far less disintegrated) portion (Köfels, Sect. 2.4); or the bifurcation into a right and a left lobe (Val Pola, Sect. 2.5). Between (and in some cases after) such spectacular scenarios there are plenty of situations favouring collisions, especially in the sometimes long tracks of rockslides: no reasonably sharp horizontal or vertical change of direction can be imagined without a large number of collisions which may be external (between the mass and immobile obstacles) as well as internal (between the elements of the mass).

To conceive an idea of the progress of disintegration by collisions, two *fundamental problems* have to be considered. The first deals with the question how far first fracture exerts an influence upon the chance of the fragments to survive in the further course of motion, the second aims at possibilities to get at quantitative information concerning the disintegrating capacity of collisions. In both cases it will be observed once again that quite general quantitative statements can effectively contribute to deduce well-specified qualitative conclusions.

The details of the first question were treated by Erismann (1988, 263) in the study already mentioned under Heading 4.1. In short, the results can be summarised as follows. Between the *mode of first fracture and subsistence* of the fragments in the subsequent percussions of a descent there is a close causal connection. The chance to survive is definitely better for block-shaped debris than for relatively thin slabs. The reason is almost trivial: typically, a slab means low resistance against bending (and also buckling in certain cases): the lever arms at which forces may act are long in comparison with thickness. And in the downhill ride of a slightly disintegrated mass rather large (and therefore bulky) slabs necessarily undergo frequent occasions of being bent, be it dynamically or statically.

So it becomes plausible that, after first fracture, mainly slab-shaped fragments – as generated by *delamination or transversal shearing* – are candidates for further disintegration. This result, stressed by the occasionally inherent instability of fragments resulting from delamination (Sect. 4.1) and by residual stresses within the material, is in agreement with field evidence: large slabs are by far less frequent in the deposits of rockslides than blocks of comparable dimensions.

It might be objected that slab-shaped fragments are generated mainly by fractures along weak planes in the material and that, as a consequence, subsequent fractures at right angles to these planes encounter the highest resistance inherent in a given material. This argument is correct in principle. It does, however, not sufficiently account for the physical laws according to which bending fracture occurs at a force proportional to the square of thickness and inversely proportional to the length of the acting lever arm. The combination of both geometric parameters easily can create a situation of *dramatically increased stress* exceeding the strength even of a resistant (or, in terms of fracture mechanics, an essentially sound, i.e. crack-free) material. Only little more than the reduction of thickness by a factor three results in a reduction of the critical force by a power of ten!

The statement thus established is a most interesting one: although prediction of granulometric details seems to be hopeless even with the use of super-computers, the main direction of weak planes often can be determined "ante eventum". So the general character of debris in the early phases of a rockslide is not necessarily beyond the reach of an expert. And this knowledge can be helpful in *forecasting the general character of an event* and, to a certain extent, also the range threatened by a catastrophe. This last remark is true in spite of the fact that the influence of disintegration upon reach will be revealed as smaller than might be expected at first sight (Sect. 5.1).

Up to this point, in the present section much has been said about bending fracture of slab-shaped fragments and very little about the conventional and/or fracture mechanical character of fractures due to the impact of a collision. So, before passing to the second fundamental question, some remarks should be made in this context. It was taken for granted that readers would bear in mind both aspects, *conventional and fracture mechanical.* In connection with bending fracture it cannot be excluded that (in case of a crack-free material as parenthetically insinuated in the second-to-last paragraph here-above) conventional strength sometimes is the determining factor. For other than slab-shaped fragments such uncertainty hardly exists. Within the elements of a mass, collisions generate shock waves which, in their turn, build up transient local stress concentrations, boosted by interference with reflected waves, by residual stresses, and/or by the vicinity of pre-existing cracks. This fact, together with the natural brittleness of the materials in question, is an argument in favour of a dominant fracture mechanical aspect. In other words, it has to be taken into account that critical K values as used in Eq. 3.1 and 3.2 are substantially reduced under impact conditions. Hence the importance of pre-existing weak zones, in particular sharp-edged cracks or highly developed parallel fabric of the material (e.g. in schist), hardly can be overestimated, and dominance of fracture mechanical disintegration is practically granted with few exceptions.

*

Probably the best (if not the unique) physically plausible approach to the question of the *disintegrating capacity of collisions* consists in determining the locations of highest energy dissipation. Despite the indisputable scatter of individual impacts there is a good chance of effective averaging: if, for instance, a number of blocks, different in size, shape, and internal cohesion, succeed each other in colliding with one and the same massive obstacle, the energy dissipated per square metre of new fracture-generated surface (or whatever an adequate definition of disintegrating capacity may be) certainly will cover a wide band; but it is equally certain that the average obtained from, say, one hundred such collisions will differ far less from that of another hundred. And, as the energy required to obtain fracture necessarily is part of the energy dissipated in an impact, there is definitely a substantial positive correlation between both.

By the trick of using *energy dissipation as an indicator* of disintegrating capacity it is made possible to draw certain conclusions on a quantitative basis (well understood, not a very precise one, but adequate for the purpose). Consider two mass points m_1 and m_2, approaching each other from opposite sides at different pre-collision velocities Δu_1 and Δu_2 (Fig. 4.4), linked with the respective masses by the equations

$$\Delta u_1 = \Delta u \frac{m_2}{m_1 + m_2} \quad \text{and} \quad \Delta u_2 = \Delta u \frac{m_1}{m_1 + m_2} \tag{4.3}$$

where $\Delta u = \Delta u_1 + \Delta u_2$. Owing to the trivial symmetry of the momenta $m_1 \Delta u_1 = m_2 \Delta u_2$ the centre of gravity of the system $(m_1 + m_2)$ is immobile during displacement. Now let the masses collide. According to the laws of impact, said centre of gravity also then remains in place, irrespective the value of the coefficient of elastic restitution $0 \leq J \leq 1$. As a consequence, the respective post-collision velocities

$$\Delta v_1 = J\Delta u \frac{m_2}{m_1 + m_2} \quad \text{and} \quad \Delta v_2 = J\Delta u \frac{m_1}{m_1 + m_2} \tag{4.4}$$

are obtained, i.e. the collision takes place as if each mass were banging against an immobile wall (represented in Fig. 4.4 by the axis A–A).

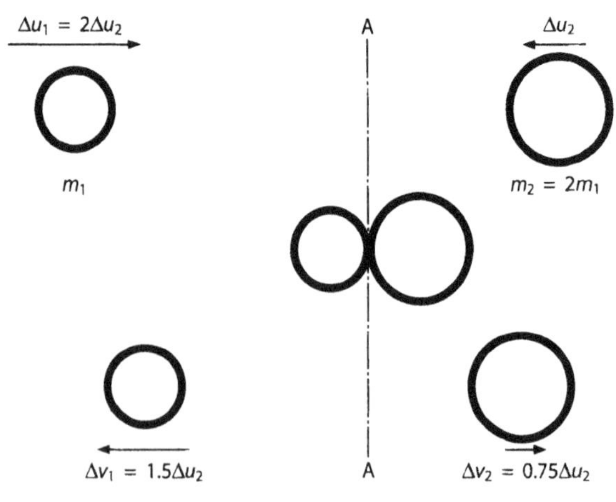

Fig. 4.4. Collision of two masses m_1 and m_2 with equal momenta. From top to bottom: situations at time $-\Delta t$ before collision (velocities Δu_1 and Δu_2); in the moment of collision; at time $+\Delta t$ after collision (velocities Δv_1 and Δv_2). Coefficient of restitution: 0.75. Lengths of *arrows* are proportional to respective velocities. During the entire process the centre of gravity of the system remains in axis *A–A*. Arbitrary systems are obtained by displacing the entire system at a constant additional velocity (sketch by Erismann)

The idealised model, by the way, does *not exclude fracture* of one (or both) of the colliding particles. Such an issue, vital in the present context, means nothing but a certain reduction of J and due consideration of the fact that Eq. 4.4 in case of fracture yield a velocity valid for the common centre of gravity of the fragments on the respective side. Said reduction of J is not a dramatic one as long as the fragments rebound within a narrow cone: the pure energy of separation (Erismann et al. 1977, 89) normally is substantially lower than the involved elastic and plastic energy. In case of an explosion-like fracture, however, it is worthwhile to investigate the energetic background more in detail as residual energy stored in a ready-to-explode particle might play a non-negligible part.

Also in another important point the presented model gives, at first sight, a false impression of being very special and therefore of little use for dealing with more general collision problems. In reality, however, it becomes a powerful tool as soon as the system $(m_1 + m_2)$ is recognised as an inertial system, i.e. a system that does not suffer any internal modifications if any velocity u, constant in value and direction, is added. The mathematical astuteness hidden behind using symmetry of momenta consists in the fact that the energy dissipated in the collision,

$$\Delta W = \frac{1-J^2}{2} \Delta^2 u \frac{m_1 m_2}{m_1 + m_2} \tag{4.5}$$

is independent of the globally added velocity u. So, if, besides m_1 and m_2, the real pre-collision velocities u_1 and u_2 are given, the entire deduction presented here-above may be forgotten, and nothing but the simple equations

$$u = \frac{u_1 m_1 + u_2 m_2}{m_1 + m_2} \tag{4.6}$$

$$\Delta u = u_2 - u_1 \tag{4.7}$$

and

$$v_1 = u + J\Delta u \frac{m_2}{m_1 + m_2} \quad \text{and} \quad v_2 = u - J\Delta u \frac{m_1}{m_1 + m_2} \tag{4.8}$$

have to be applied with Eq. 4.5 to obtain the complete set of real *post-collision parameters*, above all ΔW as a yardstick for effective collisions.

*

If there were only the optics of this chapter, the ado with mathematical formulae might be considered as exaggerated. But in the further course of reading (Sect. 5.1, 5.7) it will emerge that especially v_1 and v_2 – marginal in the present context – are excessively useful for the understanding of various motional mechanisms. Still also the conclusions concerning dynamic disintegration are of a certain importance. In first instance *locations with a high probability of fractures by impact* can be spotted. As the dissipated energy is used as indicator, a simple analysis of Eq. 4.5 suffices: maxima of disintegration can be expected where ΔW per unit of involved mass reaches maximal values.

The configuration of the masses as presented in Eq. 4.5 makes clear that the smaller mass (e.g. m_1) has a certain importance: the mass-relevant term cannot trespass the range $0.5 m_1 \leq m_1 m_2 / (m_1 + m_2) \leq m_1$, corresponding to the range of the larger mass given by $m_1 \leq m_2 \leq \infty$. So, if referred to m_1, the mass-dependent span of energy dissipation does not exceed a ratio of 1:2. The *influence of the masses* (expressed in Joules per kg of m_1) is moderate. Yet it can be observed that the risk of being smashed is higher if the counterpart m_2 is very large than if both masses are approximately equal.

The *influence of velocities*, represented in Eq. 4.5 only by their difference Δu is definitely more important. On the one hand, Δu can vary between zero and very high values, on the other hand it is squared. As a consequence, massive disintegration has to be expected primarily in locations where substantial differences of velocity between colliding rocks are possible. Little will be found within the core of a more or less disintegrated mass where all particles forcibly move at approximately equal velocities. Remarkably well preserved pieces of timber found in the very midst of rock debris – and often used for radiocarbon dating (Sect. 3.2) – are a matter of physical plausibility, not one of extraordinarily good luck. The other extreme is the frontal collision with an immobile obstacle, especially if this obstacle is sufficiently heavy and well-founded to remain in place after the impact. A gigantic example of such an ideal disintegration by collision can be observed in the Maurach gorge of Köfels (Sect. 2.4) where probably a substantial portion of a cubic kilometre of moderately disintegrated rock was converted into gravel-sized and smaller particles.

*

To visualise the *effect of collisions upon motion* (and therewith also disintegration) a very simple (and very useful, though extremely improbable) model of consecutive collisions was calculated using the above-presented equations (Fig. 4.5). Imagine a group of equal blocks moving at equal velocity strictly in line like motor-cyclists in a parade. The most distal one hits an obstacle and loses some of its velocity, but not its direction. In little more than half a second (assuming the conditions shown in the figure) a series of collisions passes from the leading to the trailing end of the group. In each collision energy is dissipated. Nevertheless the total momentum $\Sigma v_i m_i$ (where i is the ordinal number of each block), and obviously also the average velocity v_m of the group remains constant.

At first sight the fact might appear contradictory: how can energy be dissipated without influencing the average velocity? The best way to obtain a comprehensible answer consists in considering the simplest possible system, namely two moving particles of equal masses. Obviously the total kinetic energy of this system is proportional to the sum of the two squared velocities. Now compare two possible configurations. In the first both particles move at equal velocities of, say, 4 units of velocity. Then the averaged velocity is, of course, also 4, and the kinetic energy is proportional to $16 + 16 = 32$ (in squared velocity units). In the second configuration, the particles move at velocities of 2 and 6 units respectively. Again an average of 4 units results, but the kinetic energy turns out to be proportional to $4 + 36 = 40$. So the kinetic energy is revealed as consisting of two components, one proportional to twice the square of the averaged velocity (32 units), the other to the sum of the squared divergencies with respect to the averaged velocity ($4 + 4 = 8$). These components may be denominated as *external and internal kinetic energy*.

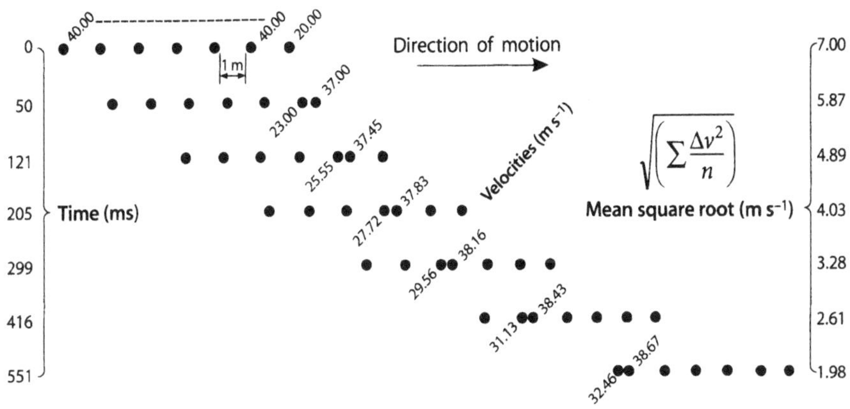

Fig. 4.5. Effect of collisions on motion. From top to bottom: seven successive positions in which seven particles (*black circles*) collide. Assumptions used to optimise transparency: equal masses of all particles; exactly in-line motion; equal initial distances between particles; equal initial velocities (40 m s^{-1}), except for most distal particle having been braked by an obstacle (20 m s^{-1}); collisions assumed as unique cause of changes in velocity (tolerable for half a second of duration); coefficient of restitution (0.7) equal in all collisions. The mean square roots of divergencies Δv with respect to the averaged velocity (n: number of particles) show the trend, in each collision, to equalise velocities, thus reducing the energy dissipated in subsequent collisions and, therewith, the amount of disintegration by collisions (sketch by Erismann)

The column "Mean square root" at the right side of Fig. 4.5 is nothing but a *quantitative expression of the internal kinetic energy.* It represents the mathematically correct weighed average of the differences $\Delta v_i = v_i - v_m$ (where v_m is the average velocity). And this value decreases drastically from 7.00 to 1.98 m s^{-1}, i.e. by a factor 3.54, and thus demonstrates the main effect of collisions within a more or less disintegrated mass: by equalising individual velocities, each collision between the particles of a disintegrated mass reduces the kinetic energy of the mass without loss of average velocity (i.e. of momentum).

It will be observed that the *geomorphologically significant results* of the above-presented mathematical considerations already have been anticipated in a postulate closing the second paragraph of the present section: "*... no reasonably sharp horizontal or vertical change of direction can be imagined without a large number of collisions... which may be external... as well as internal...*". And the locations of disintegration enumerated in the introduction to Sect. 4.1 (third paragraph) are also in agreement with the theoretical claim for vigorous decelerations due to immobile counterparts (abutments, landing areas, "crash barriers", etc.). This fact demonstrates that the locations of disintegration are, as a whole, well estimated just by using common sense. Only one important correction is contributed by mathematical analysis: the fact that little internal disintegration goes on as long as no external effects occur. And the rapid decrease of the mean square root in Fig. 4.5 shows that even external effects, if involving only one (or few) of the particles, fade out after a remarkably low number of collisions.

*

The highest degree of disintegration, as far as grain size is considered, is accomplished by *friction*, in first instance externally (between the mass and the ground), but to a non-negligible extent also internally (between the particles of the mass). The last-mentioned mechanism is mainly concentrated near the ground where massively decelerated particles are forced to interact with the rest of the moving mass. Their adjustment to the velocity of the surrounding debris occurs not only by collisions (as might be suggested by Fig. 4.5) but, mainly under the everlasting action of gravity, by frictional contacts as well. The large quantities of fine material produced by friction in the course of almost the entire descent (and ranging from sand to silt) play an eminent role in certain tribological hypotheses (Sect. 5.6). This material is, in addition, the reason for the impressive dust clouds inevitably accompanying rockfalls and rockslides – an odious obstacle for those rare observers of such events who were lucky enough not to be forced to run for their lives.

Anticipating the more complete discussions in various sections of Chap. 5 (5.1, 5.4, 5.7) one particular aspect of the *mechanisms of friction* between rocks will be looked at hereafter: it would be difficult to describe facts clearly without having done so. Disintegration by friction is essentially a succession of countless failures by local overstressing of asperities and free particles (Heilmann and Rigney 1981, 197–206; Rigney et al. 1984). Stresses, necessarily resulting from forced distortions, may be built up in various manners, in first instance by frontal or oblique collisions under conditions excluding the possibility of giving way. Free particles near the ground may be forced into a short rotating motion (Sect. 5.3; Eisbacher 1979, 328–329, 331), in many cases ending in a crushing process. Whatever the preceding mechanisms may have looked like, after failure the elastic energy accumulated in zones of high stress no longer can act in a well-defined direction and therefore is condemned to conversion into heat.

This conversion occurs by further frictional contacts and/or collisions on a reduced scale. So the compound mechanism of friction and disintegration, in descending from scale to scale, displays, on the one hand, a remarkably *fractal character* and, on the other hand, it reveals close connections with collisions and crushing. And it is nothing but natural that a finer and finer granulation more and more concentrates high gradients of velocity – and therewith high energy dissipation – in narrow spaces as shown in the "sand bed" of Fig. 2.26b. Bare surfaces having been treated in such a manner look as if they had been polished in an immense grinding machine (Fig. 4.6). In large crystalline masses like those of Langtang and Köfels the final stage can be that of rock fusion (Fig. 2.25, 2.26a).

*

Disintegration, besides its part in determining the character of motion, has various consequences. Particularly remarkable is the influence upon the *grading* of deposits. It is an erroneous opinion to believe that the sieve effect alone is responsible for the fact that grading "coarse on top" is the rule in rockslide debris. Of course, small particles are able to fall through narrow passages where larger ones are arrested; and large particles, in accomplishing angular motions, can bury and compress smaller ones as large wheels do. But it has to be taken into account that at least three of the four discussed mechanisms of disintegration prevailingly occur in the lower quarters of a mass. Crushing as well as friction depend on high compressing forces; and effective collisions require high relative velocities as mainly encountered at the bottom of the mass when colliding with asperities of the ground. And even static fracture, the fourth mechanism, shows in certain configurations a tendency in favour of near-to-bottom disposal of small particles: in the most probable form of bending (Fig. 4.2a) any smaller

Fig. 4.6. "Polished" surfaces in the tracks of the rockslides of Köfels (**a**) and Flims (**b**). Stylos give an idea of the remarkably fine "machining". In Photo **b** straightness over many metres is evident, in **a** it was made sure by aiming at the next similar surfaces. Note similarity with Fig. 2.25 and 2.26b (Photo **a** by courtesy of H. Heuberger; **b** by Abele)

by-products of fracture have a good chance to fall into the resulting triangular void; and failures following first delamination (Fig. 4.2c), be it in further delamination or in bending, are probably not independent of the overthrust load. Hence debris grading coarse on top is not only a result of having been shaken in motion: it is, to a substantial extent, the result of disintegration being particularly intensive near the bottom of a mass. Exceptions from this rule (Fig. 2.10; Sect. 5.7) have to be considered as anomalies and justify a closer investigation.

By the way, the above-mentioned mechanisms do not necessarily apply to all large blocks found on top of smaller-sized distal debris: in certain cases it is not excluded that *single boulders*, having been released later than the rest of the mass, overtake it in its motion (and possibly also after its standstill), riding (and possibly also rolling), so to say, on its back. For the moment, it is left open how far a somewhat similar problem, namely the visual post-eventum impression that large boulders had been "*floating*" *on small-sized debris*, belongs into the present context. The phenomenon originally was expressed by Griggs (1922, 142), half a century later adopted by Hsü (1975, 136), and commented in connection with certain observations by Eisbacher (1979, 330). The respective formulation by Griggs runs as follows: "*... rocks... did not slide into their present place but were rather floated into position, buoyed up by the mass of finer debris...*". For certain reasons this question will be considered under Heading 5.3.

Chapter 5

Mechanisms of Displacement

None of the stones is free, they all are linked to form a unique whole in the speeding stream and bound to remain in the order in which they started for the common rush.

Albert Heim (original in German)

5.1 Coherent and Disintegrated Motion

Before studying in more detail the role of energy dissipation in mechanisms like Coulumbian friction or lubrication, it is useful to look at coherent and disintegrated motion from a more general point of view, aiming at energetic differences depending on nothing but the degree of cohesion, irrespective of the mechanisms acting in a particular case. In doing so, inevitably *three fundamental problems* are encountered, and each of them can be formulated as a question.

1. What are the physically relevant features of a disintegrated moving mass as compared with those of a coherent one?
2. How far can such features influence the reach of an event, be it generally or in specified circumstances?
3. What is needed to describe the state of disintegration along the track of an event in such a manner that easy application and satisfactory quality of the results are granted for analysis and prediction?

In the following, by far the largest effort will be required to find answers to the numerous sub-questions implied in Question 1.

*

The difference between a coherent and a disintegrated mass might be seen in analogy to that between a solid body and a liquid. This point of view is suggested by another analogy: the motion of particles within an enormous heap speeding downhill sometimes has been compared with the *motion of molecules* in a flowing liquid. The motto at the head of the present chapter (Heim 1932, 105) sounds – whether rightly or not, is another question – like an implicit negation of such an analogy which, in fact, cannot be accepted without previous consideration of the physical circumstances.

In a disintegrated mass gravitation is a dominant principle. It generates a vertical compression which, irrespective of superposed dynamic effects, creates a vertical mean compressive stress increasing from top to bottom. The tendency of the mass to spread horizontally is nothing but a consequence of this ubiquitous stress, acting – continuously or by impact – upon oblique surfaces of particles. A necessary consequence of this process (Fig. 5.1) is a mean horizontal stress, normally growing as a function of vertical stress and in its turn – via friction – generating an additional resistance against vertical shearing. So the mean horizontal stress, in spite of depending on its vertical counterpart, fundamentally differs from it; yet in both cases the resistance against shearing deformations depends on depth. These two facts can be summarised in the

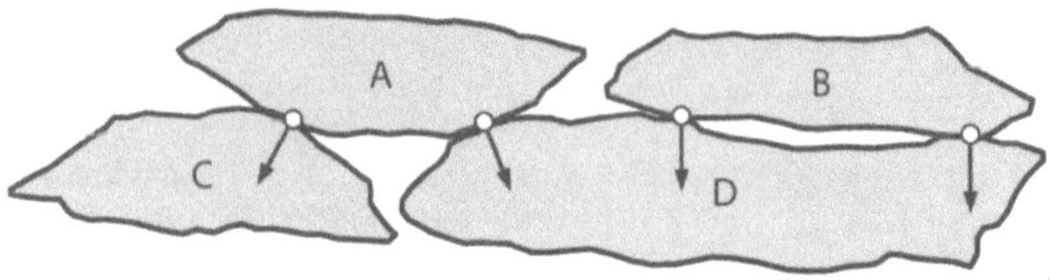

Fig. 5.1. Generation of horizontal stress in disintegrated mass. The weight of block *A* acts on oblique surfaces of blocks *C* and *D*, thus tending to push them apart. Block *B*, exerting more or less vertical forces upon *D*, remains neutral with respect to horizontal stress. Friction and the general tendency to assume positions approximately parallel to the ground reduce the mean horizontal stress as compared with the mean vertical stress (sketch by Erismann)

statement that in its behaviour *a disintegrated mass is neither isotropic nor homogeneous.* Speaking in terms of a liquid where compressive stress corresponds to pressure: neither the quality of pressure as a scalar (direction-independent) value nor the almost exclusive dependence of shearing resistance upon the local shearing velocity gradient (and material parameters) is granted – two basic physical principles of liquids are far from being fulfilled.

More than that: the anisotropy of a disintegrated mass is particularly manifest in its behaviour under *stress by extension.* While a liquid remains a continuum as long as static pressure is able to keep it together, debris, owing to friction, is far more reluctant to fill voids. As a matter of fact, a disintegrated mass may almost be considered as coherent under compression and incoherent under extension. This is true in any given direction. For instance, when moving, in descent, on a decreasing slope where the proximal particles of a mass suffer more accelerating forces than the distal ones, longitudinal cohesion is established, and the mass moves almost as if it were coherent; contrarily, in descending on an increasing slope, a disintegrated mass loses cohesion and tends to be torn to pieces (and to spread in longitudinal sense), thus resembling rather independent blocks (or clusters thereof). It is well understood that in a run-up, where gravitational forces turn from acceleration to deceleration, the above-used attributes "increasing" and "decreasing" have to be castled to obtain correct statements.

In such instances the mentioned *analogy with a liquid* turns out to be a very limited one, and it is a justified question, how far the comparison makes sense at all. The fact that a disintegrated mass, like a liquid, passes rather easily through narrow clearances and curves, obviously is a somewhat meagre argument. Anyhow, the reluctance becomes plausible that was expressed in Sect. 1.2 against the use of terms like "flow" and "stream" in connection with moving debris.

*

The difference is sharply accentuated as soon as, in addition to the above-mentioned phenomenological features, *energetic aspects* are focused. Here the preliminary work done in Sect. 4.2 will be of great use. In particular, the trick to begin the analysis of a collision between two mass points by introducing symmetry of momenta and then to extend the range of validity by adding one and the same constant velocity u to both considered velocities, now can be generalised in more than one respect.

A first step of generalisation is to consider what happens if the vector of the added velocity u is not aligned with those of Δu_1 and Δu_2 as assumed in Fig. 4.4 and the

Eq. 4.3, 4.4, and 4.5. Now in the mechanics of inertial systems no conditions are stipulated in this respect, so that the simple answer is: in the relative displacement between the mass points m_1 and m_2 and in the required energy ΔW there is no change, whatever the direction and amount of u may be. As the collision of two mass points can be considered as a fairly good approximation to the frontal collision of two extended masses, the last statement means nothing less than that the concept of external and internal kinetic energy is *valid for any frontal collision* which may occur between particles in a moving disintegrated mass.

Perhaps two *easy-to-grasp examples* are helpful for better understanding. If the vectors of Δu_1 and Δu_2 are perpendicular to that of u, it is directly evident that a collision of the two masses, in spite of dissipating energy, has no influence upon their momentum in the direction of u. And sports based on collisions (like tennis) are – besides alien effects like wind or glaring sunshine – independent of the geographic location and orientation of the court (the unique valid source of error, the Coriolis effect, imperceptible for our senses, is due exactly to the fact that the vector of u on a tennis court is not steady: it accomplishes a circle once in twenty-four hours...).

This first step is still a small one as not many collisions can be considered as reasonably "frontal" i.e. such as to produce little more than reflection in the axis of the original relative displacement (this axis, of course, being continuously displaced in function of the additional velocity u). Normally collisions are more complicated: friction, in occurring at the surfaces of two excentrically colliding bodies, in principle produces a reduction of the relative velocity and an angular acceleration. This gives the opportunity to reveal the more general physical law that stands behind all possible contacts between two particles of a moving mass. As a matter of fact, such contacts, whatever their character may be, owing to the principle of action and reaction, necessarily call for a perfectly symmetrical exchange of forces: at any moment a force F acts upon the point of contact of one body and an exactly opposite force $-F$ upon the respective point of the other body. This symmetry of forces automatically means also a symmetry of momenta which are nothing but integrals of forces over time. Thus, the general statement – of which the simple case presented under Heading 4.2 was only a very special example – is true, that, in spite of dissipating kinetic energy by impact or friction, all possible *internal contacts between particles cannot influence the momentum* of a moving disintegrated mass. To achieve such influence, the action of external forces is unconditionally necessary.

It would, nevertheless, be an error to assume that in such instances the losses of internal kinetic energy were not dissipated at the expense of the total kinetic energy of the mass. One should not forget the trivial fact that, by exception of rare or negligible cases (e.g. residual stresses; volcanic activity, Voight et al. 1983), there exists no other source able to furnish kinetic energy (no matter whether external or internal) than the potential energy of the descending mass. In other words: *the fate of the internal kinetic energy* is not sealed in the moment of dissipation (be it by collision, friction, or what not); it is sealed in the moment when, by external forces, internal kinetic energy is generated, i.e. when internal differences in velocity are built up which later can result in dissipation of energy.

There is no doubt that effects connected with internal kinetic energy stand behind the most important cases of energy dissipation within a moving mass. Yet it is possi-

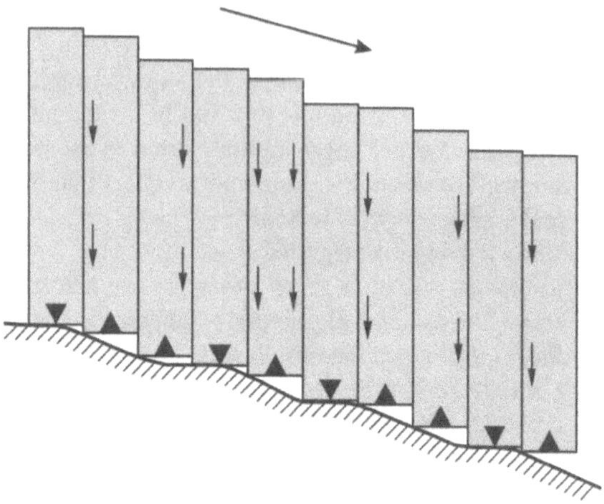

Fig. 5.2. Model demonstrating static energy dissipation within a disintegrated mass moving on undulated ground. Approximately vertical groups of particles (idealised by slabs) moving downward with respect to neighbouring elements are marked by arrows near the mutual contact surfaces. Changes of load upon the ground are marked by triangles pointing up (reduction of load) or down (increase of load). So the total effect – reduction of driving force – becomes evident: the descending (i.e. driving) portions are apparently lighter than those moving horizontally (sketch from Erismann 1979)

ble to imagine circumstances in which the generation of internal displacements and the subsequent loss of energy occur *on an essentially static basis*. In one of his early studies (1979, 25–26), Erismann developed a simple model of this kind (Fig. 5.2). It makes it plausible that, on undulated ground (one of many possible circumstances evoking relative displacements within a mass), internal friction induces a nonhomogeneous load distribution: owing to the mean horizontal stress postulated here-above the parts with reduced slope support an overproportional share of weight while those with higher slope are unburdened to a certain extent. Most of the mass, so to say, slides on a reduced slope, and a part of the vertical travel takes place in an approximately vertical motion of approximately slice-shaped mass portions, i.e. without producing a driving effect. Once more, this example stands for the general rule that internal friction, whenever it occurs, marks a loss of driving energy and thus reduces the reach of an event.

In closing this excursion into problems of energy dissipation a further remark has to be made with respect to the *motto at the head of the present chapter*. This motto is the expression of repeated observations made by Heim (e.g. 1932, 105, s. also Hsü 1978) and confirmed by others (e.g. Fig. 2.21; Shreve 1968a, 29; Eisbacher 1979b, 324, 331) according to which a moving disintegrated mass remains "in shape" to a remarkably high degree. Obviously the motto is in perfect agreement with the considerations made under Heading 4.2 about the tendency of collisions to equalise the velocities of particles. Heim's preceding statements, however, cannot pass uncommented. The description of a faster particle hitting a slower one from behind contains the sentences: "*It spurts and strikes one that… moves in its way. Just as much kinetic energy as the first had in excess of the second, it transfers in the collision to the second and loses it for itself and remains behind…*" (translation by Erismann). These (and further) formulations show that Heim's self-critical remarks about his having forgotten, at an age of 83, most of his physical knowledge (1932, 143) were made with a good reason: when using the German word "lebendige Kraft" (i.e. "living force", an antiquated expression for kinetic energy, by the way correctly applied on page 95), he must have had in mind the momentum for which the sentence, as the considerations of Sect. 4.2 show, would make sense: in a collision the total momentum is conserved, the kinetic energy undergoes a reduction. Even the Grand Old Men of alpine geology were not immune to errors...

*

Many *other physically relevant features* of disintegrated displacement are far more trivial than the somewhat intricate problem of internal kinetic energy. Yet in some cases coherent motion (and the comparison with it) needs a more differentiated consideration than a simple statement of non-existence as sufficient for questions of internal kinetic energy.

It is, for instance, a trivial fact (by the way already mentioned in this section) that *narrow clearances* like the passage shown in Fig. 5.3a, may in certain cases arrest a partly disintegrated mass consisting of elements which, if passing individually, would encounter no serious obstacle. Especially wedge-shaped blocks and/or a passage converging in the direction of displacement can thereby play the part of efficient brakes. A striking example of an expected and released rockslide that finally did not take place at all is mentioned by Heim (1932, 158–162): the mass, essentially coherent by lack of opportunity to disintegrate, was arrested by the wedge effect of solid converging lateral rocks after a practically insignificant travel! Contrarily to such a lucky outcome, a curve with steep lateral walls, if unable to let through a coherent mass, may instead evoke first fracture by bending and thus free the way (Fig. 5.3b), of course at the expense of a certain additional frictional energy lost in the contact surfaces between the mass and the lateral slopes. What will happen in a particular case, depends on the local configuration of parameters.

Here it is not intended to discuss in detail the mechanisms subsumed under the collective name of "*size effects*" which might as well (or even better) be designated as thickness effects. As a matter of fact, the examples discussed more in detail under Heading 6.2 (rather trivial) and 5.5 (less trivial) refer to mechanisms working in first instance owing to thickness: lateral extent can be regarded as a parameter of minor importance if within the bulk of the mass thickness and velocity do not vary dramatically. Now thickness remains as well as constant in coherent motion while a disintegrated mass, when moving on rather plane ground, tends to lose thickness by lateral spreading, and the potential energy thus dissipated is lost for forward motion.

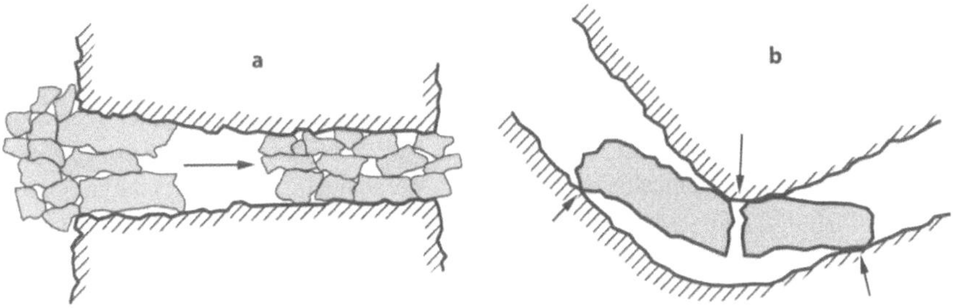

Fig. 5.3. Narrow clearances. **a** Wedge effect at entrance of gorge stopping cluster of blocks which easily would have passed one by one; arrow shows direction of motion. **b** First fracture of long block allowing the fragments to pass through narrow curve; arrows show bending (and friction-generating) forces before fracture (sketch by Erismann)

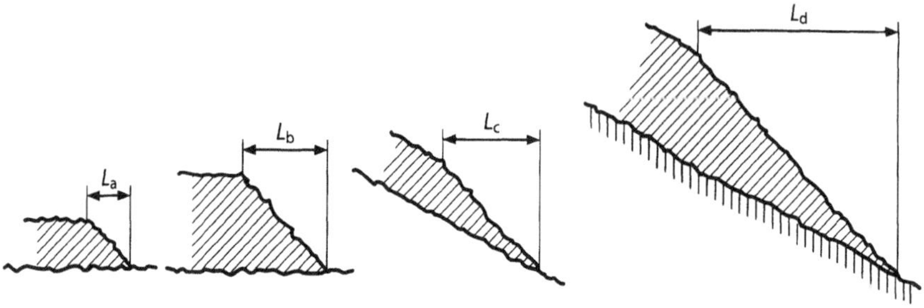

Fig. 5.4. Length L of distal talus or cone consisting of debris depends, besides the angle of rest (equal for all represented cases), on the material's thickness (and thus can be increased by channelling): $L_a < L_b$ and $L_c < L_d$. Further increase is possible on inclined ground: $L_a < L_c$ and $L_b < L_d$ (sketch by Erismann)

It is equally obvious that only a disintegrated mass is able, by the end of its travel, to generate a significant *distal talus or cone* covering an additional area (a curious variant was analysed by Voight 1973, 114–117: wedge-shaped portions of relatively plastic material, in losing thickness, drove apart neighbouring blocks). This peculiarity of disintegrated motion is especially effective if the mass moves along a narrow, but not impedingly narrow valley: after having profited from size effects due to high material thickness preserved by lateral confinement (the so-called "*channelling effect*", Eisbacher 1979b, Fig. 21–23, 25, 26; Evans et al. 1989, 436), by the end of motion a talus can develop from a higher level, thus reaching farther on a given slope (Fig. 5.4). It is obvious that such talus, if developed on a descending slope, assumes a particularly great length. So, besides the general fact that a mass moving in coherent state has no possibility to acquire drive energy by losing thickness (a partial counterweight to the above-mentioned loss of drive energy due to lateral spreading), there may exist circumstances enhancing this effect.

*

Eventually, the question has to be asked how a coherent and how a disintegrated mass deals with a local asperity encountered on its track. Besides the – never excluded – possibility of the mass being disintegrated by the obstacle and the phenomenon of scraping or ploughing (Fig. 5.5c) which will be treated in Sect. 5.4, there are two fundamentally different issues: either *the mass cuts away the obstacle or it runs over it.* In a particular case it will, of course, take the line of least resistance. And this line is different for coherent and disintegrated masses. In case of a reasonably solid obstacle, cutting away (Fig. 5.5a) means exerting a very strong hit. If coherent, a mass is able to concentrate its entire inertia in one single stroke while in a disintegrated mass substantial internal losses hardly can be avoided. The opposite refers to running over: a coherent mass has to be integrally lifted over the obstacle, a disintegrated one has a chance of finding its way with local internal displacements of particles (Fig. 5.5b). So, despite the importance of details for assessing a particular case, the general statement can be made that coherent masses rather tend to cut away an asperity while disintegrated masses are better suited to run over it. From a somewhat different point of view, the discussion started in this paragraph will be resumed in Sect. 5.4.

It goes without saying that, at least in principle, the situation presented in the last paragraph might as well be *turned upside down*, the asperity being in the moving mass

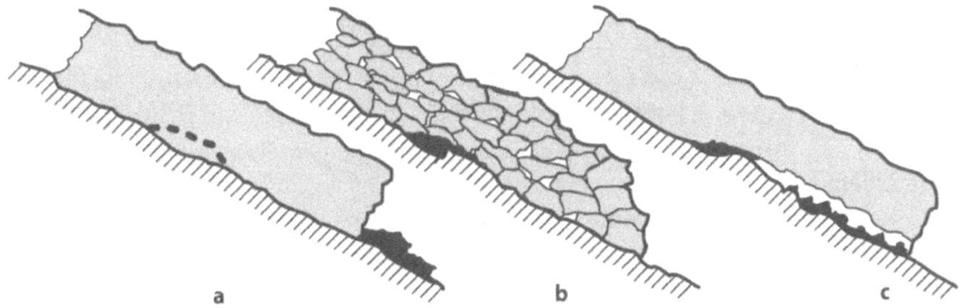

Fig. 5.5. Moving mass encountering solid asperity of the ground. **a** Coherent mass tends to cut away the asperity. **b** Disintegrated mass tends to give way by running over the obstacle. **c** Scraping or ploughing (according to size) (to be discussed in Sect. 5.4). Areas stained black: fresh fragments. In principle, the asperity might also be part of the moving mass. Consult text for particularities and probability of such inversions (sketch by Erismann)

and the cutting or giving-way mechanisms in the ground. Certain differences exist in such inversions (for instance, the freedom of forming a hump in running over an obstacle is better than that of rearranging the particles in a talus in order to give way to a moving asperity). In any case, the configuration of Fig. 5.5a and b is the most probable one for two reasons: a disintegrated ground is not very frequent, and the smaller sliding surface of the moving mass normally will be planified faster than that of the track.

*

The consequences with respect to the *reach of an event* – as postulated in Question 2 – can be drawn from the discussed mechanisms in comparatively short statements. In any case, it has to be borne in mind that, as a disintegrated rock mass is neither homogeneous nor isotropic, it is inopportune to use analogies with a liquid when comparing it with a coherent mass.

11. When being longitudinally compressed (e.g. in a descent with decreasing slope or a run-up with an increasing one), the differences between disintegrated and coherent motion are minimised by dynamically forced coherence. It can be expected that this is also true, at least to a certain extent, with respect to reach.
12. All energy-dissipating contacts in which exclusively moving particles of a disintegrated mass are involved (no matter whether mainly by collision or by friction), go at the expense of the (potential) gravitational driving energy. These losses of "internal" kinetic energy (due to differences in velocity and/or direction between particles) do not exist in a coherent mass and thus favour it with respect to far reach.
13. In a disintegrated mass moving on undulated ground (even if velocity is too low to make dynamic effects relevant) a part of the gravitational driving energy is directly lost by internal relative motion. A coherent mass bridges the undulations and thus has an advantage of reach.
14. A disintegrated mass easier passes through narrow clearances than a coherent one. This is valid both for gorges and curves. Even if forced into a partial volte-face, it may not lose its entire kinetic energy.
15. Where size (in principle: thickness) effects (Sect. 5.6, 6.3) are acting, a coherent mass has a better chance to reach far than a disintegrated one. This difference is

particularly acute if the disintegrated mass loses much thickness (and thus potential energy) by lateral spreading on plane ground.

16. The effects of Item 15 can be counteracted (perhaps even overcompensated) if the disintegrated mass is laterally confined by channelling walls and if, in coming to a rest, it can form a distal talus within a channel-shaped, downward-sloping confinement.
17. Coherent masses rather tend to cut away local obstacles while disintegrated ones tend to run over them. Cutting means track-making as the obstacle is more or less annihilated. In disintegrated motion, a certain amount of track-making is also possible by straightening the access to the run-over (area stained black in Fig. 5.5b). Quantitative comparisons of particular cases are very difficult (if not hopeless).

When considering the discordant criteria presented in the above items, it becomes clear that a farther-reaching mode cannot be ascertained a priori. Too much depends on the local conditions of a particular event, and quantification easily remains wishful thinking. Despite such uncertainty, Items 11–17 can be found useful when *considered simultaneously with geomorphological features* of an event. Thus, if for a particular portion of the track the state of disintegration is known, the probability of valuable conclusions is definitely improved: at least a hint for a reasonable subdivision into zones of low, medium, and high energy dissipation (as proposed by Nicoletti and Sorriso-Valvo 1991) might result (Sect. 6.4). And this is not a negligible asset.

The term "reach" requires, perhaps, a comment. As used here-above, it refers to the overall horizontal distance covered by an event (for further details s. Sect. 6.2). If, however, the total *area threatened* is at stake, it has to be taken into account that this area, as far as the circumstances allow lateral spreading, inevitably will be larger for disintegrated than for coherent displacement. In spite of this trivial fact it is normally justified to focus practical work on distance rather than area: the crucial question in rockslide prediction is to estimate how far a mass will reach, and it is – after due consideration of the topographic features and keeping in mind such lessons as taught by the catastrophes of Vaiont (Sect. 2.6) or Huascarán (Sect. 2.7) – nothing but common sense to account for lateral spreading by declaring as endangered reasonable zones on both sides of the presumptive main track.

*

Up to this point the two attributes "coherent" and "disintegrated" have been treated as if there were no intermediate states. This is obviously a simplification. In real events, disintegration may occur step by step at successive locations as well as it can be concentrated in a single one. And even in this last-mentioned case the location of disintegration may in itself be extended over hundreds of metres. If a continuous *description of coherence* were imperative, the use of a special parameter (expressed for instance by the average size of particles or by more sophisticated granulometric data) would have to be envisaged – not a very promising perspective in connection with the postulate of easy application, not to speak of the difficulties in predicting granulometry (as mentioned in Sect. 4.2). In such instances Question 3 leads to a discussion about the acceptability of simplifications.

Now the most radical simplification of a continuous value is reduction to a single-bit description which, in the present case, would mean the establishment of a well-defined state as a *limit between coherence and disintegration*. This approach is by no means unreasonable: from the statements 11–17 it appears probable that the differences are not as dramatic as might be guessed at first sight; furthermore many locations of disintegration, as already stated in Sect. 4.1, are detected rather easily even if nothing but a longitudinal section of a track is available. So the question remains how far disintegration should be advanced to be considered as valid. Once more, Items 11–17 can be used as a basis.

Item 11 only describes the state of best similarity and thus can be considered as marginal in the present context. Items 12 and 13 essentially contain a claim for the possibility of relative displacements within a disintegrated mass whereby the proportionalities in the basic equations for collisions and friction (Sect. 4.2, 5.4) show that, at least for a first approximation, it is of little importance whether many small or few large particles are contacting each other. So the range within which a mass can be considered as disintegrated is remarkably wide. An intentionally imprecise description by "*largest dimensions of most particles <30% of thickness of mass*" is proposed as a tentative approach. Item 14 postulates easiness of passing through narrow clearances, a condition for which the said tentative description can be used. The lateral spreading of Item 15 often occurs in the run-out phase of a rockslide. If the mass up to this point is still coherent, its being disintegrated "in extremis" is of little importance. Item 16 describes a configuration in which the reach of a disintegrated mass approximately equals that of a coherent one so that the particular state is almost irrelevant. Only Item 17, as will be demonstrated under Heading 5.4, cannot be satisfied either with the formula accepted for Items 12, 13, and 14, or with low differences between the two states as expected for 11 and 16; or with a trivial situation as stated for 15. In fact, the difference between cutting away and running over an obstacle is not a matter allowing expression by simple linear equations.

In spite of this drawback and the obvious imprecision, it appears reasonable to accept as a working tool the above tentative relation between the sizes of particles and that of the mass (Fig. 4.2 shows fragments unacceptable in a disintegrated mass; Fig. 5.5b gives an idea of what should be the maximum tolerated particle size). Therefrom it can be concluded that in the normal course of things static fracture, in particular if generated by bending or non-repeated delamination, may be regarded as insufficient while disintegration by reasonably intense impact promises adequate particle sizes. Hence the *first occasion of effective disintegration by impact* may be considered as the valid locus of transition from coherent to disintegrated state.

In a publication essentially based on quantification, the background of the above definition appears, perhaps, as somewhat woolly. To check this impression, in Table 5.1 the five key events viewed in Sect. 4.1 for locations of disintegration by impact are considered in the light of the definition. The result shows that, if sufficient topographic information is available, the uncertainty is not as serious as might have been assumed.

Comments are required only in two cases. A slight uncertainty exists for Pandemonium Creek as a sufficient disintegration, though probable, is perhaps not granted at the end of the quasi falling motion. Fortunately the influence of the signalled alternative (landing after bounce) is limited as the track between the two possible loci of transition is well-channelled and thus excludes a dramatic difference between coherent and disintegrated motion. Val Pola teaches a lesson: here the longitudinal section does not suffice to show the true locus of transition, and the contour lines and/or Fig. 2.27 and 2.28 have to be consulted to see the possibility of lateral collisions between par-

Table 5.1. Presumed locations of first disintegration by impact in five key events

Figure	Event	km	Character	Remarks
2.4	Pandemonium	0.4	Landing (quasi-fall)	Alternative: km 2.3 (bounce)
2.18	Köfels	4.1	Collision (steep slope)	Full stop of bottom part only
2.30	Val Pola	0.8	Collision (lateral)	Converging semi-falling motion
2.36	Vaiont	1.6	Collision (steep slope)	Full stop (Sect. E–E' and F–F')
2.48	Huascarán	0.7	Landing (fall)	

tial masses moving on convergent tracks to the relatively narrow passage at km 0.8. The collision must have been violent, no matter whether a simultaneous or a successive release on both sides took place. As concerning the three other events (Blackhawk once more being excluded as done in Sect. 4.1), after a look at their longitudinal sections one involuntarily thinks: where else?

Obviously the *coarse-on-top grading* normally observed in rockslides might be discussed in this section as well as elsewhere. Yet, as one of its essential causes consists in a concentration of disintegrating mechanisms near the bottom of a mass, the respective remarks have been made in Sect. 4.2.

*

Readers expecting from the present chapter a substantial amount of physico-mathematical stuff are perhaps disappointed by the aridity of the above overture in this respect. In continuing their reading they will encounter various circumstances in which formal quantitative analysis is revealed indispensable to distinguish plausibility from absurdity even in hypotheses displaying at first sight a purely qualitative character. This chapter – as, by the bye, the entire book – is addressed to experts facing rather the practical than the theoretical side of problems, and the mathematical level is adjusted accordingly. Those who are willing to penetrate into more academic fields, may study the respective literature. Some interesting examples are picked out hereafter although – owing to the assumptions made – their immediate application to rockslides cannot be recommended unconditionally. Savage and Hutter were the initiators of many publications in the field of granular displacement ("flow"), aimed at the influence of various laws of resistance upon motion: Savage (1979); Hutter et al. (1986); Savage and Hutter (1989); Nohguchi et al. (1989); Hutter (1991); Hutter and Koch (1991). Another line of thought is the application of fractal methods to quasi-chaotic avalanches generated in sandpiles and similar models: Bak and Chen (1989); Wiesenfeld et al. (1989).

5.2 Consequences of Velocity-dependent Resistance

It may be useful to start a closer discussion of mechanisms by an investigation of the *connection between velocity and resistance* against displacement. In fact, according to the mechanism dominating in a certain phase of motion, substantial variations may be observed in the behaviour of a mass. Thus quantitative comparison of idealised case histories can be arranged so as to reveal the phenomenological characters of basic mechanisms which, in their turn, can be used to judge how far the mathematical expression of a mechanism is applicable as a model for its morphologically given reality.

In such instances it is not illegitimate that, in setting up the *mental experiment* described hereafter, the physical background of the involved mechanisms will be completely disregarded: a mechanism will be defined by nothing but the causal connection between velocity and resistance (Fig. 5.6a) which necessarily is a constituent part of (positive or negative) acceleration. In other words: the mechanism will be considered as a black box (for details s. Ashby 1961). Accordingly each mechanism will be designated merely by a reference letter (N, P, Q, R) and not by a physical description. Examples of possible physical configurations, however, will be presented in certain cases for easier understanding.

For obvious reasons three of the four considered *mechanisms chosen for comparison* belong to those usually encountered in connection with mechanical motion.

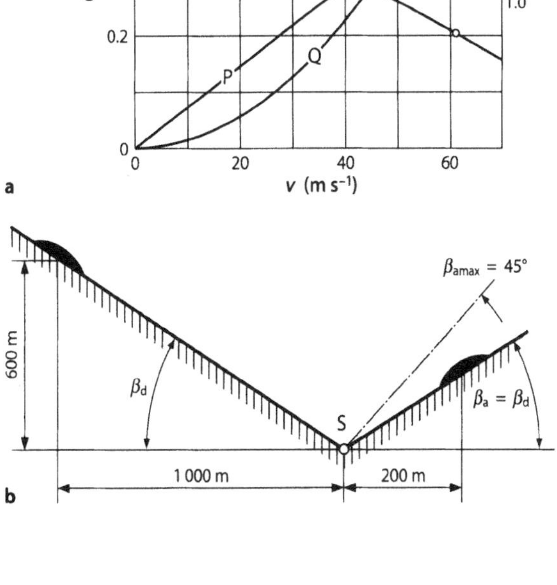

Fig. 5.6. Mental experiment with (**a**) functions $c_m f_m(v)$ expressing velocity-dependent resistance of mechanisms N, P, Q, R. Proportionalities of resistances are as follows: for N to cosine of slope angle; for P to velocity; for Q to square of velocity; for R to $1 - v / v_c$ (where v_c is constant). Constant coefficients c_m are adjusted so as to obtain, in the represented (slope-symmetrical) configuration (**b**), equal run-up heights with all mechanisms. Resulting velocities in point S are marked by *circles* (**a**). To compare mechanisms, run-up slope $\tan\beta_a$ is varied from 0.1 to 1.0. For results s. Fig. 5.7 (sketch by Erismann)

1. Mechanism N ("Not dependent"): resistance independent from velocity and proportional to the component of gravity exerted by the mass upon the ground (example: constant coefficient of Coulomb's friction),
2. Mechanism P ("Proportional"): velocity-proportional resistance (example: motion on a laminar layer of viscous fluid),
3. Mechanism Q ("Quadratic"): resistance proportional to the square of velocity (example: motion on a turbulent layer of watery fluid).

The fourth mechanism is introduced to account for the fact that under certain conditions resistance may decrease with increasing velocity (Sect. 5.5). For the sake of simplicity a linear function is assumed.

4. Mechanism R ("Reverse"): resistance proportional to $(1 - v / v_c)$, where v_c is a constant reference velocity.

On the basis of these definitions it is assumed that four masses, each according to one of the defined mechanisms, move across a V-shaped valley as represented in Fig. 5.6b. To obtain unambiguous results, three *important assumptions* are made regardless of their uncertain practical plausibility: in each case (11) the respective mechanism remains invariably effective during the entire displacement; (12) only a negligible amount of energy is lost in the transition from descent to run-up; and (13) resisting mechanisms are mainly concentrated near the bottom of the mass.

*

Obviously any of the considered mechanisms can be described by the *differential equation*

$$a = \pm g \sin\beta - c_m f_m(v) \quad (5.1)$$

where a is the acceleration along the slope, g gravitational acceleration, β the slope angle, c_m a constant (not necessarily dimensionless) coefficient, and $f_m(v)$ the function of velocity v attributed to the respective mechanism; the index m (suppressed in Fig. 5.6) stands for the appropriate reference letter N, P, Q, or R. Besides existing elementary solutions (especially for mechanisms N and P), numerical step-by-step integration is easily obtained by using the trivial set of equations

$$v = \int a \, dt \quad (5.2)$$

$$v_x = v \cos\beta \quad (5.3)$$

$$x = \int v_x dt \quad (5.4)$$

where t is time, v_x the horizontal component of v, and x the horizontal distance from the starting point of the respective phase (descent or run-up). Provided that β is regarded as positive for both phases, the sign of the first term on the right side of Eq. 5.1 is positive for descent (index "d") and negative for run-up (index "a" for "ascent").

To obtain a *standardised basis for comparison*, the four constant values c_m are adjusted in such a manner that the four run-ups reach equal heights if $\tan\beta_d = \tan\beta_a = 0.6$

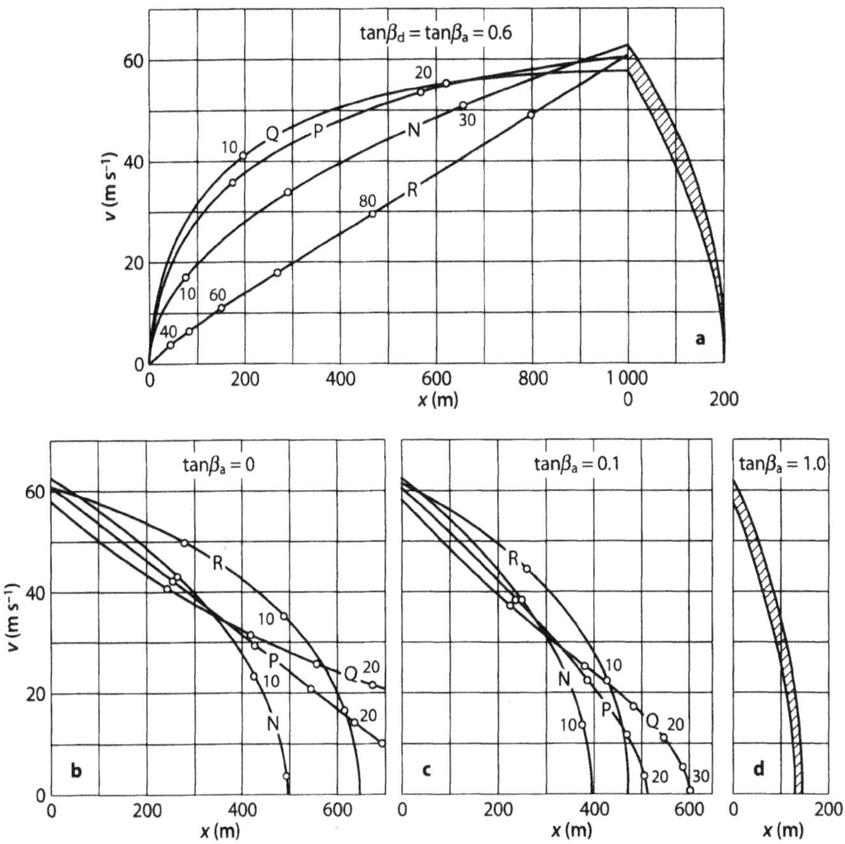

Fig. 5.7. Velocity v versus horizontal distance x in experiment according to Fig. 5.6. **a** Descent and run-up on symmetrical slopes. Coefficients c_m yield equal run-up heights. **b**, **c**, **d** Run-ups on other slopes (initial velocity as shown in **a**). Hatched areas: curves are too densely packed for individual representation. Circles with numbers 10, 20, etc.: time marks, every 10 s in descent, every 5 s in run-up (sketch by Erismann)

(equal slope angles of approximately 31° on both sides of the valley). For mechanism N, it appears useful to use $c_N = g$, whereby a coefficient of friction $\mu = f_N(v) / \cos\beta = 0.4$ is assumed. Besides c_m, the respective velocity v_{mS} in point S (Fig. 5.6b) is used as initial condition for run-up calculations.

The *comparison proper* consists in looking what happens if run-ups with various slope angles are calculated on the basis of the initial conditions described here-above. The results are shown in Fig. 5.7 as graphs of velocity versus horizontal distance.

For further *discussion* mechanism N, mathematically the simplest and certainly the most familiar in practical application (Sect. 6.3), is used as a reference basis. Accordingly, as far as quantitative results are involved hereafter, the figures of mechanism N are considered as representing 100%.

*

Of course there are several more or less *trivial results*. The narrow bands of $v(x)$ curves in steep run-ups ($\tan\beta = 0.6$ and 1.0), not exceeding a width of about 40 m in

x-direction, are due to the high ratio of gravitational to dissipated energy and, with $\tan\beta_a = 0.6$, by the forced identity of total distance run. Equally obvious is the fact that, in comparison with the other mechanisms, N has the shortest reach for low and the farthest for high β values: the resistance being proportional to the component of weight exerted upon the ground, a reduction on steep slopes necessarily occurs. Finally, the striking resemblance of mechanisms P and Q needs little explanation as both are characterised by reduced resistance at low velocities so that initial acceleration is high in descent and final deceleration low in ascent. A striking difference between P and Q will turn up hereafter.

*

In three cases, however, it is justified to speak of non-trivial and *practically significant results.*

21. In spite of being very different in other respects, all mechanisms attain approximately equal velocity maxima: as compared with mechanism N, the values of Q, P, and R range between 92 and 97%. This is mainly due to the fact that, to get from energy to velocity, the extraction of a square root is required. Anyhow, this "good-tempered" behaviour of velocity confirms a statement made by Erismann in one of his early studies (1980, 271) in connection with the energy line method which is nothing but a correct mathematical expression of mechanism N (Sect. 6.3): in many cases the easy-to-obtain maximal *velocity of mechanism N is good enough* as a first approximation even if dealing with other (or unknown) mechanisms.
22. The excessive duration of descent for mechanism R (253%) is partly due to the choice of parameters, deliberately made to show effects as clearly as possible (for $v = 0$ the resistance almost equals the driving component of weight). Anyhow, mechanism R, in descending, spends more time than N for both descent and run-up. Symmetrically thereto, in spite of requiring only a moderate time for descending (78 and 75% respectively), P and Q, in running up on mild slopes, are so time-consuming (184 and 238% for $\tan\beta_a = 0.1$) that their total time requirement may sensibly exceed that of mechanism N (131 and 146% for $\tan\beta_a = 0.1$). This fact can be important if, for instance, owing to the existence of appropriate seismic records, the duration of an event is known: in comparison with a simple type N evaluation *extra time may point to a non-N mechanism,* mainly to R (to P or Q only in combination with a run-up on a gentle slope), or to a substantial variation of μ in the course of displacement.
23. For mechanisms P and Q a low (or zero) slope in the second phase means impressing excesses of reach (166% and ∞ for $\tan\beta_a = 0$; 129 and 153% for $\tan\beta_a = 0.1$, respectively). The phenomenon is explained by the absence of resistance at zero velocity and will be commented in more detail below. As a first practical consequence, to be observed on mild or negative slopes, mechanisms *P and Q are incompatible with rapid arresting.* In this context it is interesting that the best descriptions of directly observed slides exclude dominance of P or Q for the run-outs of two famous events. Howe's comment (1909, 52, description of Frank slide) runs as follows: "*... cessation of movement appears to have been remarkably sudden...*"; Heim (1932, 111, description of Elm event) uses more dramatic words: "*... all of a sudden, the entire mass of rock grew stiff, as if by magic...*" (translation by Erismann).

It is, as a general rule, not intended to amuse readers with more or less trivial *mathematical curiosities.* In connection with mechanisms P and Q, however, there is a good reason to make an exception. For horizontal motion the first term on the right side of Eq. 5.1 obviously disappears, the variables become separable, and the elementary solutions thus obtained can, for instance, be written in terms of velocities:

$$v = v_{\mathrm{PS}} - c_{\mathrm{P}}x \tag{5.5}$$

for P and

$$\mathrm{LN}(v) = \mathrm{LN}(v_{\mathrm{QS}}) - c_{\mathrm{Q}}x \tag{5.6}$$

for Q. Here v_{PS} and v_{QS} are initial conditions of running up, according to the definition of v_{mS} here-above. Distance x is counted from point S (s. Fig. 5.6b). Hence, for mechanism P, velocity (though an exponentially decreasing function of time) is a linear function of distance x so that $v(x)$ is represented by a straight line crossing the abscissa axis at $x = 832$ m (curve P in Fig. 5.7b), a point, however, that the mass theoretically will reach only after having moved to the end of time! Mechanism Q is somewhat "better off": it also moves untiringly, but with the expectation of getting beyond any finite bound (about eight km it will make in 24 hours, ten km in 365 days)… Is it necessary to say that only perfect mechanisms, P and Q in pure culture, might be able to function as the above calculations suggest?

In spite of their seeming absurdity such results throw a light on an important practical question: in running out on near-to-horizontal ground a substantial role (or even a dominance) of a mechanism similar to type *P or Q necessarily means long run-out* and gentle deceleration, or, in other words, the danger of reach being underestimated in forecasting.

Of course, this peculiarity of mechanisms P and Q can be illustrated equally well without using Eq. 5.5 and 5.6. It suffices, for instance, to solve Eq. 5.1 for zero acceleration, assuming a certain negative slope. For $\tan\beta_{\mathrm{a}} = -0.1$, to quote an example, both mechanisms yield well-defined velocities: 13.4 m s^{-1} for P and 25.9 m s^{-1} for Q, and the character of the differential equation makes it clear that these velocities may be approached asymptotically both from above or from below. The approximately constant-speed displacement in the "bobsleigh run" *of Pandemonium Creek* (Evans et al. 1989) probably can be considered as a showhorse in this context. In fact, various circumstances favoured at least a partially Q-type motion: the mass brought with it a certain amount of water taken up in the preceding descent on a glacier, thus warranting a substantial "bow wave" (Item 5 in Fig. 2.7); steep slopes on both sides, in addition to reducing lateral water escape, allowed for a part of this water to be "recycled" when overrun by the mass; and the slope, though gentle, was sufficient to perpetuate the mechanism over a considerable distance.

So far, as remarked at the head of this section, the mechanisms were considered as being concentrated in the bottom-most part of the mass, and it is not intended here to go into details of internal displacements as discussed under Heading 4.2. Yet a *side-glance at the internal kinetic energy* may be useful. The general trend of internal contacts by collisions and/or friction, namely equalisation of velocities between the particles of a disintegrated mass, effectively reduces the energy available for further dissipation. So internal losses similar to those of type Q, as signalled by Eq. 4.5, remain small as long as no dramatic changes of motional direction occur (in Pandemonium Creek Valley obviously the meanders were smooth enough). More in this context will be communicated later, especially in Sect. 5.5 and 5.7

Once the context of field evidence being introduced into this basically theoretical section, an assumption expressed by Voight et al. (1983, 261–269) should be addressed. Their statement that, essentially, the event of *Mount St. Helens was of the P type*, is by no means in contradiction with the above considerations. On the contrary, this statement is perfectly plausible in view of the exceptional circumstances typical for a slide in a strongly volcanic environment (locally excessive temperature, melting ice, etc.).

*

If there is any "philosophical" significance in the present section, it consists in the demonstration that even apparently crazy mental experiments, in spite of being based on obviously unrealistic assumptions, can yield results of practical value.

5.3 Falling, Rolling, Bouncing

Falling, rolling, bouncing – three mechanisms so *closely linked to each other* that separate discussion would be a difficult task: a single boulder sliding on steep ground will sooner or later start rolling if not impeded by unfavourable conditions; rapid rolling of an irregularly shaped body on uneven ground is impossible without the bounces required to bridge the intervals between the – necessarily more or less protruding – points of contact; and a bounce in a steep descent, especially in its final (asymptotically vertical) phase is nothing but an almost perfect process of falling. So it is quite natural to approach the complex as a coherent system, picking out the particular mechanisms according to their physical nature.

*

The remarks made in Sect. 2.1 with respect to the phenomenon of falling are intended rather to describe the difference between rockfalls and rockslides than to *define the term of falling*. Perhaps a physically useful basis for the purpose is the criterion that acceleration should not be substantially inferior to that of free fall or, in other words, that friction should clearly be dominated by gravitation. If Coulomb's friction is assumed as a working hypothesis for the initial motion, it becomes obvious from Fig. 5.8a that acceleration a and gravitational acceleration g are linked by

$$\frac{a}{g} = \sin\beta - \mu\cos\beta \tag{5.7}$$

where β is the slope angle and μ the coefficient of friction. In the particularly interesting range $0.2 < \mu < 0.4$ the figures of Table 5.2 are obtained.

If, in order to be accepted as dominant, a force should clearly exceed the sum of all other forces and, accordingly, $a/g > 0.60$ is postulated, the occurrence of falling may, roughly speaking, be situated *above the slope range between 45° and 55°* (see bold types in the table). It is needless to say that this statement suffers the imperfection of all threshold values arbitrarily applied to continuous phenomena.

*

It is quite a long time ago since mankind – by inventing the wheel or (perhaps even earlier) by creating the first roller bearing, a couple of more or less cylindrical sticks placed across the track of a heavy slide – found out the economy of force and energy resulting from replacement of a sliding by a *rolling motion*. The experience of millennia has cemented this knowledge to such a degree that nowadays, in imagining the displacement of a large slab or a heap of blocks, one involuntarily looks out for some rolling mechanism. It is, however, necessary to remember the various effects of geometry, material, and size in order to get at a realistic idea of possibilities and limits.

Roughly speaking, rolling begins as soon as, on the one hand, sliding of a body is excluded by friction (and interlocking, if any) and, on the other hand, the vertical projection of its centre of gravity is situated outside the polygon circumscribed to the area of contact with the ground. These conditions essentially are identical with those for toppling: rolling is nothing but repeated toppling, the postulate of repetition, however, implying another class of shapes. Here the *role of geometry* is obvious. Figure 5.8b shows an easy-to-follow example: regular polygonal prisms (*n:* number of sides) will

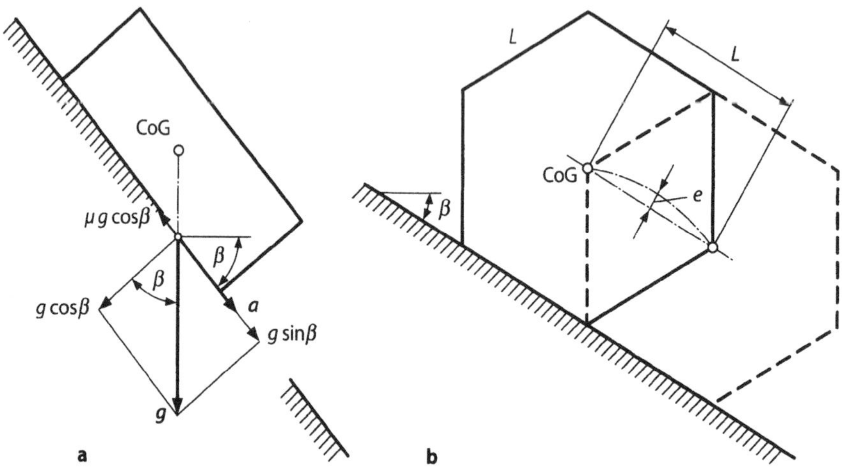

Fig. 5.8. On the verge of falling (**a**) and rolling (**b**). Falling is arbitrarily defined by $a > 0.6\,g$ (*a:* acceleration along slope; *g:* acceleration of gravity). In the particular case a coefficient of friction $\mu = 0.258$ and a slope angle $\beta = 50°$ yield this critical value (Eq. 5.7). Rolling is compulsively given by a non-sliding motion. In the particular case a regular hexagonal prism starts rolling on a slope angle $\beta > 30°$ provided that sliding is excluded by $\mu > 0.58$. The critical position of centre of gravity *CoG* is vertical above edge of contact surface. *L:* length of side = "step" length from one side to the next; *e:* deviation of *CoG* (sketch by Erismann)

Table 5.2. Falling. Comparison of acceleration with that of free fall. Quotient a/g (*a:* acceleration on slope; *g:* gravitational acceleration; β: slope angle; μ: coefficient of friction) according to Fig. 5.8a and Eq. 5.7. Italic types signal $a/g > 0.60$

β (°)	45	50	55	60
a/g ($\mu = 0.2$)	0.57	*0.64*	*0.70*	*0.77*
a/g ($\mu = 0.4$)	0.42	0.51	0.59	*0.67*

start rolling if slope angle β exceeds $180/n$ and the coefficient of friction μ exceeds $\tan\beta$. Lower friction may lead to sliding in a similar manner as the towers (and skiers) of Fig. 3.14. Examples in the range $4 < n < 16$ are shown in Table 5.3. The limitation to $\mu > 0.20$ does not mean that, once started, a prism will not, owing to its angular kinetic energy, be able to roll on, even in spite of lower friction. In fact, as mentioned in the first paragraph of this section, an appropriately shaped single boulder on steep ground will start rolling almost inevitably: sooner or later it will encounter a local obstacle that will initiate a sufficient spin.

The last line of Table 5.3 shows the deviation e of the prism's centre of gravity in its circular motion around one of its edges. This up-and-down movement is referred to the side length L, identical to the horizontal travel between reposing on two consecutive sides. So e/L is an indicator for the local *unevenness of slow rolling* motion. Of course these figures are valid only as long as the downward acceleration remains below g. If this limit is trespassed, a short bounce is inevitable and the amount of e/L

Table 5.3. Rolling conditions of regular polygonal prisms on plane slopes. *n:* Number of sides; *β:* minimum slope angle required to start rolling; $\mu = \tan\beta$: minimum coefficient of friction required to start rolling on slope angle *β*; *e:* deviation of centre of gravity with respect to a straight course; *L:* side length of polygon = longitudinal displacement when rotating from one side to the next. Parameter e/L gives an idea of the locally uneven course in slow motion

n	4	6	8	10	12	14	16
β (°)	45.0	30.0	22.5	18.0	15.0	12.9	11.2
μ	1.00	0.58	0.41	0.32	0.27	0.23	0.20
e/L (%)	20.7	13.4	9.9	7.9	6.6	5.6	4.9

is reduced. The respective critical velocity $v = \sqrt{(g R)}$, as containing the radius *R* of the circle circumscribed to the polygon, turns out to be size-dependent.

In analogy to other cases, quantitative evaluation of utterly improbable circumstances improves the easiness in understanding far more complicated configurations as encountered in reality. And it is needless to say that other mathematically defined geometries (e.g. bodies with elliptic section) might be used as well as polygonal prisms. The results obtained would seemingly be completely different from those presented above. Yet, as long as nothing but a single body is considered, at least four *relevant facts* would emerge, whatever approach might be chosen.

1. To a high degree the possibility of rolling depends on the shape of a body. The ability to roll asymptotically improves with the approximation to a circular section with the centre of gravity in its geometric centre.
2. Once having been started, rolling may continue even under conditions (i.e. configurations of *β* and *μ*) excluding the start of rolling.
3. Even on perfectly plane ground non-circular rolling bodies necessarily start bouncing at a critical velocity.
4. The critical velocity mentioned under Item 3, besides the shape of the body, depends on its size: for geometrically similar bodies it is proportional to the square root of the linear dimensions.

*

The process of rolling becomes more complicated as soon as instead of one single body *a multitude of different bodies* are involved. Once more, an extremely idealised model will demonstrate a set of problems which, far beyond the mere didactic aspect, will reveal some fundamental physical rules of disintegrated motion.

Despite their unrealistic shape, cylindrical bodies are chosen for the initial approach (Fig. 5.9). Consider at first a set of equal-sized cylinders, arrayed in several rows and kept at a small distance from each other by more than unrealistic means: something like the cage of a *roller bearing* (Fig. 5.9a). On a plane slope such a working model will move with great ease, the cylinders (or rollers) of each row rotating in a sense opposite to that of both rows above and below. Now cancel the first unrealistic condition by allowing various diameters. In a particular case only a set of four rollers is considered (Fig. 5.9b). Check the function by tracing the sense of rotation dictated to the

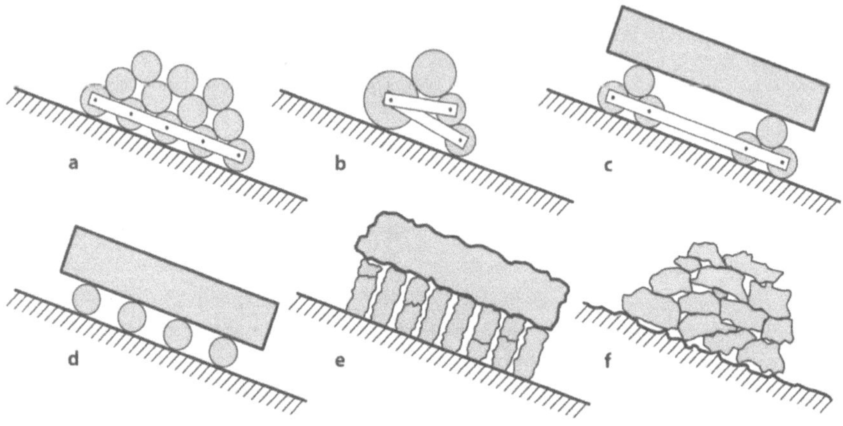

Fig. 5.9. Basic problems of roller bearings. **a** Idealised multi-plane array of rollers, easy to move (note "cage" keeping the array together); **b** blocked set of four rollers: leading and trailing rollers tend to drive the rollers of the set in opposite senses; motion is impossible without sliding in one of the contact points; **c** blocked coherent mass on even number of roller planes (if bottom rollers roll correctly downhill, top rollers drive mass uphill); **d** ideal roller bearing with single (= odd number) plane of rollers; **e** principle of functioning short-travel roller bearing as observed by Eisbacher (in toppling, rollers give the mass an effective start for further – sliding – motion); **f** everyday reality of disintegrated mass, reluctant against rolling displacement (sketch by Erismann)

rollers by the ground: if starting the process by the leading roller, the trailing roller is forced into the wrong sense (and, of course, vice versa). So the entire system "does not know what to do". As an inevitable consequence, it is either blocked or it can move only if sliding occurs somewhere. An analogous situation is observed if a large mass is supported by an even number of roller rows (Fig. 5.9c): again the rollers are in a dilemma so that the system is either blocked or forced into sliding. In case of an odd number of rows rolling would be unhindered and, in particular, a single-row roller bearing would function without the mentioned "more than unrealistic" cage as no forces would drive apart the rollers (Fig. 5.9d).

It is probably physical necessity rather than pure chance that the unique case of an *actually working roller bearing* found in the literature (Eisbacher 1979, 329, 331) belongs to the last-mentioned type (Fig. 5.9e). It is well understood that in the particular configuration the "rollers", as being relatively high and narrow, approximately rectangular bodies, after release (by an earthquake in the particular case), only could enable the mass to start motion despite a remarkably low slope angle of 13–20°. As soon as they had toppled, the reduction of resistance was lost and further motion was reduced to other mechanisms. The normal fate even of perfect roller bearings (including Fig. 5.9d) is, to a certain degree, similar: the overburden mass necessarily overtakes the rollers so that a continuous bearing function depends on the availability of further rollers moving ready for use ahead of the mass.

In turning to less exceptional models, cylindrical (or quasi-cylindrical) rollers have to be replaced by more normal clasts (Fig. 5.9f). Owing to their shape, the tendency to start rolling, as a general rule, is dramatically lower than that of cylinders. Being kept together by nothing but lateral contacts with each other or with external confinements, they are subjected to a certain lateral pressure opposed, via friction, to rotation. In addition, vertical overthrust load tends to force their largest diameter into a position rather parallel than perpendicular to the ground (Fig. 5.1). Finally, if a mass of a certain thickness, disintegrated in its lower portion, has to be considered, the probability is approximately equal for force-transmitting paths with an even or an odd number of contacts. All these arguments show that the *real probability of rolling motion is low* in comparison with that of the above-presented working models.

*

Before closing the discussion about rolling, several *particular aspects* have to be pointed out.

So far, a more or less plane track has tacitly been assumed. In the dramatic ups and downs of real motion even the seemingly ubiquitous effect of gravity is superposed by violent accelerations resulting in variations of intensity, and sometimes also sign. Thus rotary motions have a chance to be initiated, though only in the short periods of reduced compression. So *bouncing can contribute to generate rotation.* According to the circumstances, the result, when looked at over a somewhat longer time, may vary between a substantial angular displacement and nothing (if an initiated rotation is annihilated in the subsequent phase of compression). In the debris of Köfels – to quote a spectacular example – a large boulder with traces of glaciation is known (Heuberger 1994, 291). These traces give sufficient clues for a partial reconstruction of its kinematic history: it has undergone rotations on at least two of the three possible axes in space. But even such impressive phenomena normally have little to do with rolling as a means of locomotion.

In speaking of large boulders the occurrence of their *"floating" on smaller debris* (Sect. 4.2) should not be omitted. Certain reasons for grading coarse on top have already been mentioned, but the excessive phenomenon of large and accordingly heavy pieces of rock right on top of the rest requires a more consistent explanation, at least in cases where the trivial hypothesis of such boulders being latecomers with respect to the bulk of the mass turns out to be impracticable (Sect. 4.2). First of all, mere buoyancy (which might be assumed as a proposed mechanism from the wording used by Griggs 1922, 142, as cited in Sect. 4.2) cannot give a satisfactory reason. In fact, a large coherent body has, in general, a higher density than the surrounding disintegrated mass (with voids between its particles). In other words: if buoyancy were the dominating mechanism, the large boulder would sink instead of being floated! Now imagine a boulder, submersed in smaller particles, seesawing to and fro in its tendency to roll. The end going up pushes the particles on top of it upward and, to a certain extent, laterally. An effect of slackening in this region may occur, especially near the surface of the mass. On the other hand, below the said end an increased void is generated, and possibly some of the surrounding particles will, under the influence of lateral compression (Fig. 5.1), find their way into this space. So when the boulder seesaws back, it encounters a slightly more elevated support. What follows, depends on the circumstances: if the support remains low enough and the seesawing (so to say, oscillatory) energy is sufficient, a step in a slow, step-by-step climbing results; if not, the process is arrested – and the boulder has to wait for a better opportunity.

But once a large boulder of appropriate shape has reached the privileged position on top, it may profit thereof and start *rolling on the back of the disintegrated mass.* It is not excluded that some of the rather frequent boulders observed among the most distal deposits of various rockslides are not latecomers (as insinuated above) but "escalator passengers". Their privileged position, in the sense of economising energy, is twofold, by the reduced resistance of rolling and by travelling on "moving ground". And as the possibility of rolling on top of a moving mass is not exclusively reserved to large boulders, the rare opportunity to observe a descending rockslide may sometimes produce the misleading visual impression of an essentially rolling process.

*

There exists, finally, an absolute limit to rolling: under the load of an overburden, prospective *rollers can be crushed.* The method of calculating this phenomenon, current practice in mechanical engineering (e.g. for dimensioning roller bearings), first was applied to rockslides when an explanation was searched for the low slope angle in the case of Köfels (Erismann et al. 1977, 87–88). The authors of that study were in the comfortable position of dealing with a giant rockslide (maximal thickness in the range of 500 m). Thus, assuming 200 m as an average, they still obtained a practically prohibitive stress in spite of using the best possible geometry (s. Fig. 5.9d): a set of ideally cylindrical rollers of equal size, densely arrayed between the perfectly plane surfaces of ground and coherent mass. For lower thickness the equation determining the Hertzian pressure s_h should, in the worst case of horizontal ground, be written

$$s_h = 0.591c\sqrt{s_m E} \tag{5.8}$$

where $c > 1$ is a coefficient taking into account the imperfection of geometry. Keen readers will, by the way, have observed that s_h does not depend on the diameter of the cylinders: many slender rollers are as good as few thick. The other parameters are, despite a somewhat different notation, identical with those of the mentioned study: s_m is the average pressure upon the roller-covered surface, E is the modulus of elasticity of the material. With a density of 2500 kg m^{-3} (for a coherent overburden) and $E = 15$ GPa (for a sound material) Eq. 5.8 can directly be written as a function of thickness H, namely $s_h = 11.335\,c\sqrt{H}$ if s_h is expressed in MPa and H in metres. It is not an easy task to estimate c without assuming well-defined particular cases. Yet, in order to see that $c \approx 10$ still is a fiction, it suffices to imagine circumstances in which one of three rollers on one third of its length is quasi-perfect (and $c = 9$ means prohibitive stresses above 20 odd m of thickness!). Another approach is to replace cylinders by spheres: tentative calculations have shown that even in a perfect array a couple of metres thickness would yield dramatic overstresses. In parentheses it should be remarked that for the exceptional configuration of Fig. 5.9e the rule $c > 1$ is not valid as the "rollers", being only slices with more or less cylinder-sectorial ends, allow an abnormally dense packing. This makes sense only with such "kamikaze rollers" which, after having done their duty (i.e. helped to start the mass), leave the rest to other mechanisms.

The general impression gained from the above reflections becomes more and more transparent: if considered as a basic mechanism of locomotion and if no particular effects are at work, *rolling turns out to be far less important* than might have been assumed at first sight. It is, in fact, reduced to the displacement without overburden (be it of single blocks or of particles at the surface of a mass), to roller bearing function under particularly thin masses (hardly more than 20 m thick), and, in exceptional circumstances, to roller bearing "suicide" function under thicker masses, this function being cut after a short time (e.g. by toppling of "sliced" rollers or by continuous crushing).

By no means, however, does this result make obsolete the information expressed under Items 1–4: the limits to rolling in first instance are valid for rockslides; in *rockfalls*, on the contrary, rolling must be considered as a mechanism of high importance.

Anyhow, it is worthwhile to complement the mentioned items by a set of statements as obtained in the course of discussion.

5. Rolling is an important mechanism of locomotion in rockfalls. Its importance in rockslides, as far as not occurring at the top of a mass, is limited both in coherent and in disintegrated motion (see hereafter).
6. In both possible concepts (roller bearing under a coherent mass and disintegrated debris), irrespective of the degree of geometric perfection, blocking mechanisms are inevitable in arrays of a certain complexity. This effect is enhanced by the actual shapes of particles.
7. The absolute limit for the concept of a roller bearing under a coherent mass is given by crushing of the material. In principle this limit is independent of the roller's diameter. By exception of "kamikaze" cases of short duration, hardly a thickness of 20 m can be considered as overburden, and ball bearings are completely out of the question.
8. Rudimentary rolling motions are possible owing to shaking or bouncing on uneven ground. Their importance for the displacement of rockslides is strictly limited.
9. Large boulders acquire a location on top of smaller debris not by buoyancy (which would let them sink) but by micromechanisms occurring systematically in the process of being shaken. Once on the back of the mass, they have a chance to roll forth, overtaking at least part of the other debris.

All results thus presented are in good *agreement with field evidence* as well as with theoretical considerations. In particular the repeatedly mentioned fact that even heavily disintegrated debris tends to preserve the sequential order of clasts and, in a more general way, to remain "in shape" (Fig. 2.21; Sect. 2.3, 2.4, 2.8, 4.1, 5.1) would become questionable if rolling were one of the basic mechanisms in rockslides, especially in large ones. The rarity of rounded medium-size clasts in the deposits of rockslides is a further argument. And it is perhaps needless to say that the importance of sliding mechanisms, as treated in the two immediately following Sect. 5.4 and 5.5, is increased to the same extent as that of rolling is diminished.

*

Contrarily to rolling, *bouncing is frequent in rockslides and in rockfalls* as a necessary consequence of the attained velocities and the unevennesses of tracks. Certainly, Howe's dictum (1909, 52) that in the Frank rockslide "*... the movement of the individual fragments consisted of a succession of bounds... the movement of the mass as a whole suggested that of a viscous fluid...*" requires comments in two respects owing to the simple fact that eyewitnesses could see nothing but the surface of the moving mass. So they saw the part where bounds were unhindered by any overthrust and where a certain optical similarity with a flowing liquid did exist (s. Sect. 5.1 and 5.4). Nevertheless, any unevenness not razed off by a moving mass necessarily evokes a more or less local taking-off to a more or less extended bounce. So bouncing may be considered as one of the basic means of locomotion in rockslides, all the more in rockfalls, and it is worthwhile to have a closer look at its mechanisms.

As long as a projectile has no contact with the ground, the acting forces, being reduced to gravitation and aerodynamic effects, are comparatively unproblematic. The *aerodynamic resistance or drag* F_v of a bouncing body follows Eq. 3.5 in which, of course, the density of air (and not that of water) has to be used for ∂_1. On this basis it is easy to set up the differential equations describing motion. As in turbulent conditions (amply valid, in terms of Reynolds numbers, for rapidly moving clasts in the relevant size range from several centimetres to several metres) drag is quadratically proportional to velocity v, its horizontal and vertical components likewise are quadratically proportional to the respective components v_x and v_z. So they can be integrated separately in the form

$$\begin{aligned} \frac{dv_x}{dt} &= a_x = -v_x^2 \frac{c\partial_1 A}{2\partial_2 V} && \rightarrow \quad v_x = \int a_x dt \quad \rightarrow \quad x = \int v_x dt \\ \frac{dv_z}{dt} &= a_z = \pm v_z^2 \frac{c\partial_1 A}{2\partial_2 V} - g && \rightarrow \quad v_z = \int a_z dt \quad \rightarrow \quad z = \int v_z dt \end{aligned} \tag{5.9}$$

where the following notation is used: *t:* time; a_x, a_z: components of acceleration; *c:* drag coefficient (in certain cases two values, c_x and c_z); *A:* sectional area at right angles to vector of velocity (in certain cases A_x and A_z); ∂_2: density of rock; *V:* volume of body; *g:* gravitational acceleration; *x, z:* co-ordinates (origin: take-off point).

For the further work it is most important to estimate the *influence of the main parameters* (initial velocity v_0 and characteristic diameter *D*) upon the reach of a bounce. In Fig. 5.10 this influence is visualised for a horizontal vector of v_0, a rather normal condition for a "jumping hill" resulting in considerably far bounces (a substantial positive vertical component of velocity cannot be considered as a frequent occurrence, and a negative one dramatically shortens its reach). Assumed is an approximately spherical body ($c \approx 0.5$ when taking into account local irregularities), an altitude in the range of 2000 m ($\partial_1 \approx 1.0$ kg m^{-3}), $\partial_2 = 2500$ kg m^{-3}, and $g = 9.81$ m s^{-2}; v_0 is chosen in a normal (50 m s^{-1}) and in an excessively high (100 m s^{-1}) range, both combined with diameters *D* of 0.01, 0.1, and 1.0 m. As a reference (and more than that, as will be demonstrated), the respective trajectories in vacuum are shown.

A first important result is immediately conceivable: even at velocities around 100 m s^{-1} for boulders larger than metre-sized *air drag is almost negligible* so that vacuum trajectories may be used. This is a very substantial advantage as instead of solving step by step a set of non-linear differential Eq. 5.9, the straight-forward solutions

$$\begin{aligned} \Delta x &= x - x_0 = v_{x0}\, t \\ \Delta z &= z - z_0 = v_{z0}\, t - 0.5\, g\, t^2 \end{aligned} \tag{5.10}$$

readily yield the co-ordinates for any desired point of a trajectory (take-off conditions are marked by the index "o"). In addition, except for utterly improbable cases (some examples will be mentioned hereafter) calculations of reach are, in comparison with those of other rockslide parameters, remarkably accurate. Reach is slightly overestimated so that, in prediction, errors are on the safe side.

Besides trajectories as shown in Fig. 5.10, also those attaining the *farthest possible reach*, as resulting from a certain (ascending) angle β_1 of the initial velocity vector,

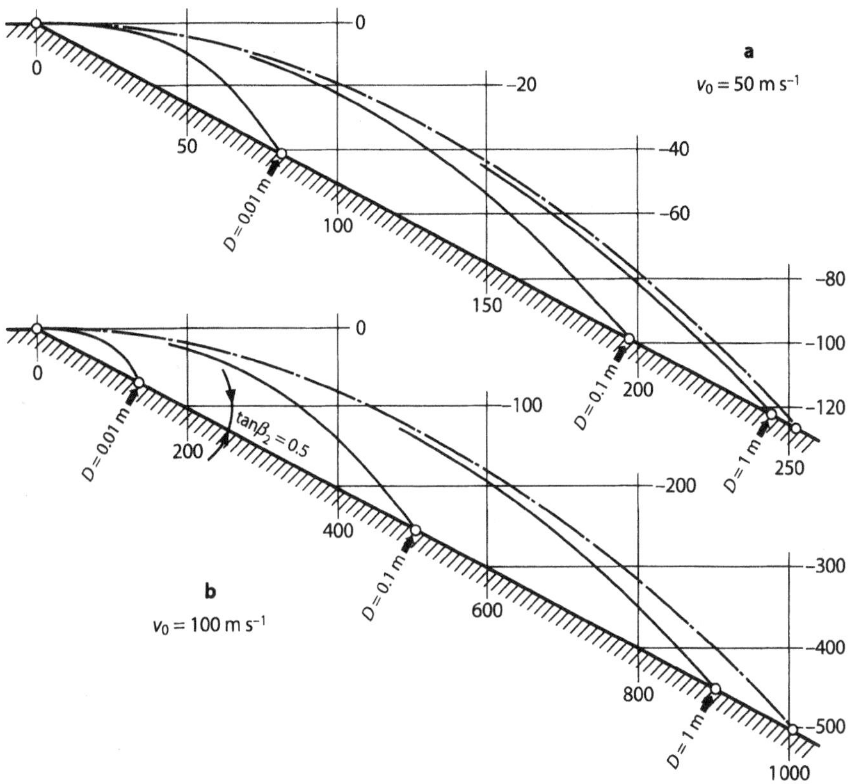

Fig. 5.10. Influence of aerodynamic drag (altitude about 2 000 m) upon trajectories of bouncing blocks (diameter D). Vacuum trajectories (*dots* and *dashes*) serve as a reference. Horizontal initial velocity v_0 is 50 m s^{-1} (**a**) and 100 m s^{-1} (**b**). Distances and relative altitudes are in metres (note: **a** and **b** are at different scales). A certain scatter of parameters has to be taken into account especially for $D < 1$ m (sketch by Erismann)

are of particular interest. In a general form, the reach x of a vacuum trajectory is expressed by

$$x = 2\frac{v^2}{g}\left(\sin\beta_1\cos\beta_1 + \cos^2\beta_1\tan\beta_2\right) \tag{5.11}$$

where β_2 is the (descending) slope angle. The equation

$$\beta_1 = 45° - \frac{\beta_2}{2} \tag{5.12}$$

yields the condition of ultimate reach. To quote an example: replacing $\beta_1 = 0$ by $\beta_1 = 31.72°$ in the case of Fig. 5.10 increases the reach of both vacuum trajectories by almost 62%. Caution is, therefore, recommended in prediction if the take-off circumstances do not definitely exclude a positive vertical component of velocity.

To pave the way for a reasonable approach to practical applications a critical review of *tolerated simplifications* may be useful.

11. Neglect of compressibility effects in the air is justified as long as Mach numbers (i.e. quotients velocity/velocity of sound) remain essentially below unity. For a velocity of 100 m s^{-1} this is definitely the case.
12. As far as bouncing velocities are sufficiently elevated, the influence of wind normally can be neglected for metre-sized and larger blocks (by exception of heavy storms).
13. The reserve made in Sect. 3.3 with respect to the shape-dependent drag coefficient *c* is equally valid in the present context. The resulting uncertainty, however, is confined to small particles.
14. Slow rotation in bouncing (most probable around the transversal y-axis and stable if occurring around one of the body's main axes of inertia) in first instance induces averaging of both *A* and *c* which, in such instances, automatically are equal in the directions *x* and *z*.
15. Aerodynamic lift, supporting the boomerang-like (i.e. curved) flight of a slab-shaped particle, gyroscopically stabilised by fast rotation around a quasi-vertical axis, hardly can be considered as an important factor since the conditions leading to such an escapade obviously are rare enough.
16. In case of sufficiently fast rotation a non-negligible Magnus effect may be observed: owing to the boundary layer of air being swept along by the rotating surface, an aerodynamic force acting at right angles to both the vectors of velocity and spin is generated. As a rule (i.e. in case of forward rotation around the *y*-axis), the resulting force is a negative lift and reduces the reach of the bounce (the resulting error is on the safe side in case of risk prediction). Rotation around an essentially vertical axis may result in a non-negligible lateral deviation; if so, both start or landing azimuths cannot be extrapolated over the entire aerial travel.

So an accurate quantitative treatment as claimed above turns out to be valid within the limits dictated by Items 11–16, and the essential results referring to aerial motion can be expressed by two *conclusions*.

17. Even for velocities around 100 m s^{-1} it is justified to base predictions on vacuum trajectories. This is valid for large single blocks as well as for entire masses moving coherently or, if disintegrated, in not too loosely packed state (allowing the formation of a common boundary layer of air carried along with the mass or parts thereof).
18. In loose disintegrated masses large particles bounce farther than smaller ones. A tendency for a certain two-dimensional gradation (large grains concentrated in distal regions and on top) is the result. For celestial bodies the density of the atmosphere (if any) has to be accounted for.

*

So far, bouncing was considered as an individual phenomenon beginning by a take-off from a jumping hill and ending with a landing. Very often, however, the phenomenon will not be terminated by such landing and *rebounding* will take place, an energy-recuperating mechanism definitely less appropriate for accurate calculation than the bounce proper. The problem lies in the complexity of landing: an irregularly shaped, possibly rotating body hits a ground that may be naked rock (approximately plane or uneven) as well as scree or unconsolidated soil. So it is nothing but natural that the coefficients obtained from laboratory or field experiments range within an extremely wide band and the correct choice for prediction may be a difficult task.

To obtain, from the known incidence data of a landing, quantitatively satisfactory *initial conditions for a subsequent rebounding* take-off, at least two basic parameters must be taken into account in a plane model (already simplified by neglecting lateral effects). The first is the coefficient of elastic restitution as already used in Eq. 4.8, in the present context expressed by

$$J = \frac{v_P}{-u_P} \tag{5.13}$$

where u_P is the component of velocity perpendicular to the slope in landing and v_P the respective component in starting to rebound (Fig. 5.11a). The second parameter basically is the coefficient of friction μ of a non-rotating body in a rebounding process. Assuming that Coulomb's rule is valid during the impact, the loss in velocity can be deduced from the loss in momentum which in its turn is the product of μ with the change of momentum perpendicular to the ground. Written in terms of velocities (if the mass is considered as remaining constant and thus can be left aside) this argument yields

$$v_L = u_L - \mu(v_P - u_P) \tag{5.14}$$

where, in analogy to the notation in Eq. 5.13, u_L and v_L are the pre-impact and after-impact longitudinal components of velocity. Of course, as u_P is negative, the bracket

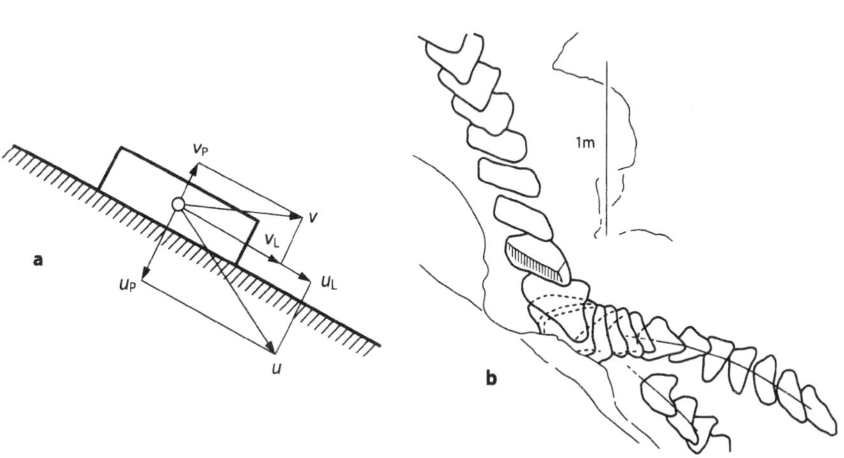

Fig. 5.11. Theory and reality in rebounding. **a** Non-rotating (or point-like) body in the moment of collision with the ground. *Circle:* centre of gravity; u, v: velocities before and after collision; u_L, u_P, v_L, v_P: longitudinal (index L) and perpendicular (index P) components of u and v. Note that both components are reduced, longitudinally by a frictional shock during collision, perpendicularly by non-elastic losses in the material. **b** Real clast, split into two by the impact. Despite the obvious differences the theoretical approach, if adjusted in certain respects and reasonably interpreted, yields results useful also for practical work (sketches: **a** by Erismann; **b** after Bozzolo 1987, 28, copy from serial photographs taken during experimental rockfall)

on the right side is the sum of the two involved absolute values. If the bouncing body is rotating, μ has to be readjusted appropriately and even can, in case of a circumferential velocity exceeding u_L, acquire negative values (the body being driven forth by the angular kinetic energy).

Here a short excursion into the problems of *repeated rebounding* is useful. In looking out for a description of falling which would be physically more plausible than that deduced from Eq. 5.7, the possibility of a continuous sequence of congruent rebounds on a plane slope was investigated. The result would not be worth even the abridged description that will follow if there were not one point of particular interest. Three criteria of congruence were postulated as initial conditions for two subsequent impacts: equal values of u_P, equal values of u_L, and the collision with the ground. The three equations set up on this basis were formulated to express the space of time Δt between two consecutive impacts using J, μ, the slope angle β, and velocity u_P. The three results

$$\Delta t = \underset{\text{A}}{-u_P \frac{J+1}{g\cos\beta}} = \underset{\text{B}}{-u_P \mu \frac{J+1}{g\sin\beta}} = \underset{\text{C}}{-2u_P \frac{J}{g\cos\beta}} \tag{5.15}$$

when equated in pairs, yielded $\mu = \tan\beta$ (from A = B in Eq. 5.15) and $J = 1$ (from A = C). The interesting point lies in $\mu = \tan\beta$: it makes evident that travelling through the air is, in all, by no means more economic than travelling on the ground, at least, as long as Coulomb's rule holds good. Of course, an airborne body, besides the (more or less negligible) aerodynamic drag, does not dissipate any energy; but in landing it is forced to pay back every Joule it had been saving (s. also comments to Fig. 5.18).

To avoid misunderstandings, two remarks have to be made in this context. On the one hand the conclusions drawn from Eq. 5.15 do not exclude *rebounding on extremely smooth slopes*, i.e. in case of $\mu > \tan\beta$. A – probably rare, but not impossible – issue in case of high restitution is the following one: some normal bounces may take place, and each will cover the distance given by its initial conditions. At a certain landing, however, the body will "declare itself bankrupt" in terms of motional energy: in longitudinal sense it will come to a rest before having made use of the entire braking potential given by the impact, and, of course, before taking off for the next bounce. Now on slightly inclined ground this take-off will be somewhat inclined so that a minute displacement will occur in rebounding. If restitution is good enough for a series of such rebounds, a slowing down, bounce-by-bounce creeping will be the result. Theoretically this motion never will stop, in reality it certainly will be arrested by (so far neglected) non-linear effects.

On the other hand, also *serial bouncing as a current phenomenon* in rockslides and especially in rockfalls is by no means excluded. The condition to obtain, for instance, the above-postulated congruent trajectories consists in replacing the plane track by a conveniently undulated one. In such instances $J = 1$ no longer is a condition inasmuch as part of the energy required in perpendicular direction can be taken from longitudinal motion. The necessary consequence is that this energy is made available by providing $\mu < \tan\beta$. Exactly this has been done in an interesting mental experiment which will be presented in more detail under Item 24 hereafter. For the moment it may suffice to remark that on steep slopes continuous bouncing is as normal as sliding and rolling (in certain cases even more).

Behind the simple model as used above – treatment of a particle as a mass point and accounting for rotation by adjustment of the coefficient of friction – stands a far more complicated reality. As mentioned above, the ground may consist of any material; the shape of a (possibly rotating) clast as well as small unevennesses in the ground (below the size perceptible in a map) can result in dramatic modifications of the geometry (Fig. 5.11b); the phenomenon as a whole can be significantly three-dimensional; there may occur collisions with other particles of the rockfall or immobile obstacles. So in prediction the determination of *realistic rebounding parameters* requires great care both in estimating average values and scatter bands. In addition, details of formulation have to be taken into account: figures presented in the literature may refer to components of velocity (as in the present section) or to the total velocity; information may be expressed in terms of energy instead of velocity, thus being proportional to the square of the respective velocity; eventually, the energy lost may be used in place of the energy restituted.

It would exceed the frame of this book if all hypotheses concerning the complex of falling, rebounding, rolling, and also sliding (considered as a whole or in partial mechanisms) would be presented and commented in extenso. Still it is considered worthwhile to exemplify different *philosophies of approach* by extracting from appropriate investigations the respective lines of thought. Of course, the choice of examples mainly was made with regard to the method used to attack a problem rather than the general quality of a study. As a consequence other valuable publications remained uncited (e.g. Azimi et al. 1982; Hacar Benitez et al. 1977; Habib 1976; Heierli et al. 1985).

21. Statham (1979), in the endeavour to establish a simple quantitative description of the size effect observed (large clasts reach farther than smaller ones), consciously renounced to present a physical explanation and concentrated work on the establishment of a handy algorithm expressing the observed effect with satisfactory accuracy – a typical black-box approach,
22. Broili (1974) took advantage of the rare possibility to set up a series of full scale experimental rockfalls at the very site of a large catastrophic event that, in the outskirts of Lecco (Northern Italy), at the foot of Monte San Martino had killed eight persons in 1969. Cameras located along the track gave the basis for extensive evaluation; in hitting a scree after a blast-released, quasi-vertical fall of 210 m, the clasts lost most of their energy; yet the remaining 14–25% sufficed to generate rebounds up to 200 m long, followed (on a slope decreasing from 30–25°) by gradual transition to rolling. An interesting detail: small particles rebounded farther than larger ones, but finally were left behind in rolling.
23. Bozzolo (1987; et al. 1988), in addition to a careful analysis of the existing models, investigated the role of details: cameras were placed so close to the tracks of relatively small, though still realistic experimental rockfalls that (besides the occasional destruction of a camera by a capricious clast) in certain cases three-dimensional histories of individual impacts could be established, showing, for example, that the rotation of clasts normally was too slow to grant a purely rolling contact with the ground.
24. Campbell (1989) demonstrated in a two-dimensional, computer-based mental experiment that under favourable conditions ($\beta = 40°$, $J = 0.9$, regularly uneven ground, no resistance in longitudinal sense) a few disc-shaped particles near the

ground not only rebound continuously, but also maintain a larger number of further discs in a state of levitation. For similar conditions this approach is a good example for the extraction of useful knowledge from a simple quantitative model. Only in one point a certain criticism cannot remain untold: the pretence to have proposed the "*... only mechanism... that holds promise of accounting for all the features attributed to long runout landslides...*" (p. 664) sounds too optimistic in view of the extremely favourable conditions chosen and the fact that long run-out qualities mainly should work in the range $0 < \beta < 10°$ (s. Sect. 5.5).

*

The fact that the mathematical treatment of bounces is one of the few problems in rockslide dynamics in which the results obtained excel by an accuracy unknown in questions of friction or strength, may be considered as a good reason to crown a mainly theoretical discussion by the analysis of extremely spectacular real *giant bounces*. As a matter of fact, in the Huascarán event (Sect. 2.7) boulders of up to 65 tons are assumed to have accomplished aerial travels exceeding 1 000 m, in particular cases even 4 000 m. Therewith a particularly fascinating field for the application of Eq. 5.10 to 5.12 is opened. Under Heading 2.7 the phenomenon has already been considered, with due reference to the important publications by Körner (1983, 97–99) and Plafker and Ericksen (1978, 288–306), but as far as this was possible without using mathematical means. So it will be possible to reduce the following discussion to the indispensable line of thought.

First of all, it should be recalled that in the event of 1970 the mass, contrarily to that of 1962, was not confined, between km 5 and 8 of Fig. 2.48, by Quebrada Armapampa but inundated a large area. So it makes sense to investigate this area for *possible take-off points* not considered in the literature. Luckily enough, the excellent map (scale 1 : 25 000, contour interval 20 m) annexed to the book containing Körner's study (ed. Patzelt 1983) makes it possible to do the work with reasonable accuracy (Fig. 5.12).

Four sections were laid at locations offering a total of nine more or less appropriate "jumping hills". From each of these points two trajectories were aimed at the periphery of the village Huashau (partly damaged by boulders and spattered by mud). According to Plafker and Ericksen the landing incidence could be assumed as being between 15° and 30°. Hence a solution of Eq. 5.10 was seemingly impeded by the fact that *landing instead of take-off conditions* were known. Yet, owing to the perfect parabolic symmetry of vacuum trajectories, "reverse aiming" could be applied: the equations were solved as if the projectiles were shot up the hill, from landing to take-off. So, after calculating the parameters emanating directly from landing incidence and Eq. 5.10, namely total time and landing velocity, it was easy to get at the initial conditions by applying said equation appropriately.

To assess the plausibility of the 18 trajectories thus obtained, it is useful to compare the velocities required for take-off (v_{15} or v_{30}, the indices referring to the respective angles of incidence at landing) with those after the preceding descent. Thereby, in spite of being highly improbable, an absolute *maximum of physically imaginable velocity*, distinctly higher than that obtained in Sect. 2.7, was admitted. The altitude of 3 800 m was attained by the centre of gravity of the mass after a displacement of about 4 400 m in horizontal and about 2 100 m in vertical direction (Fig. 2.48: km 0–4.2 and altitude 5 900–3 800 m). If, anticipating the deductions of Sect. 6.3 and assuming an

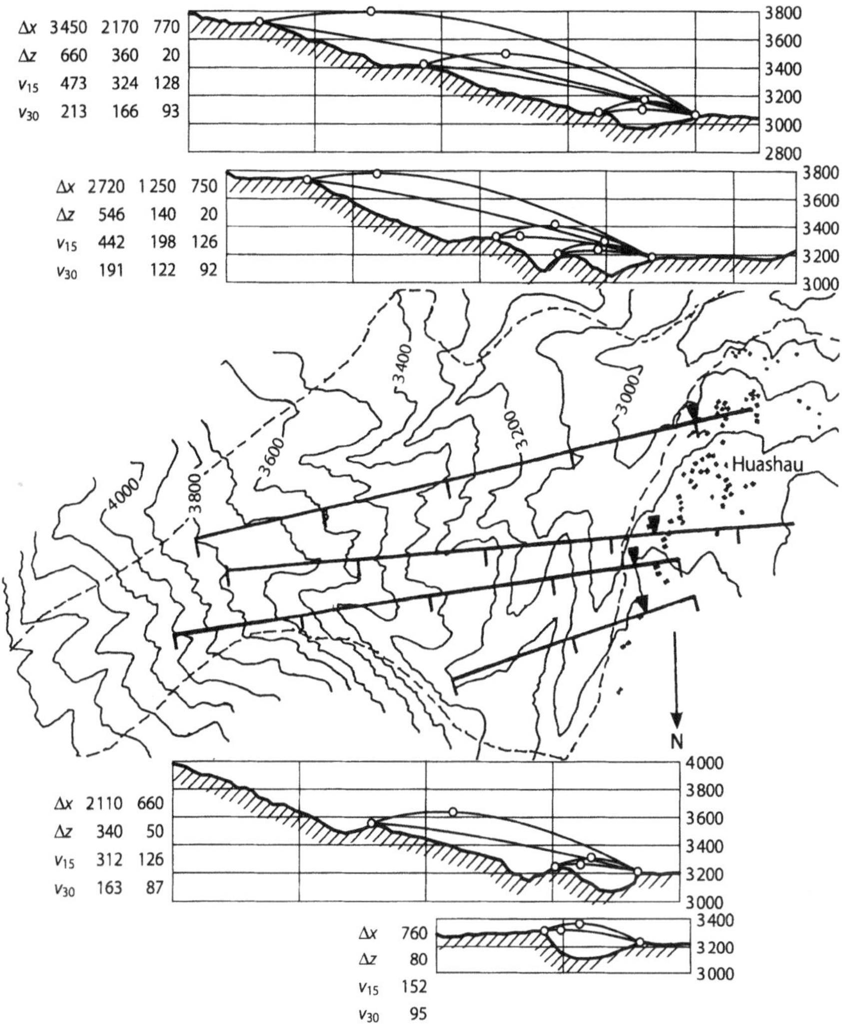

Fig. 5.12. Map showing locations of possible giant bounces half-way in the Huascarán event (complete map in Fig. 2.48). Corresponding sections are aligned thereto. *Dashed lines:* area overrun by debris. *Circles:* possible points of take-off, culmination, and landing (periphery of Huashau). *Triangles:* landing points. Δx, Δz: horizontal and vertical bounce distances (m). v_{15}, v_{30}: take-off velocities (m s^{-1}) correlated with landing incidence of 15 or 30°. Horizontal division: 1 km (sketch by Erismann)

incredibly gentle energy line with $\mu = 0.15$ (averaged) despite the massive energy loss in the transition from quasi-vertical falling to sliding on a reduced slope, the potential energy at the considered point is given by a level difference of 2 100–660 = 1 440 m corresponding to a velocity of 168 m s^{-1}. Taking into account the assumptions made (neglect of drag, limitation of reach to the periphery of Huashau), it becomes evident from the calculated results (Fig. 5.12) that the largest bounces of 3 450 and 2 720 m could not have occurred, scarcely those of 2 110 and 2 170 m. Here an additional factor has to be considered: at 168 m s^{-1} the aerodynamic drag is more than 2.8 times that at 100 m s^{-1} so that, in spite of the density reduced by a factor 3:4 when passing from 2 000 m to the range of 3 000–4 000 m, the drag is still more than twice that of Fig. 5.10b. In such instances it becomes obvious that no bounces in the range of 2 000 m length can have been accomplished; and the last critical arguments – that there is no guaranty for the required positive take-off angle and that in the velocity range of 200–400 m s^{-1} simplification of Item 11 no longer holds good – may be left in the quiver.

In Sect. 2.7 it already has been shown that acceleration by the Venturi effect in a funnel is not a practicable way to make plausible exorbitant velocities. A further explanation might be found in Sect. 5.7 under the key word "*kicking*". In fact, a comparatively small block may be accelerated to a remarkably high velocity by the impact of a larger one. However, the size of numerous boulders found in and around Huashau excludes the faintest hope for the coming true of this hypothesis.

The more arguments are considered, the clearer it appears that opportunities of *shorter bounces* merit a closer inspection. And a look at Fig. 5.12 shows, as it is, that there are plenty of possible jumping hills allowing a bombardment of Huashau from distances not exceeding 1 500 and even 1 000 m. So the velocities required may range (by one exception in which an almost straight trajectory of 1 250 m is the condition for a landing incidence of 15°) between approximately 90 and 130 m s^{-1}.

Is it illogical, in such instances, to consider the *total absence of "super-bounces"* beyond 1 500 m and of velocities beyond 130 m s^{-1} as the most probable solution?

5.4 Unlubricated Sliding

In current engineering practice unlubricated friction between coherent sliding bodies normally is treated according to *Coulomb's rule*. In other words: it is postulated that the resistance opposed to a sliding motion be proportional to the compressive force acting at right angles to the contact surface. Thus the factor of proportionality μ (the coefficient of friction) is assumed to be constant, i.e. depending only on the characteristics of the contact surfaces (material, rugosity, etc.), but independent of load, relative velocity, and time (or relative distance travelled). An exception characterised by a substantial increase of resistance (a ratio 2 : 1 not being considered as exceptionally high) is tolerated for zero velocity.

This "*zero-velocity effect*" may have different causes, ranging from interlocking asperities (Sect. 3.2) to the absence of a spontaneous lubrication by an almost ubiquitous microscopic layer of water. It goes without saying that the first-mentioned mechanism has a good chance to prevail in connection with surfaces of rock: as soon as even a minute amount of kinetic energy is available, the moving mass acquires, in addition to the forces driving a body ahead, an increased ability to deal with all sorts of obstacles. Some of the rare rockslide masses released on soft slopes excel in being provided with a ready-for-use "prefabricated" track like those of Gros Ventre and Rossberg (Sect. 3.1; Alden 1928, 347–349; Heim 1932, 156–159).

*

In resuming the discussion of Coulomb's rule, it is essential to face the fact that the mentioned proportionality, though resting on a solid basis of experience, for a long time had no theoretical background reducing it to more fundamental physical principles. This made questionable its application to new circumstances. Therefore a short excursion into some *general problems of tribology* may be useful. Most of the research in this field is centred around technological problems as encountered in the development of bearings, guiding facilities, brakes, etc., i.e. devices designed on the basis of well-machined metallic or plastic materials in which asperities are described in microns and not in centimetres or metres as in case of rock. One might, perhaps, be tempted to try a fractal approach in order to transpose results from a microscopic to a giant scale. It has, however, to be taken into account that such an approach works only if the basic mechanisms remain similar and – at least in principle – only have to be adjusted to an appropriate scale. The above-mentioned example of a similar effect being due to macro-interlocking and poor micro-lubrication shows that caution is required: it has to be made sure that similarity refers to the mechanisms as well as to their perceptible effects.

In such instances several tribological publications have been combed through for information promising a useful application in understanding the background of Coulomb's rule (if at all valid) and of other mechanisms occurring in unlubricated sliding of rock.

That two congruent blocks, when sliding side-by-side along equal tracks, will undergo twice the resistance of a single one, is nothing but trivial. In this particular context the message of Coulomb's rule can be expressed by saying that the total resistance remains equal if the two blocks are moved on top of one another instead of side-by-side. Somehow, *similar conditions of resistance must exist* to fulfil this postulate. The basic idea of an explanation for technical materials consists in the assumption that, as

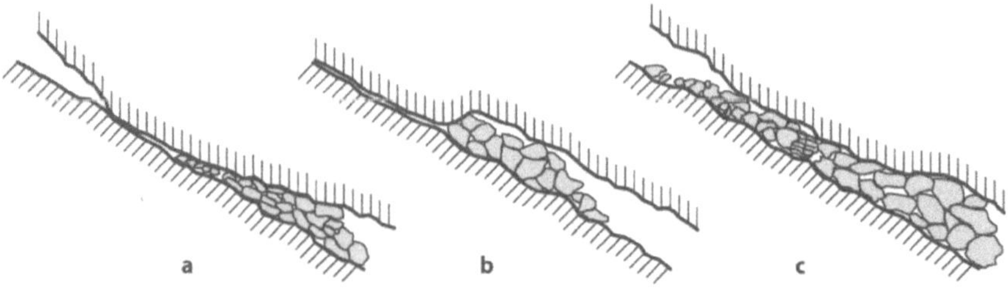

Fig. 5.13. Effects of crushing. Contact points of asperities, both on the ground and the moving mass, are crushed when reaching material's critical stress. Hereby they are adjusted to their counterface (**a, b**); stress is near to critical. As long as subsequent sliding lasts, track-making by smoothing of surfaces occurs. Any fragments caught between ground and mass may be crushed to sand by the wedging effect of a very acute angle (**a**); they also can be pushed forth in case if a near-to-right angle (**b**); crushing even can occur where space, though far from allowing direct ground-to-mass contact, narrows down sufficiently to force a load-bearing role upon fragments (**c**). Typical fracture by transverse extension under compression is shown in horizontally hatched piece (**c**). All mechanisms may occur on various scales, thus not being confined to coherent state of mass (sketch by Erismann)

no surface can be made perfectly smooth, contacts always consist of asperities, locally smoothed down to contact patches by deformation under compressive stress. So an increased load in first instance means more and larger patches and not a higher stress. This idea probably was first introduced by Bowden and Tabor (1964, 1973) who considered μ as a function of shear strength and hardness. It was applied by many authors (e.g. Czichos 1968, 1971; Heilmann and Rigney 1981). A compact formulation is found in a summary by Rigney et al. (1984, 195): "*... deformation changes the near-surface microstructure in ways which make the material unstable to local shear. This in turn produces transfer of pieces of deformed material which are... mixed with counterface material... to produce ultrafine-grained material.*" Now rapid deformations of rock cannot be expected to attain an order of magnitude as in metals or plastics. But in rock there is an equivalent mechanism distributing a given force over an automatically adjusted area consisting of contact patches under a clearly limited stress. This mechanism is crushing.

In fact, as soon as the compressive stress exceeds the strength of the material, crushing – as presented in Eq. 5.8 for the case of cylindrical rollers – locally destroys the coherence of the rock, the fragments partly are driven aside, and the geometries of the two involved surfaces acquire a better reciprocal adjustment (Fig. 5.13). The result is a contact patch with near-to-crushing stress distributed more or less evenly. Of course, this process, owing to the continuous relative displacements in a rockslide, is highly variable so that said patches perpetually change size, shape, and number. It is not excluded that this fact even improves the integral result by the averaging effect of incessant parameter variation. Anyhow, it becomes plausible that in rock-on-rock sliding the simple mechanism of suppressing stress maxima by *crushing creates relatively even frictional conditions*. This is a strong argument (though not an unquestionable proof) at least for an approximate validity of Coulomb's rule.

*

To acquire a somewhat more detailed idea of what happens between the ground and a moving rock mass, it is opportune to recall the collisions treated in Sect. 4.2 and the modes of dealing with obstacles discussed under Heading 5.1.

In fact, two of the modes shown in Fig. 5.5 entail an acquisition of additional material by abrasion, and all three may result in acceleration of certain involved masses. Mode a means abrasion of asperities and their acceleration, as a rule, from zero to the velocity of the mass. Mode c turns out to be very similar if inverted (as mentioned in Sect. 5.1) to a system with an asperity moving with the mass and scraping the ground: again there is first a (rather rough) abrasion, then acceleration. Mode b is somewhat different not only by the fact that no additional mass is acquired: the local jumping hill of the obstacle slightly decelerates and deviates a series of particles, thus shooting them, so to say, into the disintegrated mass where the obvious incongruence of velocity vectors (in length and direction) with the surrounding portions of the mass rapidly is annihilated by an immediate sequence of collisions (if the angular difference is small, the situation is very similar to that of the mental experiment of Fig. 4.5). Finally any loose material trapped somehow between ground and mass may (not must) be detached from one of the surfaces at the expense of kinetic energy, and it may (not must) be accelerated up to the velocity of the moving mass, for instance in a situation as shown in Fig. 5.13b. Anyhow, it becomes clear that at least two mechanisms have to be considered in the seemingly so simple field of unlubricated sliding: *dissipation of energy by disintegration and by acceleration.*

*

The basic parameter for determining the *energy of disintegration* is the area of separation. Seen from this angle it is obvious that a more or less clear cut as can be expected for mode a (Fig. 5.5) cannot dissipate a large energy, especially not in a brittle and possibly stratified material not exempt of residual stresses. So the interest is concentrated in first instance on crushing and similar mechanisms in which parts of a mass are disintegrated to small fragments thus yielding a large total area of separation. If, for example, a volume ΔV is disintegrated into cubes with a side length ΔL, said area is $3\Delta V / \Delta L$. The energy required to effectuate the separation is not easy to determine. In their comments to the consequences of the large collision having occurred in the event of Köfels (Sect. 2.4), Erismann et al. (1977, 82, 89–91), have used the specific energy of separation G that can be deduced from the fracture toughness (Sect. 3.2) and the elasticity of a material. For Köfels this energy was found to be in the range of 100 N m^{-1}. Of course, kinetic energy cannot crush rock without losses which, as known from the poor efficiency of cement mills and other technical installations, are far higher than the net energy of separation; so G has to be multiplied by a factor of "ineffectiveness" $k \gg 1$ to obtain the energy per unit of separated area. Eventually, it appears useful to refer the energy to the volume V of the moving mass. So the specific energy of disintegration

$$\frac{\Delta W}{V} = 3Gk\frac{\Delta V}{V\Delta L} \tag{5.16}$$

is obtained. Essentially this lost energy is independent of velocity.

For commenting on the equation it is useful to consider the availability of the *parameters*. G can be determined as a parameter of the material. $\Delta V / V \approx \Delta m / m$, the relative amount of disintegrated mass, and ΔL, so to say the degree of disintegration, are the main parameters which in some cases can approximately be estimated post eventum (a forecast is practically hopeless). The remaining parameter, dimensionless k, is definitely the weakest link in a chain not excelling in well-known figures. A first determination was attempted in the just-mentioned publication about Köfels (p. 91), and an ineffectiveness $167 > k > 125$ (efficiency below 1%) was estimated. In the frontal collision as observed for Köfels the situation was, of course, not identical with the one considered here, and certain facts (frictional energy losses not separated from crushing in the evaluation; residual kinetic energy) point to a lower k in the present context. The question, however, is whether "lower" means a division by 1.5 or by 15. All in all, predictions with the present knowledge of parameters are more than risky, and a realistic experimental determination is not an easy task.

Despite this serious deficiency, the value of Eq. 5.16, owing to its transparency, is higher than might be expected. At least a tentative estimation of a loss described as "very high" may perhaps be dared. With $G = 100$ N m^{-1}, $k = 100$, and $\Delta V / V = 0.01$ (a definitely high figure for the material between the sliding surfaces) $\Delta W / V = 300 / \Delta L$ results for the specific energy. For normal rock in cubes of 10 mm this corresponds to a lift of about 1.2 m against gravity, for cubes of 1 mm ten times more, etc. So the important statement can be expressed that a significant loss in kinetic energy requires a *rare coincidence of parameters*: a very high value of $\Delta V / V$ and a very low one of ΔL.

To complement the line of thought, it is useful to throw a side-glance at the *granulometric aspects*. Above all, it is obvious that a mass consisting of equal-sized cubes is not a realistic model as sizes necessarily vary within a considerable range. Asperities not are watchmaker's tools generating fine grain right from the start. On the contrary, most of the fragments are broken out in not too small portions of varying sizes. Their subsequent fate depends on the individual situation: if they are caught in a wedgelike space preceding one of the patches of contact, the chance of being milled to fine debris is high (Fig. 5.13a); if, contrarily, they are pushed along by non-wedgelike elements (e.g. owing to the almost right angle at the head of a patch, Fig. 5.13b), they may subsist in their original form throughout the entire travel; a third possibility is getting into a space too narrow to preserve them from being strongly compressed so that they are forced to play either the part of a transient patch locally subjected to crushing (Fig. 5.13c) or that of a short-lived bearing roller in the sense of the remarks made under Heading 5.3 (though not necessarily as stilted as shown in Fig. 5.9e for a particular case). Anyhow, the rock volume crushed to fine debris necessarily is substantially smaller than that broken off the respective surface. On the other hand the fact has to be taken into account that severely disintegrated portions are the main suppliers of area of separation.

*

The main difference between the mechanisms according to modes a and c in Fig. 5.5 lies in the solidity of asperities. In mode a they are cut away more or less in a single stroke; in mode c they subsist and generate a groove as long as allowed by the circumstances. The higher amounts of involved material and energy in comparison with mode a are immediately obvious. So, if aiming at explanations for the coefficients of friction known from field evidence, there is no getting round the *mechanism of scraping* which necessarily is frequent and thus challenging to attempt an estimate of its possible decelerating potential.

Though evasive to exact calculation owing to the lack of detailed knowledge of mechanisms and parameters, scraping still offers the *possibility of quantitative approach*. Figure 5.14 shows a simplified model. As, of course, a scraping asperity is far

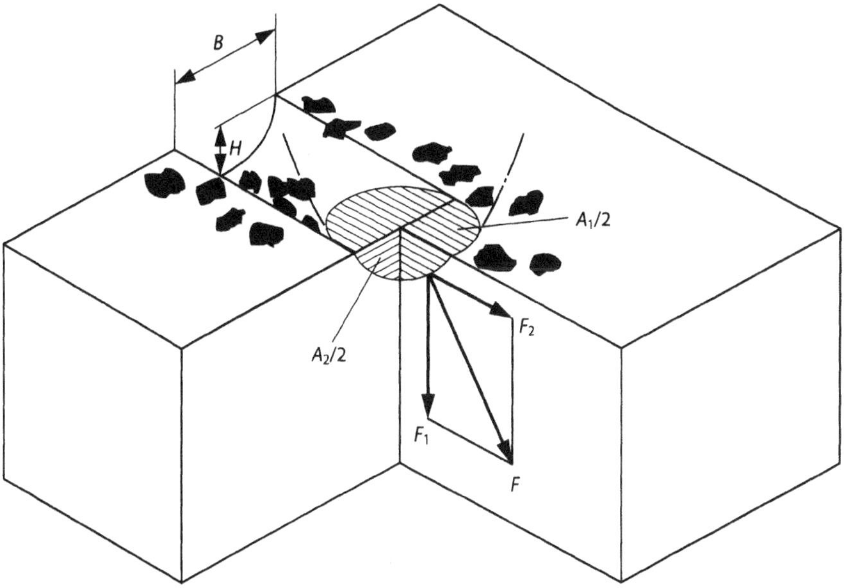

Fig. 5.14. Schematic view of resistant asperity scraping surface of weaker ground. Asperity (*hatched*) is a paraboloid, in figure cut at ground level for better visibility (extension marked by *dots* and *dashes*). For dimensions of asperity see groove behind it (*B:* breadth: *H:* depth). $A_1/2$: ground-parallel half-area (back half of asperity, being located in groove, transmits no forces); $A_2/2$: cross-sectional half-area. *F:* total force. For two reasons component F_1, perpendicular to ground, is larger than forward-acting F_2: material is kept together by surrounding rock, and A_1 is larger than A_2. *Black dots:* fragments of ground (for better visibility only a few are shown) (sketch by Erismann)

from having the shape of a plough or a file-tooth, it is represented by a segment of a rotational paraboloid, forced to generate in its counterface a groove of width B and depth H. The resulting fragments (some of which are shown in the figure) are pushed into the (by assumption, not by necessity, sufficiently extended) space between ground and mass; several may find their way to the groove. To obtain a maximum of decelerating force it is further assumed that the entire overburden exclusively is supported by various contact surfaces of mode c.

The components F_1 (at right angles to the ground) and F_2 (parallel to the ground) of the total force F exerted by the overburden must create stresses which, at least locally, exceed the strength of the material. Such strength, however, is not equal in the two directions - not only owing to the (most probably surface-parallel) stratification of the rock but also owing the freedom of removal of fragments. This freedom is more effectively barred for F_1 (coherence with unloaded half-segment behind; larger distance to surface) so that a somewhat higher critical stress can be expected. In addition, half-circular area A_1 in the plane of the surface is larger than sectional area A_2 at right angles thereto (normally $H < B/2$ may be expected). So it becomes obvious that, after integration of stress over the respective area, definitely F_1 *turns out to be larger than* F_2, yet without exceeding its order of magnitude.

The resulting total amount of braking forces may cover a wide range. For the individual contact a very rough estimate of the "local coefficient of friction" F_2/F_1 just has been formulated. Thereby also μ, as being the average of all involved contacts $\Sigma F_{2i} / \Sigma F_{1i}$ (i = ordinal number of contact), must be in the order of magnitude of, but definitely below, unity. In other words: scraping as dominant mechanism suggests a rather high coefficient of friction. Contrarily to crushing, it amply can suffice to account for *friction as known from field evidence.* It is not pretended, however, that therewith the unique candidate for the purpose is presented.

In parentheses it should, perhaps, be recalled that simplified models never can cover the *diversity encountered in nature.* Here an example in the context of scraping: if the material consists of a relatively soft matrix with hard inclusions, breaking out such an inclusion means nothing but paving the way for the next inclusion to start scraping after abrasion of some material of the matrix. And, of course, no intermediate modes between a and c (Fig. 5.5), as dictated by the participating materials, can be excluded.

*

Energy lost by accelerating additional masses (Fig. 5.5a and inverted c) is easily analysed when applying Eq. 4.5 to 4.8 with the required adjustments to be compatible with Eq. 5.16. For acceleration from zero to the accordingly reduced velocity of the mass and no elastic restitution (J = 0, as compulsory for the additional mass Δm being absorbed by the mass m) the velocity after a collision is

$$v = u \frac{m}{m + \Delta m} \approx u \frac{V}{V + \Delta V} \tag{5.17}$$

and the dissipated energy

$$\Delta W = \frac{u^2}{2} \frac{m \Delta m}{m + \Delta m} \approx \frac{u^2}{2} \frac{V \Delta m}{V + \Delta V} \approx \frac{u^2}{2} \Delta m \tag{5.18}$$

where u is the velocity of the mass before the collision, m and V the moving mass and its volume, Δm and ΔV the additional mass and its volume. In principle the last equation needs no comment: it simply makes clear that the *dissipated energy is proportional to the square of the velocity* of the mass. Comments are only required to the approximations following the right side of the equation. The first is based on the assumption that the densities of V and ΔV are approximately equal; the second is valid if ΔV in a rough approximation can be neglected in the denominator. Finally it should be remarked that – especially in mode c – ΔV is not the result of one single collision; but the mathematical character of the equations allows merging of several collisions as long as no other essential influences intervene.

*

The case history of a fragment in *detour-generating mode b* (Fig. 5.5) can be reduced to the following two phases: (1) change of velocity and direction of the fragment by oblique collision with an unyielding obstacle and (2) multitude of direct and indirect collisions within the disintegrated mass, finally establishing an approximately equalised velocity. According to the arguments of Sect. 4.2, phase 1 applies an "injection" of external kinetic energy, partly opposed to that of the mass and thus reducing it. Phase 2 does nothing but reduce the internal kinetic energy. As both phases – in particular the essential phase 1 – follow the momentum theorem, it needs no further explanation to show that from this phase a *loss of energy proportional to the square of velocity* must result.

The *amount of this loss* depends on the particular circumstances of the series of collisions occurring while the mass passes the obstacle. And the number and variety of physically relevant parameters are large enough to qualify the calculation as a hopeless gamble. Still once more a reasonable qualitative estimate of a maximum is not excluded. It must, in fact, be considered as an extreme case if the mentioned "injections" over the whole length of a mass occur perpendicularly to the velocity vector of the mass (normally the formation of a softer jumping hill by tailbacked fragments must be expected as insinuated by the stained area in Fig. 5.5b). In such instances the particles colliding with the obstacle are re-accelerated from zero to the velocity of the mass as if they were external particles hit by the mass. This means that Eq. 5.17 and 5.18 may be applied (obviously, to be correct, $V + \Delta V$ should be replaced by V, and V by $V - \Delta V$ if ΔV is the volume involved). And in extreme cases – e.g. a mass reduced in thickness by spreading and several consecutive obstacles covering its entire width – ΔV, being more than once the entire layer of colliding fragments, may represent a percentage of the mass that no longer can be considered as ranging near to negligibility. The combination of spreading and uneven ground turns out to be an efficient brake: let ΔV be 0.1 V, and the loss in velocity will be 10%, that in (external) kinetic energy 19%.

*

The consequences of processes according to modi a, b, c (Fig. 5.5) are not confined to disintegration and braking. There is, for instance, the complete eradication of the relevant asperities. There may also, as an intermediate state, be the saturation with fragments of the space between coherent or quasi-coherent mass and ground whereby fragments support more of the overburden than the remaining rudiments of asperities (if any). Figure 5.13c can be interpreted as the beginning of this process that finally leads to layers of finer and finer debris, within which a sliding surface is established according to the principle of least resistance. Anyhow, there is a general tendency of smoothing or, in other words, of *track-making*. Impressive field evidence is readily available. Where in Fig. 2.25, 2.26b, or 4.6 can be found asperities apt for forced displacement or able to force parts of a mass into a detour?

It is by no means insignificant that the events from which the just-mentioned photos were taken, Flims, Langtang, and Köfels, belong to the cubic-kilometre class. In fact, track-making underlies a strong *size effect*. A large mass, as usually being thicker than a smaller one, exerts a higher average pressure upon the ground, thus generating more and larger patches of contact which play the part of a carpenter's bench or a grinding machine. And a large mass, as usually being longer than a smaller one, extends the work over a longer distance.

Once the key word "size effect" having emerged, the further size effects occurring in unlubricated sliding should be enumerated. So it will be observed that both in Eq. 5.16 and 5.18 the mass-specific dissipated energy $\Delta W / m$ is proportional to the specific involved mass $\Delta m / m$. Now in a surface of given unevenness the volume available to act as Δm is given by number and size of "workable" asperities. And this maximal volume is proportional to the area of the surface. This means that, when assuming equal unevenness and complete abrasion of asperities, m grows with the cube of linear extension, Δm only with its square. This "*area-to-volume*" rule already has been commented in connection with release. While, among the examples presented in Sect. 3.2, rather the comparison of elephant and daddy longlegs was appropriate, this time it is that of skiff and crew of eight. Of course, as in the example the rule is not the unique criterion (owing to wave development, Reynolds number, presence of a coxswain), also in a rockslide other effects cannot be neglected (to quote only that of the immediately preceding paragraphs: size effect by track-making). And, to close the comment on this ubiquitous rule, its applicability to obstacles generating detours is as good as in case of dealing with additional masses.

*

All in all, the considerations so far made, in spite of inherent deficiencies in the knowledge of mechanisms and, above all, parameters, clearly show some fundamental possibilities and limits of interaction between the ground and a descending mass. The *consequences* are obvious.

11. Provided that rock-on-rock sliding mainly occurs on a contact surface consisting of a multitude of protruding patches continuously readjusted by crushing (and relative displacement) and thus compressed by a near-to-critical stress, Coulomb's rule makes sense without being proved conclusively.
12. A reasonably accurate calculation of the energy lost by the mechanisms commented hereafter under Items 13, 14, and 15 is not possible owing mainly to deficient knowledge of parameters.
13. Energy lost in crushing or scraping off mass portions is essentially independent of velocity. If originated by crushing, it hardly can exert a noticeable braking effect; if originated by scraping, it can account for a substantial part of the total resistance against motion.
14. Energy lost in accelerating crushed and scraped off mass portions is essentially proportional to the square of velocity. It hardly can exert a noticeable braking effect.
15. Energy lost by obstacles forcing parts of a mass into a detour is essentially proportional to the square of velocity. It can account for a substantial part of the total resistance against motion.
16. There exists a general tendency to track-making by disintegration and an universal presence of size effects reducing resistance against motion (and thus increasing velocity and reach) in large events.

*

Probably the most important result of the present section consists in the statement that, as long as no further mechanisms have to be taken into account, *Coulomb's rule gives a reasonable basis* for a quantitative treatment of unlubricated sliding. If expressions like "makes sense" have been used instead of sheer affirmations, it was done in the intention to stress the fact that the statement cannot be considered as being proved for rock in a scientifically unquestionable manner. Especially the setting up of laboratory-scale experiments meriting the qualification of being realistic in the optics of a rockslide is an extremely difficult task. And experiments conceived (and useful) for other purposes seldom are conclusive in the context of this book. A typical example is a study by Lajtai and Gadi 1989, who investigated the dramatic increase of μ on initially polished granite surfaces during a short-stroke to-and-fro motion.

Nevertheless the *practical suitability* of Coulomb's rule is backed up by more than the frequently negligible contribution of velocity-bound resistances. Strangely enough, it is an intrinsic deficiency (if this term is acceptable for a quality opposed to easy quantification) of unlubricated sliding that improves the chances of Coulomb's rule. This deficiency lies in the fact that, in spite of being quasi-independent of velocity, resistance is not constant, be it owing to frequent changes of physically relevant conditions, be it owing to the track-making effect that grows under constant conditions. So, even in cases where unlubricated sliding definitely is the dominant mechanism, μ has to be considered as a function of distance, if not of velocity. As a consequence, calculations anyway should be carried out after division of the track into appropriate sections, and any complementary information should be accepted as helpful to establish reliable reference data. An important example is the evaluation of bounces to determine velocity at the location of take-off (Sect. 2.7, 5.3). And such complementary information, as will be shown hereafter, can be a means to simplify the differential equations of motion.

When using nothing but Coulomb's rule, the *basic differential equation of displacement* – which yields velocity in a first and distance in a second integration over time – is applied in its simplest form

$$a = g(\sin\beta - \mu\cos\beta) \tag{5.19}$$

where a is acceleration, g gravitational acceleration, β the slope angle, and μ the coefficient of friction. A resistance influenced by non-negligible accelerations (be it by incorporating additional masses, be it by making detours) depends, in addition, on the square of velocity v and a coefficient c_v expressing the circumstances of acceleration (as given, for instance, by the accelerated mass per unit of time and the amount of its acceleration). Normally in the resulting equation

$$a = g(\sin\beta - \mu\cos\beta) - c_v v^2 \tag{5.20}$$

the second term on the right side, though not negligible, is definitely smaller than the first, and the equation has to be solved only over a restricted time. If in such a case the second term is substituted by an increased value of μ approximately accounting for its presumed average, the resulting errors, expressed in terms of velocity and distance, very often will be tolerable. It is well understood, by the way, that the differences between the curves labelled N and Q in Fig. 5.6 and 5.7 represent an incomparably higher degree of diversity: N follows Coulomb's rule while Q stands for an exclusively quadratic dependence on velocity, not a partial one as in Eq. 5.20.

In the *literature* various steps in the insinuated direction do exist. Körner (1976, 235–237, 1977, 1983, 96–101), assuming analogies with snow avalanches and considering the fundamental work of Voellmy (1955) as well as studies by other authors (e.g. Salm 1966; Scheller 1970; Scheidegger 1973), excluded as negligible a further (v-proportional) term in the differential equation, and thus obtained a result analogous to Eq. 5.20. Among those who adopted Körner's method, Nicoletti and Sorriso-Valvo (1991, 1372–1373), in discussing the effect of geomorphic shape upon the reach of rockslides, introduced the useful notions of low, medium, and high energy conditions which definitely merit an extensive application. A detailed assessment of the problems thus raised would exceed the frame of the present section, but the complex of questions will be resumed in the light of practical aspects in Chap. 6. So, for the moment, the above arguments may suffice.

Nevertheless, one particular example appears to be of some use in the present context. It demonstrates, at the end of this section, a fundamental principle that directly can be deduced from Coulomb's rule. In considering possible analogies between the motion of rock masses and liquids, one of the authors (Erismann 1979, 20–21) investigated the implications of a *mode of displacement resembling laminar flow*. It is a tempting task to repeat the respective argument without the use of equations, just on the basis of a simple illustration (Fig. 5.15).

Assume the descent, on a slope, of a long (virtually "infinite") mass consisting of many layers. To fulfil the essential conditions of laminar flow, each layer should move somewhat faster than that immediately below it, and the distribution of velocity should be a quadratic one, having its apex at the ground (in the following this distribution is insignificant and therefore not respected in Fig. 5.15). Now let the coefficient of friction at each sliding surface be μ, by one single exception where the friction is slightly reduced to $\mu - \Delta\mu$. Finally assume, for the sake of simplicity, that the driving forces exceed the braking forces so that motion is accelerated throughout the mass. As long as $\Delta\mu$ is nought, an account of the frictional forces (driving from above, braking from below) yields for each layer a braking total exactly equalling its frictional braking force when descending alone directly on the ground. Together with the gravitational driving force of the layer the resulting acceleration a is equal for all layers. Keen readers certainly have observed that this statement is valid irrespective of the thickness of a layer. If $\Delta\mu$ is introduced, the layer i_+ on top of the "privileged" sliding surface acquires a slightly increased acceleration $a + \Delta a$, the layer i_- below it an accordingly reduced one, $a - \Delta a$ (a_{01}, Fig. 5.15). Consequently, sooner or later layer i_+ will equal the velocity of the next layer i_{++} on its top, i_- that of i_{--} below its bottom (v_1, Fig. 5.15). Once moving at equal velocity, each pair of layers will stick together as any difference of

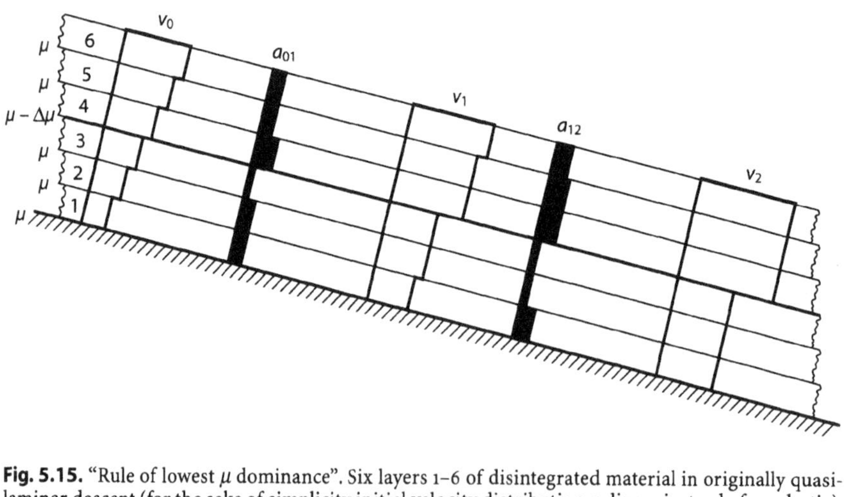

Fig. 5.15. "Rule of lowest μ dominance". Six layers 1–6 of disintegrated material in originally quasi-laminar descent (for the sake of simplicity initial velocity distribution v_0 linear instead of quadratic). Sliding surfaces have equal coefficients of friction μ, except $\mu - \Delta\mu$ at half-thickness (*bold line*). Resulting acceleration a_{01} is reduced for layer 3, increased for layer 4 until velocities equal those of 2 and 5 respectively (velocity distribution v_1). Now pairs of layers 2+3 and 4+5 are exposed to $\mu - \Delta\mu$ (accelerations a_{12}) so that finally only two blocks of layers (velocity distribution v_2) are formed (for comments s. text) (sketch by Erismann)

velocity immediately would entail correcting frictional forces. And as increase and reduction of accelerations have been acknowledged as independent of layer thickness, both pairs of layers will continue the game (a_{12}, Fig. 5.15). So the final issue is only a question of time: *two blocks of coherently moving layers*, separated by the sliding surface with the reduced coefficient $\mu - \Delta\mu$ (v_2, Fig. 5.15).

From this almost trivial deduction it becomes obvious that, as a stable mechanism within a disintegrated moving mass, the possibility of *laminar sliding is reduced to absurdity*. This result is in agreement with the various observations of field evidence confirming preservation of the initial sequence of materials in rockslides as expressed in the motto of this chapter and commented under Heading 5.1: laminar sliding necessarily would entail a reversal of sequence.

Of course, this "*rule of lowest μ dominance*" may appear somewhat theoretical when confronted with the seemingly chaotic processes going on in a rockslide. Yet a tendency in the postulated sense is inevitable as long as Coulomb's rule is dominant in a moving mass. And how could a surface acquire such impressing privileges as shown in Fig. 2.25 and 2.26b if there were no powerful rules of dominance?

In parentheses it might be added that the local development of a quasi-laminar disintegrated motion is nothing but natural in certain circumstances. A striking example is shown for Köfels (Fig. 2.20), and similar mechanisms are at least probable for Flims and other large events having passed across a valley (Schneider et al. 1999).

5.5 Lubrication

Lubrication is essentially a technical term. It describes the reduction of frictional resistance by introducing a third medium (the lubricant) between the surfaces of two bodies displaced with respect to each other.

This definition offers the advantage of leaving open the technological details of the lubricant and the very mechanism of economising frictional energy. In fact, *technical bearings* – in many cases rotational – display a wide variety of different solutions: the lubricant may be a liquid of high or low viscosity (e.g. oil or water), a gas (air or other), a near-to-liquid or otherwise particular solid (grease, graphite), and it may be kept in place either by its own coherence (oil, grease) or by means provided ad hoc (hydraulic pressure, sealings). In addition, a lubricant may modify its character in service (reduction of viscosity due to frictional heat) or even be generated from at least one of the involved bodies by such a modification (self-lubricating materials).

The hypothetical possibilities of *lubrication in rock motion* are almost as many-sided as in technical applications. In considering the proposals hitherto made for lubricants (water, snow and ice, mud, clay, air, dust, etc.) and for their genesis (from external sources or, by abrasion or frictional heat, from the rock involved) one is tempted to use "Water" for liquid, "Air" for gas, "Fire" for generation by heat, and "Earth" for solid and thus to subdivide the four most important groups of mechanisms according to the *four classic elements* postulated by Empedocles.

Before going into the details of these mechanisms, it is useful to recall the fundamental *preconditions of effective lubrication*, starting again from a technological viewpoint. One of the main problems in optimising a technical lubricant consists in the inherently conflicting requirements as concerning viscosity: on the one hand, viscosity should be low to minimise the resistance in the relative displacement between the involved bodies; on the other hand, it should be high to reduce its rate of escape from the gap between the bodies to locations of lower hydraulic pressure. Besides the choice of the lubricant, various technical means are used to overcome the problem. They range from sophisticated handling of the indispensable pressure gradients within the system (example of non-rotating bearing: Erismann 1989, 278–280) to forced lubrication by a pressurised lubricant (example with air as lubricant: Hovercraft vehicle), often in combination with appropriate sealings.

Hereafter an additional observation will turn out to be useful. In certain cases of forced lubrication the lubricant does not entirely separate the two bodies and thus bear the entire load. In such *load-sharing forced lubrication* most of the load is supported by the pressurised lubricant, the rest by surface-to-surface contacts. In parentheses it should be remarked that the advantage drawn from this hybrid technology consists in simple positioning and sealing obtained at the expense of slightly increased friction.

As a consequence, assessing natural lubricating systems means the study of *lubricant supply and escape* and, if possible, the establishment of a quantitative balance between both. The term "supply" stands for the entire mechanism (or set of mechanisms) required to locate the lubricant in the free space between the bodies.

Lubrication is particularly important in cases of long run-out, i.e. in connection with low positive (or even negative) slope angles. Therefore in the present section equations, as far as connected with slope, are simplified by formulation for *horizontal track*. Errors committed in practical use are of marginal importance on soft slopes.

*

Not many key-words are more frequently used in connection with downhill mass movements than "*Water*". Sometimes it seems that – in spite of Terzaghi's unmistakable warnings (1950, 91: "... *water in contact with many common minerals... acts as an anti-lubricant...*"; 94: "... *in no event can the slide be explained by the 'lubricating effect' of water...*") – the role of water as an efficient lubricant is taken for granted without considering in detail its particular qualities in this field. In the present book, besides Sect. 2.2 which will be commented hereafter, the importance of water so far mainly has been stressed in the contexts of destructive power (Sect. 2.5, 2.6, 2.7) and release (Sect. 3.4), and it will be treated in further detail in connection with fluidisation, creeping, and secondary effects (Sect. 5.6, 5.7, 7.1, 7.2).

One of the mentioned particular qualities consists in a very *low viscosity* of 0.00134 kg m^{-1} s^{-1}. With a good reason "watery" is a common attribute for low-viscosity liquids. In accordance with the remarks made here-above about the preconditions of lubrication, low viscosity means that water, if not a practically irrelevant layer of minute thickness is considered, will, on the one hand, leave almost unhindered the relative displacement of the involved bodies, but that, on the other hand, it will escape very quickly under the overthrust load if not maintained in place or supplied in sufficient quantity. So dealing with water as a lubricant in rockslides mainly consists in detecting circumstances under which water has a chance to contribute in supporting the overburden load during a sufficient time to exert a perceptible influence upon the effective coefficient of friction.

The *most promising mechanism* in this context has been proposed by Abele (1984, 1991a, 1991b, 1997b). It is not intended to repeat the (mainly phenomenological) description and comments given in Sect. 2.2 and visualised in Fig. 2.7. Only the wording of the crucial sentences (1997, 2) should be presented in a compressed form: "... *when moving over a water-saturated valley fill..., a rockslide mass exerts... a compressive effort upon the fill. The rapidity of motion prevents the pore water from escaping unpressurized and forces it to support part of the rockslide's weight...*" In the immediately following sentences the balance between pressurising and escape is stressed as the essential problem, and the analogy to forced lubrication of technical bearings is mentioned.

As usual, straightforward *quantification* quickly is revealed as barred by insufficient knowledge of parameters. Nevertheless some quantitatively accessible models can help to assess the basic idea.

The *volume of potentially available liquid* (which may be more or less pure water as well as sufficiently fluid mud) is limited by the pore volume between the larger particles of the fill. If, as a first approximation, such particles are assumed to be equal-sized ("first-order") spheres in highest density packing, the pore volume is about 26% of the total volume occupied by the fill. And if a second-order set of spheres is introduced (each tetrahedron-like pore between four first-order spheres containing the largest possible sphere), the remaining pore volume is still substantial. Now the model is obviously unrealistic: on the one hand the more irregular shape of real particles and the impossibility of a highest-density array point to a larger pore volume which, on the other hand, is reduced by the existence of smaller particles – however not below the size where they are readily transported by rapid water motion and thus have to be considered as part of the liquid. Taking into account the somewhat complicated reality, one may assume "several percent of the total volume occupied by the fill" as a reasonable (though certainly not particularly precise) estimate.

Such an amount of lubricant, ranging in metres of thickness, would warrant a perfect lubrication if its entire volume could be made available as lubricant. This would mean reduction of all pores to nought, obviously an impossible assumption. So the next step of argument concerns the *actual availability of lubricant* under the compressing action of the sliding mass. "Pumping" water to the interface between fill and slide means forcing the proper fill material into a lower position. In principle, this can occur by elastic deformation, by disintegration, by re-arrangement, and, of course, by any combination of these mechanisms.

Elastic deformation is not very promising as, in order to yield a maximum of deformation, it requires an even distribution of stress in a body. And the multitude of bodies forming a valley fill, with minute points of contact, is very far from this ideal. And even an accumulation of three absurdly unrealistic conditions as perfectly even stress distribution in the fill (e.g. consisting of vertical, coherent prismatic columns), high ratio of strength to modulus of elasticity in the material of the fill, and near-to-failure compression (requiring a slide mass several kilometres thick) would fail to result in a substantial displacement of water.

For *disintegration by crushing* the above-mentioned first-order spheres model, though not strictly realistic, does not dramatically differ from what can be expected in practice. For instance, in both cases the relative deformation of the fill, when disregarding the support by the displaced water, can be expressed by

$$\frac{\Delta H_F}{H_F} = \frac{cgH_S\partial_S}{s_F} \tag{5.21}$$

where H_F and ΔH_F are, respectively, the thickness of the fill and its reduction under load, c is a dimensionless factor given by the geometry, g gravitational acceleration, H_S the thickness of the slide, ∂_S its density, and s_F the compressive strength of the fill's material. Only the factor c is different for the model, where it has the value $\sqrt{12/\pi} = 1.1027$, and reality. Anyhow, a tentative calculation on the basis of Eq. 5.21 gives a result definitely more encouraging than that based on elastic deformation: with a slide 100 m thick a compression of 1% of the fill thickness is possible even when assuming a solid fill material ($s_F = 275$ MPa). This is probably sufficient to start forced lubrication, but it appears precarious with respect to a water supply sufficient to keep lubrication operational for a reasonable time.

Unfortunately *re-arrangement of particles* in the valley fill, probably the most promising "lubricant pump", is particularly evasive to mathematical treatment. It seems that only a computer simulation of sufficiently complex realistic configurations or large-scale experiments in a testing laboratory might provide satisfactory basic information (how far the experiences with technical means of compacting like road rollers or pulsators might be useful, is an open question). In spite of this flaw of knowledge, certain clues to the effectiveness of the mechanism do exist (Fig. 5.16): if a particle I is held at a certain elevation in a wedge-shaped gap between two neighbouring particles II and III which resist its tendency to push them aside, a massively increased vertical force F

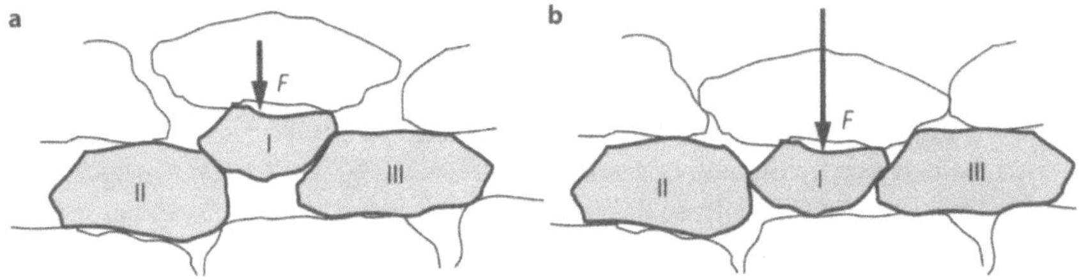

Fig. 5.16. Forced lubrication by pressurised water. "Pumping" of water by forced re-adjustment of particles. **a** Under moderate force *F*, particle I is unable to push aside particles II and III. **b** Increase of force *F* enables I to displace II and III and to assume a lowered position, followed by other particles. Reduction of interstitial space forces water to rise, pressurising it when way out is sufficiently barred (sketch by Erismann)

exerted upon particle I may overcome this resistance. And the vertical displacement of this particle may be many centimetres (against millimetres, at the best, in case of being locally crushed). So it is probable, though no proof can be presented, that the term "most promising" at the head of this paragraph is not an overstatement. It would, however, be somewhat rash to consider such arguments as a definite guarantee for a sufficient endurance of the lubricating process as a whole.

In this context it may be helpful to describe in some more detail, though only qualitatively, the *process following the compression of the fill.* Initially only part of the deformation according to Eq. 5.21 occurs owing to the supporting (i.e. lubricating) effect immediately exerted by the water. Simultaneously water starts escaping by any possible passage, allowing the slide – if no additional water is supplied – to dip, continuously compressing the fill while the water is decompressed. This process is arrested by the water's surface reaching atmospheric pressure and the fill being compressed in accordance with Eq. 5.21. In parallel thereto the lubricating effect fades out. Of course, the half-life period of this process is an important parameter correlated with the quality of lubrication. It will be observed, by the way, that, as a certain compression of the fill is a precondition for water being pressurised, the overburden never can be supported exclusively by the lubricant. Thus the lubrication not only is forced but also *load-sharing* in accordance with the above definition.

The arguments so far presented make probable Abele's hypothesis, but they do not conclusively prove it. If it is presented here as the most promising explanation for the phenomenon of long run-out, two eminently important circumstances play an essential role: in a general manner, none of the other hypotheses proposed in the literature has been found equally congruent with the geomorphological evidence and its evaluation (quantitative, as far as possible); in particular, this congruence is striking in the comparison of *overall slopes* (Fahrböschung) as demonstrated in Sect. 2.2 (Fig. 2.5).

Soon after first having observed the correlation between overall slope and water-saturated valley fill, Abele involuntarily was given an opportunity to *check the idea* in an essentially "blind" test. In the autumn of 1992, in the very first phase of co-operation between the authors of the present book, Erismann (hereafter: E) had begun the study of interpolation algorithms for the dependence of overall slope on volume. It was natural to use the book on Alpine slides written by Abele (A) which contains a graph with the overall slopes of numerous events (1974, 45). This information was used by E simply as statistical material, without looking into geomorphological or other details. Somewhat later A explained his hypothesis, and E, spontaneously taking his data sheet, began to ask questions like: "*Does event X fulfil your conditions?... and Y?... and Z?... Say nothing but Yes or No...*" The resulting list of affirmatively commented events yielded exactly the points marked by circles in Fig. 2.5. Finally A proposed to cross-check the astonishingly unambiguous result using data from American events. Unfortunately these data were not at hand. So the work was done later by E when he had taken over the bibliographical collection of his friend who no longer could enjoy the confirmation of his idea: the result was very similar to that obtained from the Alpine events.

One last point should be mentioned to complement the discussion of forced lubrication by pressurised pore water. As Fig. 2.7 shows, one of the important ways of escape is through the bow-wave 5. The "slot" connecting it to the space from which escape takes place is pushed ahead with the slide mass like the air inlet of a jet aircraft. But the *pressure encountered by the escaping water* may vary within an extremely wide range: if the slot is penetrating into unmoved water, the expression for dynamic pressure $p = \partial v^2/2$, besides velocity v, has to be used with the density of water for ∂; if instead, for any possible reason, immersion completely ceases, the density of air may become valid, i.e. the counter-pressure drops by a factor of about 800 at sea level, and 1000 at an altitude of 2000 m. The turbulent and, of course, dramatically variable nature of the bow-wave makes hopeless any guess a priori, and the average pressure only can be described as "something between zero and 0.2 MPa at 20 m s^{-1}, 1.25 MPa at 50 m s^{-1}, and 5 MPa at 100 m s^{-1}", adding the remark that the respective static water pressures correspond to depths of 2, 12.5, and 50 m. As a whole, once more a case for a well-programmed computer simulation is evident.

For the sake of historical correctness the reader's attention should be drawn to a *precursor*. Under the title of "Flowslides", Rouse (1984, 491), with reference (among others) to Bishop (1973), not only criticises the air-lubrication hypothesis, but also speaks of "*... a temporary transfer of the normal stress* (author's remark: and therewith of weight) *onto the fluid in the void space...*" and postulates (p. 497) that "*... particles forced into a dense arrangement...*" finally lead to "*... a temporary increase in pore pressure...*". These statements coincide in more than one respect with those expressed by Abele, though the geomorphological context is not linked to the hypothesis by equal correlations. For instance, it is assumed that a post-failure study of a flowslide would "*... have difficulty in differentiating it from any other fast flow...*", while Abele in the mentioned experiment immediately could identify the events fulfilling his geomorphological conditions (and turning out to have low overall slopes).

Once Abele's hypothesis having been launched, it would be an interesting task to check it with respect to its compatibility with the geomorphological circumstances of further, in particular *large events* not depending on channelling to remain thick. To quote the very showpiece: in their excellent description of the giant Saidmarreh rockslide in Iran (doubtlessly a case of long run-out), Watson and Wright (1967, 129) insinuate the importance of water. But without a closer investigation ad hoc it might be premature to draw definite conclusions.

It will have been observed by a majority of readers that, without using the terms of this section, forced lubrication by water repeatedly has been referred to *in connection with release*. The most impressing examples are found in Fig. 2.45 and 3.19. Again the analogy with technical applications is striking: forced lubrication is a common means to make bearings practically frictionless at extremely low velocities.

Further details of escaping, valid to a certain extent for group "Water", but at least to an equal degree for "Fire", will close the discussion of the last-mentioned group.

*

In view of the extensive discussion in Sect. 2.3 and an earlier study (Erismann 1979, 31–33), the lubrication belonging to the group "*Air*" could be left aside entirely if there were not, in the literature, an interesting study (Krumdiek 1984) written by an author familiar with aerodynamics and pointing to the so-called "ground effect" known especially to old glider pilots who sometimes had trouble landing without the help of aerodynamic brakes. Unfortunately this study is somewhat imprecise in the mathematical formulation of physical facts. An example: on page 540 two equations for aerodynamic lift are found, one conventional and essentially correct, the other multiplied by a non-dimensionless coefficient of ground effect, falsifying the dimension of lift from that of a force into that of an energy.

In a more general view, the *analogy between a rockslide and a wing* of an aircraft is definitely a poor one. The most obvious point in this context refers to the fact that under normal flight conditions a wing produces approximately two thirds of its lift by suction from above and one third by compression from below. Now even a very thin rockslide mass only 10 m thick requires as much as a pressure difference of 0.25 MPa between bottom and top of the airfoil, and two thirds thereof are almost 1.7 times atmospheric pressure. Rightly it might be objected that the ground effect changes the balance between sucking and pressure. But anyhow the necessary lift has to be granted by a sufficient velocity. And here the difficulty becomes serious: the velocity required for taking off is $v = \sqrt{[(2F_L)/(A\,\partial c)]}$, a formula approximately analogous to that used by Krumdiek, though with the notation adopted in the present book: F_L is lift, A wing area, ∂ density of air, c maximal lift coefficient. For the mentioned "flyweight" slide F_L/A is 0.25 MPa, ∂ about 1.25 kg m^{-3}, and c may be 3 for a wing with highly efficient flaps. The resulting velocity of 365 m s^{-1} requires a free fall in vacuum of 6 800 m before a take-off can be envisaged. And some questions – where the rockslide has its highly effective supersonic landing flaps, and what should do a slide 20 times thicker (not to speak of the size effect) – remain unanswered.

All in all the hypothesis is reduced to a pure overpressure-on-bottom mechanism as proposed by Shreve (1966, 1968a, 1968b). And the fluidisation assumed by Krumdiek as a consequence of having taken off needs no comment as taking off does not occur.

*

In passing to the key-word "*Fire*", the phenomena in question should be specified more precisely to avoid misunderstandings. In fact, a lubricant generated by fusion might with a good reason be attributed to "water" and an evaporated one to "air". The justification for the formation of a special group lies in a particularity of all heat-generated lubricants: without fail, they come into being at the locations of maximal energy dissipation. In an anthropomorphous view the statement might be made that heat provides lubrication exactly at the locations where it is most needed. And as the lubricant in all instances comes either from the bottom of the moving mass or from the top of the ground (or from both), the term "self-lubrication" is justified.

All this sounds, with somewhat modified accents, like a repetition from Sect. 2.4 where, besides the impressive example of Köfels used to demonstrate the fusion of

Table 5.4. Energetic aspects of lubricants generated by heat. "Rock" is used for Köfels gneiss. Disparity sign > means "somewhat more owing to the neglect of minor effects". Second and third lines are averaged figures referring to "model" rockslide: density of material 2500 kg per cubic metre; average thickness 100 m; vertical travel of centre of gravity 1000 m; horizontal run-out (entire energy dissipated in generating lubricant). Second line shows mass of lubricant, referred to area of sliding mass. Third line shows volume of lubricant, referred to total sliding area on ground (≈ 5 times sliding area of mass). Evaporation volume refers to water

Medium Process	Ice Melting	Rock Melting	$CaCO_3$ Dissociation	Water Evaporation	Dimension
Process energy	0.334	1.28	> 1.27	>2.24	$MJ\,kg^{-1}$
Amounts of lubricant:					
Mass/area of sliding mass	7330	1916	>1916	>1095	$kg\,m^{-2}$
Volume/area on ground	1.6	0.15	>0.15	>0.22	m

rock, other mechanisms of self-lubrication (melting of ice; evaporation of water; chemical dissociation of carbonate; dehydration of gypsum or serpentine) as well as the first publications in this field by Habib (1967, 1975) and Goguel (1969) are mentioned. Still the considerations presented hereafter will throw a complementary light upon the complex field of lubrication. They deal with both generation and escape of lubricant.

Looked at from an engineering point of view, lubrication is a means to economise energy (and wear). So it cannot be unimportant how much energy is needed to generate the lubricant. A list of the most interesting candidates may be unexpected in certain respects.

A first impression in considering Table 5.4 probably refers to the remarkable *lubricating power of melting ice* (well known from skating and probable for Sherman, Sect. 2.3). It is backed up by field evidence: among the key events of this book, Pandemonium Creek and Huascarán, both very rapid, had to move on glaciers over a non-negligible distance. It should be observed, in this context, that snow may not necessarily offer equal possibilities: its lower consistence may entail its being pushed ahead by a rock mass instead of serving as a track. Yet there exist reports of low friction observed in events on snow (McSaveney 1978; Kelly 1979).

The second particularity is given by the *energetic requirements* of various processes. Fusion and dissociation of rock are quasi equal in this respect, while evaporation of water is almost at twice this level. It has, however, to be borne in mind that the escape mechanisms for liquids and gases are different so that no rash conclusions should be drawn (s. hereafter).

Of course, Table 5.4 is not realistic in every respect: even in an event consisting of descent and horizontal run-out (as assumed for the model rockslide) not the entire decelerating energy can serve to melt rock or to evaporate water. Still the more detailed calculations carried out for Köfels (Erismann et al. 1977, 96–99) show that only a small portion of heat is lost by conduction. So, at least in the case of a mainly coherent mass (as that of Köfels must have been), the volumes presented in the last line of the table so to say as "lubricant carpets" covering most of the ground, possibly not are dramatically overestimated. And the weak point, as already observed for the group "Water", once more is found in the mechanism of escaping which, on the one hand, cannot be estimated with reasonable (though moderate) accuracy like generation and, on the other hand, means both a loss of lubricant proper and one of *heat carried away by convection.*

In such instances the interest in *escaping mechanisms* is further increased, and four questions have to be considered. One refers to the possibility of a size effect, the second to the behaviour of various lubricants when being forced through narrow passages, the third to the geometry of such passages, and the fourth to quantification of load-sharing lubrication. The third and the fourth are those relevant for group "Water" as well.

*

The first question yields a straightforward quantitative answer. The only condition to be respected is that the rate of lubricant generation – like the energy available for the purpose – be approximately proportional to the thickness of the mass. Now compare two similar masses, one of which is twice as thick as the other. Obviously their relation of lubricant generation per unit of sliding area will be 1 : 2. The larger event, however, will not offer, per unit of area, more interstitial passages for escape, be it trough its thickness or laterally. On the contrary, the total path will be longer, the degree of compaction higher, and time probably more limited by faster motion. And to accelerate the lubricant to double velocity of escape, more than twice the pressure is required. In other words: the lubricant balance of generation versus escape is established at a pressure higher than proportional to thickness. No matter, whether the lubrication turns out to be load-sharing or not, a positive *size effect is granted.* Nowadays, however, self-lubrication no longer would be presented as the only physically plausible mechanism known to entail a size effect – as done by one of the authors 20 years ago (Erismann 1979, 43, Item c): too manifold are the examples quoted in this book. Still the existence of extra-terrestrial events with long run-out underlines the importance of self-lubrication (Lucchitta 1978).

*

The second question can, at least partially, be answered in a satisfactory manner, though quantification is confined to relative statements. When moving through a sequence of narrow passages (certainly the usual configuration, especially in a disintegrated mass), under normal conditions the *resistance of a gas is higher than that of a liquid.* Under the conditions of generation at the bottom even of a large slide, gas density is below that of water (not to speak of fused rock). And even if the opposite were the case, after several passages the necessary reduction of pressure and density soon would invert the relation (as the density of the liquid is almost independent of pressure). And, of course, lower density means more volume, and more volume means more resistance against escaping. In addition, the resistance of a gas is increased by the necessity to reach the sonic velocity in most of the passages. In fact, a direct quantitative comparison of fused and dissociated rock yielded a definite superiority of CO_2 as a lubricant (Erismann 1979, 35–43). This study might be criticised with good reasons for the necessarily ill-founded assumptions made concerning the geometry of the escape system (an example: no possibility of clogging was accounted for); but this has little to do with the relative figures obtained in the comparison of two fundamentally different lubricating systems. Anyway, it is an unlucky fact that the traces of dissociation disappear almost as quickly as those of evaporation so that only one well-documented report of field evidence is known (Hewitt 1988, 66).

Corrections to the general superiority of gaseous lubricants may be necessary if the mentioned "normal conditions" fail to be fulfilled. In particular fused rock at a temperature near to solidification may undergo a non-negligible increase of resistance due to increasing viscosity.

*

The third question, relevant both for "*Fire*" and "*Water*", as announced above, deals with the rate of escape as a function of the degree of disintegration in a mass. That an essentially coherent mass, offering in principle no ways of escape through its thickness, is a favourable precondition for self-lubrication, is easy to see. Yet also the existence of a more or less *impermeable layer* at the bottom of disintegrated masses, as assumed in Fig. 2.7, Item 4, is not an uncommon observation. How can such a layer be formed?

It is well known that in their depth many disintegrated rockslide masses consist of "*... finer grained, tightly packed, and relatively nonporous...*" material (Shreve 1966, 1640, referring to Sherman slide). The respective tendency of gradation (coarse on top), most frequent in disintegrated masses, has been commented under Heading 4.2. Now in any case the *lubricant contains solid pollution*, either carried along after generation or being part of the product of heat transformation (quartz in melted feldspar, CaO powder in CO_2 gas, etc.). So it cannot be excluded that such pollution may clog up the obviously narrow passages between relatively small particles. Especially a sufficiently viscous mixture of fine-grained material with water may act, in the extremely rapid sequence of partial mechanisms, as an excellent short-time sealing material. And fused rock, even without any additives, but ready to re-solidify at the first contact with a colder surface, may equally be effective. Figure 2.25 shows a crack in the ground below the Langtang slide, reaching relatively far, but having absorbed only a very small volume of frictionite. Similar spin-offs have been observed on both top and bottom of the sliding surface, but it seems that, at least in the particular circumstances, not much of the fused material was lost.

Partially, this fact may counterbalance the general superiority of gaseous lubricants, but it would be very *difficult to quantify* this amount, even on a purely relative basis.

*

Eventually, the fourth question refers to *load-sharing lubrication*. Its existence in the case of forced water lubrication is discussed above; for starting heat-generated self-lubrication it is simply a must. As a matter of fact, there must be in the immediate vicinity a powerful source of heat, and this source hardly can be other than frictional. In the initial state there is nothing but dry friction; when the required temperature is reached, local generation of lubricant starts, immediately followed by increase of pressure and escape; according to the circumstances an equilibrium between generation and escape is attained, concomitant with the respective degree of pressure, and hence load sharing. Now it must be taken into account that the process reduces friction in accordance with the amount of weight supported by the lubricant: the effective coefficient of friction as well as frictional heating are thus reduced.

Only in extreme cases, if at all, can the *entire weight be supported by the lubricant* alone (perhaps the continuous layer of frictionite of Fig. 2.25, more than 100 m long, is an example). In such a case friction is dramatically low, and dramatic velocities are possible. In addition, friction becomes essentially dependent on velocity. Whether even in this state a balance of generation and escape can exist in reality, depends, besides generation and escape proper, on the character of the lubricant and the conditions of cooling – once more a case for detailed investigation of various parameters, followed by large-scale computer simulation. It is, for instance, not excluded that after the development of a rather effective layer of lubricant the system becomes unstable by dominance of escape and collapses by lack of heat, thus re-starting rock-to-rock contacts. This would mean nothing but a *load-sharing "pulsating in time"* complementing the more trivial load-sharing in space.

What can be done with the means available at the time being, is to determine approximately the resulting coefficient of friction μ_{eff} in the (normally valid) parameter configuration in which the resistance in the lubricant can be neglected in comparison with that of Coulomb's friction. In such instances

$$\mu_{\text{eff}} = \mu\left(1 - \frac{p_L c}{H \partial g}\right) \tag{5.22}$$

is valid where μ is the rock-on-rock coefficient of friction, p_L the average load-bearing pressure of the lubricant, c the (dimensionless) fraction to which the sliding area is exposed to p_L, H the thickness of the mass, ∂ its density, and g gravitational acceleration.

A last remark should be made in the field regarding both "Water" and "Fire". One of the most spectacular possibilities of a descending rock mass is its motion *on top of a lava stream.* It will, however, not be treated in detail as belonging only partially to the topics of this book and requiring a considerable amount of space to be presented in a satisfactory manner. It suffices to recall that a block practically immersed in flowing magma rather has to be considered as an alien element within a viscous fluid than as a body lubricated by the fluid. For interested readers in first instance the extensive description of the Mount St. Helens event by Voight et al. (1983, 261–269) is recommended.

*

Under the heading of "*Earth*", the last group of lubricating mechanisms refers to solids as lubricants. There exist various basic mechanisms, ranging from almost perfect triviality to a degree of sophistication requiring some deeper digging to become transparent.

In fact, if a material excelling by *naturally plane surfaces* – for instance graphite or mica – is found located between two surfaces in relative displacement, only the question may be non-trivial how it could get there without being part of one of the respective bodies. The rest is simply the almost complete absence of rugosity: a fact highly interesting from the viewpoint of materials science (especially of crystallography), although nothing but a common parameter when considered in the optics of motional dynamics.

Travelling on top of a layer consisting mainly of *fine-grained rock particles and water* (clay, mud) can only be effective if the lubricant is available in large quantity and has the above-postulated more or less ideal degree of viscosity between too watery and too solid. The impressing examples of Gros Ventre and Rossberg/Goldau (Alden 1928, 347–349; Heim 1932, 71–74) are mentioned in Sect. 3.1. It would, of course, be a fascinating challenge to attempt, on the basis of the little-changed deposits of these historical events, a detailed simulation yielding the mechanism (in first instance N, or P, or, most probably, something in between) as presented at the head of Sect. 5.2. For further details s. Sect. 5.6.

As mentioned in Item 24 under Heading 5.3, perhaps the most challenging mechanism for a lubrication of the "Earth" type would consist of a number of highly energetic particles keeping the rest of a mass in levitation by "*playing volleyball*" with it. Such mechanisms have been studied in computer simulation by Dent (1985) and Campbell (1989). Both have based their work on representing a mass of debris by spherical (Dent) or disk-shaped (Campbell) bodies moving on periodically uneven ground.

Dent's work suits the requirements of lubrication particularly well: a very large particle on top is considered as overburden, the smaller ones holding it in suspense as lubricant. For reasons unknown to the author of the present section the coefficient of restitution is assumed to be unity at zero relative velocity (p. 4), thus favouring low velocities. Further the trend to rotate is artificially retarded to reduce the effect of rolling (p. 7). Contrarily to these aspects which impede a perfectly realistic approach, a point of particular interest in this model consists in the fact that a positive size effect was observed (p. 9). Subsequent analysis showed that it was due to the influence of multi-particle collisions in which tangential momentum transfer is reduced (pp. 20, 22–23). A further valuable statement refers to the phenomenon reported in Sect. 5.3 under the label of a rebounding body "declaring itself bankrupt", the braking capacity of an impact exceeding the momentum stored in the body. Here, however, rotational instead of linear motion is considered, and thus friction is tangential (pp. 14, 17).

Obviously also *Campbell's study* had to deal with the problem of rotation which cannot be solved in a perfectly satisfactory manner when using spheres or discs (p. 659). And also his solution – to let them rotate without correction –, despite the easier approach it offers to analysis, certainly does not comply with the behaviour of real debris, irregularly shaped, sharp-edged as they are and (as discussed in Sect. 5.1) tending to move with their shortest axes at right angles to the ground. On the other hand, Campbell was able to demonstrate plausibly that, under the extremely favourable (not to say impossible) conditions stressed in Item 24 of Sect. 5.3, three highly energetic particles can hold in suspense no less than 37 other particles of equal size – a remarkable performance.

And as in both studies the angle of collision turns out to be the most important parameter (Dent p. 14; Campbell p. 664), an essential *difference between simulation and reality* cannot be concealed. So both give excellent insights into various problems, but both show that a realistic simulation requires a far higher preparatory and computational work. By the way it should be observed that, according to the details of a particular set of chosen parameters, the mechanisms in both cases may be interpreted as lubrication, as fluidisation, or as something in between (would "flubrication" be the right term?). So they also will shortly be mentioned in the immediately following Sect. 5.6.

*

To check the capacity of this class of mechanisms with respect to long run-out, hereafter a simple, *easy-to-calculate model* will be used in which a realistic coefficient of restitution ($J = 0.7$) and a horizontal track (simulating a typical run-out situation) are assumed. All further assumptions are chosen to favour the well-functioning of the mechanism, even at the price, in some cases, of unrealistic conditions (Fig. 5.17). A mass m_1 moves over the ground. Both the ground and the bottom of the mass are plane by exception of the singular points mentioned hereafter. A single row of highly energetic particles, forming together a comparatively small mass m_2, achieve a to-and-fro motion between ground and mass, thus holding m_1 in levitation. This motion is

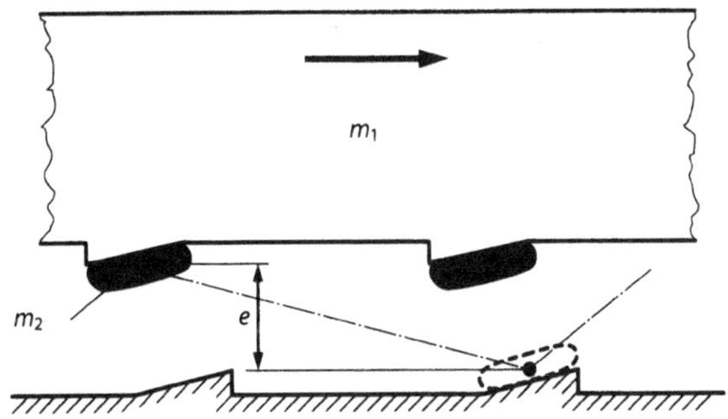

Fig. 5.17. Partial view of running out rockslide mass m_1, dynamically lubricated by many particles of m_2 (two are visible), rapidly moving in a zigzag course between m_1 and ground, thus keeping m_1 in levitation by multiple impacts from below. *Arrow:* motional direction of m_1; *dots and dashes:* path of m_2; *e:* stroke of m_2. Note the – deliberately idealised – angular adjustment of impact points in ground and m_1 granting an optimal path for each particle of m_2 (feature dropped in Fig. 5.18). Energetic estimations on this basis are definitely optimistic (sketch by Erismann)

optimised by a locally perfect individual angular adjustment of the surfaces of both m_1 and ground so as to avoid collisions between particles or undue lateral obliquities in their paths, and, above all, to carry along m_2 with m_1. Furthermore the weight of m_2 is assumed to be negligible. Thus only the energy required to obtain total restitution after each impact is accepted as a loss, and it is assumed that this energy can be taken, without any further loss, from the kinetic energy of m_1.

Let g be the gravitational acceleration, u the vertical component of the velocity vector of m_2, and Δt the time required for a complete cycle of the particles (with one impact at the ground and one at m_1). Then levitation obviously takes place if the momentum exerted by m_2 in hitting m_1 equals the action of gravitation during Δt, i.e. if

$$2\,m_2\,u = m_1\,g\,\Delta t \tag{5.23}$$

is true. When introducing the vertical stroke e of the particles, Δt can be expressed by $2e/u$, and the vertical component of velocity is written

$$u = \sqrt{ge\frac{m_1}{m_2}} \tag{5.24}$$

thus making possible the determination of the respective kinetic energy and, after multiplication by $(1 - J^2)$, its loss in a cycle. This lost energy, extended to the duration t of the run-out, easily can be transformed so as to express the respective loss in elevation

$$\Delta z = \frac{1 - J^2}{2}\,t\sqrt{\frac{gem_1}{m_2}} \tag{5.25}$$

that m_1 would undergo in losing an equivalent amount of gravitational energy.

Now introduce parameters reasonably favourable to well-functioning, e.g. $J = 0.7$, $m_1/m_2 = 20$, $e = 0.2$ m, and $t = 50$ s. The resulting figure, $\Delta z = 79.9$ m, corresponding to the kinetic energy of m_1 moving at a velocity of 39.6 m s^{-1} and dissipating said energy in an horizontal run-out distance of 990 m, gives, at first sight, a promising, though not dramatic impression. This is, however, the moment to remember that the considered loss of energy comes from *nothing but keeping operational the process of levitation* under particularly favourable circumstances. And a side-glance at the assumptions made stresses this impression, as the following – by no means exhaustive – examples will demonstrate: if, for instance, there were in m_2 two rows of particles above one another, thus giving rise to twice as many collisions, Δz would be doubled; if, as might be concluded from experimental collisions carried out with loose talus material, a velocity restitution coefficient around 0.45 (Bozzolo 1987, 22, Tab. 2.4.1) were used, an increase of Δz by a factor exceeding 3.0 would result; if the highly sophisticated individual adjustment of the points of collisions were replaced by plane surfaces, the process either would fade out, m_2 necessarily being overtaken by m_1, or a continuous supply of fresh particles (requiring to be freshly energised in the vertical) would be needed. The final result is obvious: the "volleyball" model, not excluded as a mechanism in steep slopes, is practically hopeless for the explication of the phenomenon of long run-out.

The above arguments show with particular clarity the *value of partial quantification*. If in a system a sufficient part of parameters can be estimated, and the respective partial mechanisms can be quantified sufficiently well, and if the result of such partial quantification is near the limit between function and non-function of the total mechanism, a qualitative or semi-quantitative consideration of the remaining parameters and partial mechanisms can pave the way to a reasonably well-founded yes-or-no answer. For normal-sized rockslides, however, there exists an argument beyond any doubt.

Consider in some detail the consequences of dropping the – really fantastic – assumption of individually optimised angles of collision. As the forces acting upon the particles of m_2, owing to the short duration of collisions, necessarily are far higher than in case of a roller bearing (Sect. 5.3), such particles, in the reality of not too thin a rockslide, cannot subsist in a shape allowing rotational contacts. This is equivalent to saying that the particles, in colliding with either of the surfaces (ground or m_1), undergo an incremental process of sliding. It was already insinuated in Sect. 5.3 that, to allow for serial bouncing, the coefficient of friction μ shall not exceed the slope tanß. This condition, $\mu <$ tanß, trivial for bare sliding, has a more general background. Consider the simple case of motion on horizontal ground. Of course, in order to maintain a body at a given distance from the ground, the lifting force, averaged over a reasonable time interval, shall equal the weight of the body. This is true also for a force generated by a rapid sequence of impacts (Fig. 5.18). The same applies to any force proportional to said lifting force. Hence, provided that Coulomb's rule is valid, the resulting average of frictional force within the considered time interval, irrespective the course of the succession of force peaks, must be the product of weight and coefficient of friction. In such instances the lubrication may exist, but its *effect is reduced to nought* as compared with unlubricated sliding, even if the coefficient of restitution is unity.

*

The physical and geomorphological heterogeneity of the mechanisms treated in the present section makes useful a short review of the *results obtained.*

1. Various analogies exist between technical and natural lubricating mechanisms, for instance: the fundamental balance between supply and escape of lubricant; complex systems like forced lubrication by pressurised lubricant; load sharing between

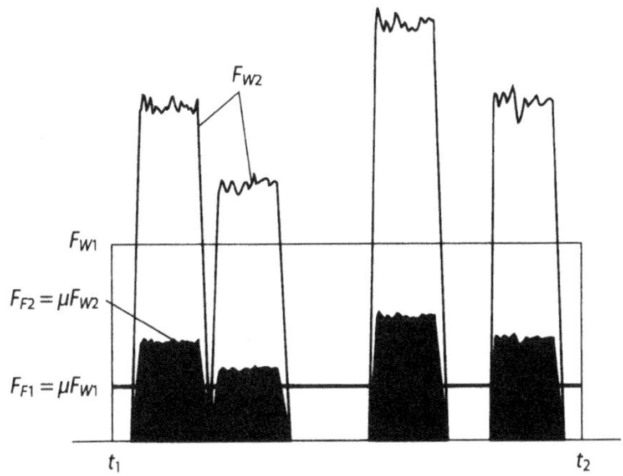

Fig. 5.18. Forces exerted by non-rolling particles of m_2 (Fig. 5.17) upon plane bottom of m_1 or plane ground. To obtain levitation of m_1 in time interval t_1–t_2, total momentum of irregular impacts (force peaks F_{W2}) must equal that of weight F_{W1}, i.e. area below F_{W2} must equal area of rectangle below F_{W1}. If Coulomb's rule (coefficient of friction μ) is valid and friction is unable to reduce to nought difference of velocities, frictional force peaks $F_{F2} = \mu F_{W2}$ are obtained (*peaks stained black*), yielding, in their turn, an average force $F_{F1} = \mu F_{W1}$. In such instances, frictional losses in pure sliding and with lubrication are equal. Note analogy to repeated rebounding commented in Sect. 5.3 (sketch by Erismann)

lubricant and solid-to-solid contacts. Only for Item 4 lubricants found in debris are reported in certain cases.

2. Water as unique lubricant is highly probable in the form of weight-pressurised forced lubrication by originally interstitial water of fills, predominantly in laterally confined valleys or large events (size effect). Load sharing is probable. Correlation with overall slope is remarkable.
3. Air as a lubricant, after physical analysis and comparison of terrestrial events with such on celestial bodies, seems to be far less effective (if at all) than was assumed for a considerable time.
4. Self-lubrication (key word: "Fire") by heat-generated lubricants (melted ice/snow or rock, chemically dissociated rock, evaporated water) is promising by high efficiency (lubrication at best suited points), despite relatively low volume of lubricant. Unique lubricants perceptible in debris: melted and dissociated rock. Size effect and load sharing are compulsive.
5. Rock and soil as lubricants (key word: "Earth") are trivial and consistent with field evidence in case of perfect surface (mica, graphite) or materials containing water (clay, mud). Non-trivial is the "volleyball" model of self-lubrication with energised particles holding the main mass in levitation. This model can work only on sufficiently steep slopes.

*

The importance of the present section lies in the fact that lubrication may entail long run-out with catastrophic consequences. So it is essential that, in spite of lacking knowledge for accurate forecasting, it was possible to make evident the physical limits of the considered mechanisms, and that the resulting statements mainly are based on quantitative reflections.

5.6 Fluidisation

The fundamental *difference between lubrication and fluidisation* lies in the location of the mechanism achieving a reduction of resistance. In case of lubrication this mechanism, irrespective its physical nature, is considered as concentrated in the immediate vicinity of the boundary between moving mass and ground; in fluidisation, on the other hand, the mechanism is active at least in a considerable portion, normally even in the entire thickness of the mass. In somewhat imprecise terms, lubrication might be considered as a quasi-bidimensional phenomenon, while fluidisation is fundamentally three-dimensional.

The main *geomorphological consequences of this difference* are obvious. In motion, a fluidised mass, like a liquid, normally displays relative displacements within its entire volume. Lubricated motion, on the contrary, entails moderate relative displacements between the particles of a disintegrated mass, thus keeping it "in shape" and preserving to a large extent the sequential order of constituent parts. This fact is a powerful clue in analysis a posteriori, excluding a majority of rockslides from the suspicion of having been essentially fluidised in their downhill ride (Sect. 2.4, 5.1, 5.4; in particular comments to Fig. 2.21, 5.15). Hereafter doubts will be expressed in a particular case on the basis of this simple though often disregarded clue.

In parentheses it should be observed that for solid lubricants, as far as they obey Coulomb's law, the *rule of lowest μ dominance* (Sect. 5.4) is valid so that, contrarily to the above definition of fluidisation, a tendency must prevail to concentrate the entire drop of velocity near one single plane.

In some cases, fluidisation can be considered as a "*multiple lubrication*", i.e. a mechanism of lubrication acting in a sandwich-like array on several planes, approximately parallel to each other and to the ground. Transition from this hybrid concept to pure fluidisation logically is given as soon as no distinct planes of lubrication can be perceived: the sandwich has been turned into a hamburger. The practical importance of this particular model may be questioned with good reasons; its didactic value, however, will soon be demonstrated.

*

As a matter of fact, the just-mentioned sandwich model is particularly useful to give a quantitative, though not in any respect a realistic idea of a mechanism of fluidisation encountered rather frequently in connection with rockslides: *fluidisation by water*.

Imagine a multiple-layer, sandwich-like mass, rather similar to that of Fig. 5.15, yet containing a layer of water between any pair of consecutive layers of rock (Fig. 5.19a). Let the respective thicknesses be e_r (rock) and e_w (water) and the densities ∂_r and ∂_w. For a first approximation assume impermeable rock layers with negligible rugosity and an extension large enough to neglect also the peripheral losses of water in the considered space of time. It should be added, in this context, that the term "negligible rugosity" in first instance means that the asperities of a surface must be small (i.e. low) in comparison with the thickness of the adjacent layer of water. Finally let e_r and e_w, when compared with the thickness H of the mass, be small enough to allow a reasonable approximation when replacing the model by a mass entirely *consisting of a Newtonian fluid* equalling it in its averaged values of density ∂ and viscosity E which obviously are given by

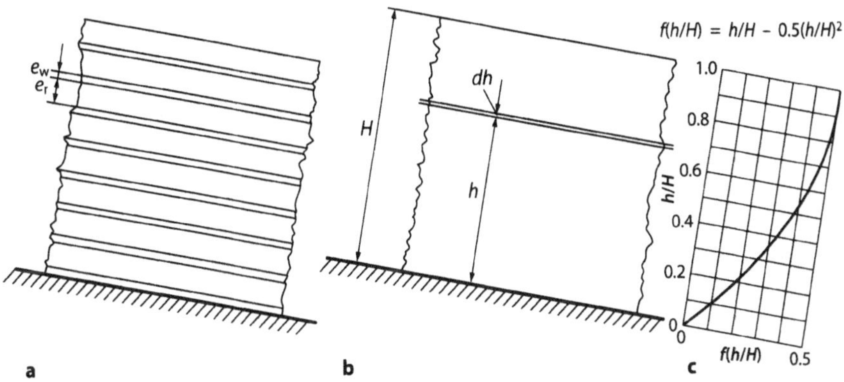

Fig. 5.19. Fluidization by water. **a** Simplified "sandwich" model facilitating quantitative understanding of viscosity-controlled motion of disintegrated water-saturated mass (e_r: thickness of impermeable rock layer; e_W: thickness of water layer). For better clearness only few layers are shown. Fundamentals are discussed for infinitely extended layers; then finite lateral extension is introduced to allow for lateral water escape. **b** Continuous viscous mass as approximated by sandwich model. Extensions (*H:* total thickness; *h:* distance from ground) are used to establish differential equations of motion (Eq. 5.28, 5.29). **c** Standardised velocity distribution (equilibrium after sufficiently long travel under constant conditions), valid for all masses moving under the rule of constant viscosity. $f(h/H)$ is proportional to velocity (sketch by Erismann)

$$\partial = \frac{\partial_r e_r + \partial_W e_W}{e_r + e_W} \tag{5.26}$$

and

$$E = E_W \frac{e_r + e_W}{e_W} \tag{5.27}$$

where $E_W = 0.00134\ \mathrm{kg\ m^{-1}\ s^{-1}}$ is the viscosity of water. This trick simplifies quantitative treatment dramatically.

In fact, for motion at *constant velocity* (i.e. the stable velocity asymptotically approached on a given slope, s. mechanism P in Sect. 5.2) the shearing stress *s* at the level *h* (Fig. 5.19b) can be expressed by the effect of gravitational acceleration *g* on the left side of

$$(H - h)g\partial \sin\beta = s = E\frac{du}{dh} \tag{5.28}$$

and, on the right side, that of viscosity *E*. β is the slope angle and *u* the velocity at level *h*. This differential equation easily can be solved to obtain velocity *u* in the formulation

$$u = f\left(\frac{h}{H}\right) gH^2 \frac{\partial}{E} \sin\beta \tag{5.29}$$

where $f(h/H) = h/H - 0.5(h/H)^2$, as being dimensionless and identical for any possible configuration of parameters, undoubtedly represents the best choice for an easy-to-use variable (Fig. 5.19c). The maximum velocity at the surface of the mass is given by $f(h/H) = 1/2$, and to obtain the mean velocity, averaged over $0 < h < H$, the average $f_m(h/H) = 1/3$ can directly be used in Eq. 5.29.

It will be observed that Eq. 5.26 and 5.27 yield unchanged results if both e_r and e_w are multiplied by one and the same dimensionless factor. This means nothing less than that the distribution of velocity described by the differential Eq. 5.28 is *independent of the absolute thicknesses* attributed to the layers as long as e_r/e_w remains unchanged (and, of course, as long as the accepted range of layer thicknesses with respect to total thickness H and to the size of asperities is respected).

Now cancel one of the obviously unrealistic assumptions of the model and give it a finite breadth B (which, owing to possible longitudinal cracks, is not necessarily the total width of the mass). This change yields a *possibility of escape* for the water (which so far was considered as trapped between the quasi-infinite layers of impermeable rock), thus enabling water layers to narrow down and to increase resistance accordingly. Now the crucial question for the well-functioning of the model lies, as easily can be concluded from Sect. 5.5, in the time available for not too seriously hindered motion of the mass.

It would go beyond the frame of this book to deduce in detail the equations showing the consequences entailed by the *laminar flow of the escaping water*. Fortunately, its final result displays, in the chosen formulation

$$\frac{de_w/dt}{e_w} = \frac{2}{3}\left(\frac{e_w}{B}\right)^2 \frac{p}{E_w} \tag{5.30}$$

a remarkable transparency. On the left side the velocity at which a water layer narrows down (de_w/dt) is referred to its thickness e_w. Thus the relative (percental) loss in layer thickness per time unit is expressed. The right side, on the one hand, is connected with the geometry both by the constant $2/3$ and the squared figures e_w (the width of the passage) and B (i.e. twice the length of the passage); on the other hand, it represents the antagonism between driving pressure p and braking viscosity E_w.

The most interesting modification of the presented model would refer to a quasi-normal, *three-dimensionally disintegrated mass* with a reasonably homogeneous granulometric distribution of particles. Certainly, the quantification of any laminar escape mechanism necessarily contains the term p/E and a dimensionless constant value c corresponding to the coefficient $2/3$ of the model. So on the right side of the equation only the influence of the term e_w/B needs to be considered in more detail. Thereby it has to be taken into account that this term is dimensionless so that any possible substitution must equally be dimensionless. Leaving open its precise character, a function $f(e_w/L)$ may be introduced for the purpose. Numerator and denominator stand for width and length, obtained by reasonable averaging from all significant passages of the mass. In addition, it is granted that both undergo a more than proportional size effect: e_w because of the high dependence of resistance on the width of passages, L because, in its centrifugal motion, the escaping water not only covers an increasing distance, but also assumes an increasing volume by the water added en route. Equation 5.30 thus can be regarded as a slightly modified special case of

$$\frac{de_w/dt}{e_w} = cf\left(\frac{e_w}{L}\right)\frac{p}{E_w} \tag{5.31}$$

With relative easiness c and $f(e_w / L)$ could be determined for a fairly regular array of particles; for more realistic configurations, however, it would be another task for a fast computer.

Fortunately, in the frame of the present book such endeavours are not absolutely necessary as the most important *generalised conclusions* can be drawn from Eq. 5.29, 5.30, and 5.31, including the accompanying arguments.

1. It is obvious in the context of Eq. 5.29 that the general behaviour of a water-saturated disintegrated mass, as long as flow is laminar and the mass is able to hold the water, resembles that of a viscous liquid.
2. In such instances necessarily there is an increase of velocity from the bottom to the top of the mass. The main consequences in a mass of finite length are reversal of sequential order of the elements and rapid loss of thickness.
3. A strong positive size effect ("small slide slower") is given by the square of thickness H in Eq. 5.29.
4. According to Eq. 5.30 and 5.31, the relative loss of water (and increase of effective viscosity) is slower in large than in small masses: driving pressure p is proportional to the linear extension, braking effect of path length (expressed by B or L) is in any case overproportional. The result is a second size effect ("large live longer").
5. The absolute extension of passage width e_w exerts a decisive (definitely overproportional) influence upon the duration of functional life.

Whenever fluidisation by water is suspected, Items 1–5 can be used to *check the probability* of fluidisation. In particular very fine-grained water-saturated materials like clay show an excellent endurance in serving as sliding tracks (here, of course, further effects like capillarity may play a certain role).

How far *valley fills* serving as a basis for pressure lubrication by water (Sect. 2.2, 5.5), if consisting of appropriate material, may partially or integrally perform an additional internal deformation like a viscous fluid, is left open. In his basic publications to the subject, Abele (1991b, 1997b) used the term "mobilisation" which might have meant such a motion below the rockslide surfing on top. The possibility of a single layer of conveniently viscous rock-and-water mix acting as a proper lubricant has already been briefly mentioned under Heading 5.5.

Finally, some remarks should be made about the limits of the described form of fluidisation by water. Obviously the bounds of viscous flow are trespassed as soon as, at increasing velocity, energy dissipation by *inertial effects* overtakes that due to laminar flow. The description and analysis of a similar transition, though under somewhat different conditions, is one of the main issues of Bagnold's remarkable work (1954, 1956). In the optics of this book, the transition marks the boundary between rockslides and fast mud flows which are considered as a mixture of rock or soil particles and water moving essentially like a low-viscosity, "watery" fluid wherein the rock particles take part, according to their size and the respective velocity of sinking, as almost plankton-like, density-raising parts of the fluid or as alien bodies, forced by weight, friction, buoyancy, hydrodynamic effects, etc. into a motion at more or less reduced velocity. The physical difference thus outlined was the reason of the reluctance expressed in Sect. 2.7 against the use of the catastrophe of Huascarán as a key event of this book.

At the other end of the scale of velocities, fluidisation by water in *creeping*, by virtue of its frequent occurrence, must be considered as an important mechanism. Innumerable are, in fact, the small events which happen, so to say, day by day all over the world. And initiation of rockslides by this kind of mechanism has already been mentioned more than once in the present book (Sect. 3.1, 5.5).

*

To the author's knowledge, Shreve (1966) was first to propose a layer of air as a lubricating mechanism in rockslides. He was, however, definitely not first to consider, in a more general sense, air as a means able to increase the reach of descending masses. In fact, somewhat more than a year earlier (in spite of having appeared in the same year, the two manuscripts had been received by the respective publishers on March 20, 1965, and September 20, 1966), Kent (1966), in comparing limited scree-forming rockfalls with the far more extended events of Saidmarreh (Harrison and Falcon 1937, 1938), Frank (Mcconnell and Brock 1904), and Madison (Sharpe 1938), came to the conclusion that the obviously more economic locomotion of the large slides was due to their capability to be *fluidised by absorbing air*. It is perhaps of a certain interest to quote the original formulation of the essential statement: "*... it seems highly probable that the high speed and high fluidity... was due to fluidization with air – the air entrapped during the first few hundred feet of fall being 'entangled' in the rock for long enough to permit the rock to flow as a liquid...*" (p. 82).

Unfortunately, very little is said about the *details of the supposed mechanism*, and, in particular, nothing about the way in which air could be trapped, compressed, and "entangled" with the disintegrated mass. Its compression, however, is explicitly mentioned in the case of Frank: "*... the phenomenon was accompanied by a violent compressed-air blast... mud and small rocks were spattered over the surface far ahead of the slide...*" (p. 81) and: "*... in some cases people were carried many tens of feet without damage before being buried...*" (p. 82).

On the basis of the arguments dedicated to air lubrication (Sect. 2.3, 5.5), it would not be difficult to reduce to absurdity the hypothesis of air fluidisation by using, mutatis mutandis, the same arguments (in fact, the crucial condition in both cases consists in the possibility to overcome gravity; and for fluidisation even the dubious chance of trapping and compressing air by a large bounce cannot be postulated). It seems, therefore, more useful to draw the reader's attention to another line of thought that equally was not analysed conclusively. As a matter of fact, it is well known that the *blast accompanying a fast rockslide* (and to a certain extent preceding it, Sect. 2.5, 2.7) can be strong enough to blow down a man, not to speak of carrying along small stones. This is perfectly plausible even for a moderately rapid event: when replacing water by air at sea level in Eq. 3.5 and assuming a velocity of, say, 50 m s^{-1}, a force approximately equalling the weight of a person is obtained – an impressive hurricane. The respective dynamic air pressure $p = 0.5\,\partial_a v^2$, however, being in the order of 1550 Pa, is just high enough to hold in suspense a square metre of rock some 5 cm thick! And little will change at twice the velocity (100 m s^{-1}): the man will be blown away like a twig, but the thickness of the rock square will not essentially exceed 20 cm...

This simple mental experiment is, perhaps, a useful demonstration for someone endeavouring to acquire a feeling of how to think in terms of rockslides...

As mentioned under Heading 2.3, the impact of Shreve's impressive hypothesis (1966, 1968a, 1968b) evoked, in the late sixties and especially in the seventies, a school of thought based on the idea of *air as a means increasing the reach* of rockslides. There were, however, certain unanswered questions encouraging scientists to look out for other solutions, even at the cost of going against the tide.

Some of these questions were raised as soon as *Lunar and Martian events* were considered more in detail (Guest 1971; Howard 1973; Luchitta 1978). It was found that long range did not necessarily have to be considered as an earth-bound (or rather: atmosphere-bound) privilege. It will be shown hereafter that in accepting the fact certain critical questions remained unconsidered. Anyhow, it could not be overlooked that remarkably low overall slopes existed without the presence of a supporting gas (the unique plausible mechanism in the eyes of many experts at the time). So it is natural that certain researchers tried to find mechanisms able to solve this enigma. Two important examples based on computer simulation (Dent 1985; Campbell 1989) already have been discussed, mainly in Sect. 5.5. There also the result of the respective analysis was summarised in Item 24. A third study will be presented hereafter. Two things are remarkable in this context: all three examples were based on straightforward mechanics; and all three appeared several years after the most-cited publication which will close the discussion of this group of mechanisms.

Under the denomination "*mechanical fluidisation*", Davies (1982, 14–15) suggested that "*… a high energy input… causes high impulsive contact pressures between… grains…*", in other words an approach to the analogy between a gas and fine-grained solid material. The experiments set up to visualise flowing (a mass consisting of sand on a longitudinally vibrating, slightly tilted bed) must, however, be considered with a certain reserve: the parameters correlated with an optically flow-like motion of the sand (amplitude 5 mm, frequency 25 Hz) result in peak accelerations of no less than ±12.6 g, an overwhelming effect reducing friction to a practically negligible role. As a consequence, conclusions referring to the motion of a rockslide mass (in which, as a rule, besides the stored potential energy no relevant source of energy is available) would be more than risky. An exception might be made in case of releasing an almost-ready-to-move mass on a soft slope by the vibrations of an earthquake: in this particular case the exciting energy per unit of mass, in spite of being far below that of the experiment, still is large as compared with a kinetic energy that is just starting at zero velocity.

Little has to be added to the comments given in Sect. 5.5 with respect to the "*volleyball" models* of Dent (1985) and Campbell (1989). For fluidisation similar basic conditions have to be fulfilled as for lubrication: in first instance the effect of gravity must be overcome if a more economic locomotion than given by mere friction shall be expected. With respect to the energetic situation, however, fluidisation is handicapped by the simple fact that, as the mechanism must penetrate the thickness of a disintegrated mass (instead of the ideal array of highly energetic particles in a, certainly very questionable, single layer as shown in Fig. 5.17), a dramatically increased number of collisions between particles is indispensable. So even the moderate chance of a possible function on steep slopes, as given in Item 5 of Sect. 5.5, must be considered as more than improbable in case of fluidisation.

Like the above-mentioned researchers also Hsü (1975), in setting up his often-cited study, stood, on the one hand, under the impression of the new results obtained by rapidly improving astronautics. So he considered the problem of long run-out in the absence of an atmosphere as a serious question. On the other hand, he was fascinated by various more or less old publications, in first instance those by Buss and Heim (1881), Heim (1932, 94–95, 104–105), and Bagnold (1954, 1956). It seems that on this basis he came to the idea that "flowing" – and therewith also long run-out capacity – is similar in fluids and long run-out rockslides, and that this similarity exceeds the trivial form of visual perception, thus being valid also in a physical sense. So the conclusion is presented that Bagnold's claim for a real fluid, in which solid particles can be dispersed, somehow must come true even in the waterless environment of a celestial body. As a final consequence it is declared that the *part of a fluid is played by dust* and that "*... one can easily fancy a stream of colliding blocks swimming with terrifying speeds in a sea of small stones and dry rock powder...*" (Hsü 1975, 135–136).

Both the choice of references and the conclusions cannot pass without comment. In fact, Heim's arguments on the subject, despite their realism in many points and their literary beauty, in other respects suffer a certain lack of physical coherence (s. the critical remarks in Sect. 5.1). And Bagnold hardly would have given his consent to the application of his results to anything differing from solid bodies dispersed in real fluids. So, in order to attribute to dust the qualities of a fluid, its *analogy to a fluid would have to be proved* in a physically plausible manner. Unfortunately this proof has remained unsaid. In fact, besides a tentative calculation of the dispersion's density Hsü (1975, 136) and the interpretation of the result as a "*... mixture of one-third dust and stones and two-thirds air or vacuum ...*", there is little information that might give a hint about the manner in which the dispersion would be introduced between ground and blocks and how it would act to reduce friction or replace it by a more economic mechanism accessible to physical analysis (in particular if two thirds of said dispersion consist of vacuum, p. 136).

There are, in addition, other points provoking contradiction. For instance, a statement immediately following the discussion of Bagnold's work and running as follows: "*... a single element... present at Elm and Blackhawk, on the Earth and on the Moon... could be the dispersion of fine debris particles between colliding blocks, ... the dense dust cloud....*" An implication of this statement is that the dust cloud is approximately equally dense on the Earth and on the Moon. This, however, is far from being granted. The *influence of aerodynamic drag*, as demonstrated at length in Sect. 5.3, enables large particles to reach dramatically farther than small ones. Now on the Moon there is practically no atmosphere and no drag. This means that large and small particles, when shot under equal initial conditions, accomplish equal almost parabolic trajectories. And in case of being accelerated by a collision, the smaller particle undergoes more acceleration than the larger one (Sect. 5.7). In other words: on a celestial body with no or almost no atmosphere small particles as a rule reach farther than others, and the cloud, instead of being kept together by drag, turns out to become flimsy and widespread. In such instances it is at least questionable whether the dust for the proposed mechanism is at all available.

This is not the unique point in which more attention should be paid to the *differences between celestial bodies*. On the one hand, in release, a reduced gravity allows, assuming equal strength of the involved materials, for a larger mass accumulation before fracture takes place; on the other hand, in descending, a reduction of gravity means reduced velocity and, for a given size of the mass, a reduced kinetic energy, thus possibly making the mass unable to cut away a given obstacle that would not subsist with a higher gravity. So the uncommented introduction of the giant Moon rockslide Tsiolkovsky in a figure showing terrestrial events might entail undue misunderstandings (Hsü 1975, 139, Fig. 7). The comparison of rockslides on different celestial bodies is more difficult than it might appear at first sight. More than that: the size of the best-known extra-terrestrial events exceeds that of the largest slides on Earth so much that the extrapolation of any size effect becomes questionable. We simply do not know whether the Tsiolkovsky slide on the Moon, with a mass six times that of Saidmarreh on the Earth, when looked at from a Lunar point of view, really is a long run-out event or not.

In spite of such disappointing results, a conclusive rejection of Hsü's idea would remain questionable if not based on physically indisputable arguments. Very probably the unique conceivable possibility of *dust as a fluidiser* would be a dynamic one whereby the dust would separate the larger particles from each other and from the ground by multiple mechanisms as illustrated in Fig. 5.17 and discussed in connection with Dent's and Campbell's studies. Now this mechanism has proved to be near to useless in lubrication, under conditions far more favourable than in the case of dust (larger dynamic particles in one or few layers); so the chance of dust as a fluid-like fluidising medium must be considered as non-existent. In fact, the only advantage that motion possibly may draw from dust, is the reduction of Coulombian friction obtained from smoothing the effective rugosity (Fig. 2.25, 2.26b; remarks on track-making in Sect. 5.4).

*

As if there were some sort of famine of interesting ideas in the field of fluidisation, a further possible mechanism was proposed by Melosh (1979, 1983), Goetz (1981), and Goetz and Melosh (1980). This time the work was the result of more than intuitive argument: it was backed up by physically well-founded considerations and presented in a form allowing to check the line of thought, equation by equation. The object was *acoustic fluidisation.*

To grasp the basic idea, it may be useful to consider a very simple model (Fig. 5.20) or to set up an equally simple experiment. Imagine a plane table, slightly tilted by books under two of its legs. A brush, standing on the bristles, is located on its top. Now the tabletop is slightly knocked with a plastic hammer. If the slope angle is well adjusted, at a certain intensity of the knocks the brush will accomplish a short sliding displacement after each knock and than come to a rest. Well-calibrated serial knocking will generate a quasi-continuous creeping motion. The process is almost trivial. Before starting the experiment, a number of bristles are in contact with the tabletop (Fig. 5.20a). They are slightly bent by the tendency of the brush to slide. Now let, for the sake of simplicity, a single transversal shock wave (climax looking down) run along the table. When passing below a row of bristles, it will unload them and let them straighten, simultaneously increasing load and bend of the other bristles, and thus allowing the brush to accomplish a minute displacement (Fig. 5.20b). The same occurs at each of the rows so that after the wave has passed the entire length of the brush, said "microsteps" have added up to a perceptible displacement (Fig. 5.20c, d).

It is obvious that in a rockslide a gamut of most vigorous acoustic signals is active, and that their superposition makes possible particularly high peaks. Melosh demonstrated that such peaks can reach values allowing momentary separation of the con-

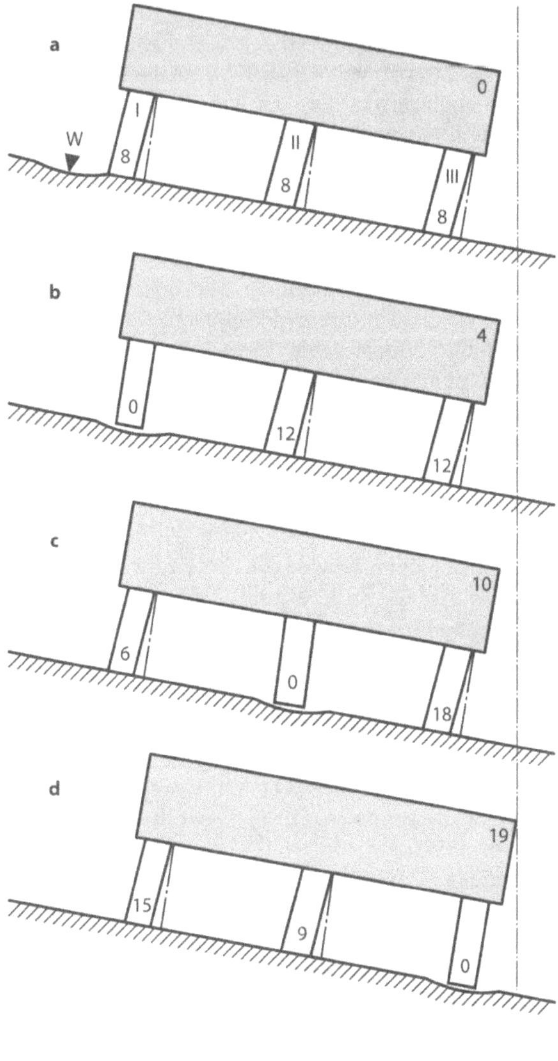

Fig. 5.20. Acoustic fluidisation, simplified model demonstrating principle. **a** Acoustic wave (symbolised by single unevenness *W*) approaches mass supported by legs *I*, *II*, *III* (elastic mainly in shear); deformations of legs are equal (s. figures at bottoms of legs and reference lines). **b**, **c**, **d** By successively reaching legs *I*, *II*, *III*, wave annihilates the respective loads, and legs are reset to no-load positions, thus increasing loads and deformations of legs remaining to support mass. Process results in displacement of mass (s. figures at top right of mass and reference line at the right) (sketch by Erismann)

tacting surfaces, each separation potentially entailing a local annihilation of shearing stress and a discharge of elastic energy, accompanied by increased stress and deformation at the remaining contact points and – as essential effect – an incremental displacement of the considered portion of the mass. More than that: it was showed that the acoustic energy necessary to "run the show" is far smaller than the frictional energy that would have to be overcome in accomplishing an equal displacement. In principle, *acoustic fluidisation is suitable for function* as a mechanism increasing reach if the configuration of parameters is favourable.

Certain conditions are trivial and easy to explain. For instance the *wavelength* of the acoustic signal must be shorter than the considered portion of the mass (if it were distinctly longer than the entire mass in Fig. 5.20, no differentiated unloading of legs one by one would be possible) and longer than the directly involved elements; in a disintegrated mass this refers to the characteristic grain size (if in Fig. 5.20 the wave were definitely shorter than the contact length of a leg, reasonable unloading again would be impossible). An energetic limitation lies in the last-mentioned condition as a sufficient vibrational energy requires, of course, a reasonably high frequency.

More problematic is the rapidity of *attenuation*. It is raised as an open question in the publication of 1979 (p. 7516). It is, however, taken for granted that propagation of acoustic waves from particle to particle functions well in view of the particles remaining in contact with one another (1983, 161). This last statement is not eo ipso certain in the dramatic motion of a rockslide, and it should not be forgotten that the main source of acoustic signals is given by collisions which would not occur if contact were continuous.

These open questions have, of course, a substantial weight; yet probably the most important problem which apparently was somewhat underestimated by the authors of the publications referred to, is a geomorphological one. In the present section (as in various others) it was stressed that in most rockslides the *sequential order of components* is preserved. This is impossible in a laminar flow in which the top of the mass continuously overtakes the lower portions, thus necessarily leading to a reversal of sequence. Now it is postulated that the pressure-dependent distribution of apparent viscosity, as experiments showed, may induce a rate of flow depending on stress (and depth) by a power of up to eight. Figure 5.21 shows a graph of this relation. Obviously the velocity gradient is far higher in the lower than in the top layers of the mass. Still about 20% of the mass takes part in this gradient. So it is easy to imagine that overtaking – and therewith also a certain distortion of the sequential order – is not avoided.

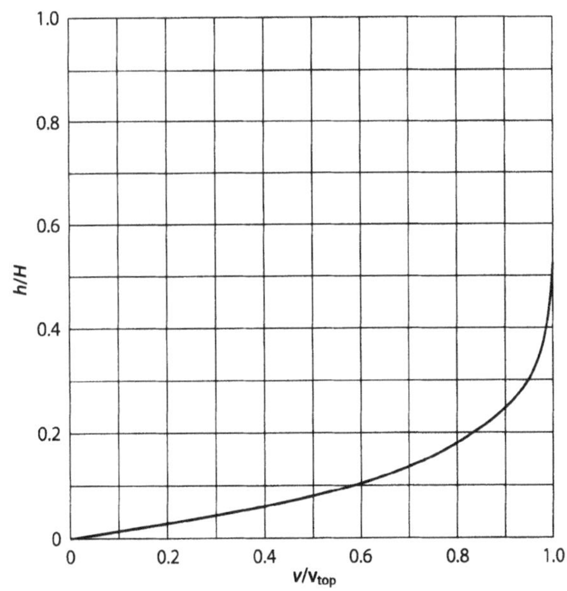

Fig. 5.21. Acoustic fluidisation, standardised representation of equilibrium velocity v (for definition s. Fig. 5.19) against distance from ground h (referred to respective values v_{top} and H on top of mass), assuming, in accordance with Melosh (1983, 162), that v is a linear function of the eighth power of h. Figure shows that near-to-ground portions of mass are definitely overtaken up to at least $h/H = 0.2$. After a certain distance, distortion of the sequential order necessarily must occur (sketch by Erismann)

The *conclusions* resulting from these arguments can be formulated as follows.

11. Acoustic fluidisation, as being physically plausible, is a mechanism not entirely excluded in rockslides.
12. Insufficient knowledge of difficult-to-determine parameters makes impossible, at the time being, a definite judgement.
13. As the prevailing mechanism in long run-out, acoustic fluidisation is excluded by the incompatibility of the laminar (though non-linear) character of the resulting flow with the frequent geomorphological occurrence of preserved sequential order of components

*

Fluidization apparently is a particularly fascinating field able to motivate the creative imagination of researchers. If the proposed solutions do not in every respect comply with the visions of their authors, this is mostly due to the initial situation from which the approach to a problem was started. Such is, more or less, the fate of all work done under the rule of trial and error.

5.7 Various Mechanisms

By no means should the title of the present section be misunderstood as meaning something like "miscellaneous trifles". On the contrary, the mechanisms discussed hereafter not only represent a substantial scientific interest: in certain conditions some of them also may play a decisive part in questions of life and death. And their being integrated to form a section is neither due to insignificance nor to a common physical basis but to two simple facts: each presentation of a mechanism, if included in the appropriate description of a key event, would appear too long; to fill a separate section, however, it would be too short. So, as a compromise, the formation of a somewhat multicoloured section was decided.

*

The systematic use of curves in an essentially horizontal plane (hereafter denominated as "*horizontal curves*") as "built-in speedometers" has been presented in the context of Pandemonium Creek (Sect. 2.2), and certain difficulties of the method (due to the dynamic behaviour of the mass) were alluded to in the case of Huascarán (Sect. 2.7). In addition, examples of sharp curves as sources of energy dissipation were mentioned in the descriptions of Pandemonium Creek (Sect. 2.2) and Val Pola (Sect. 2.5). In the literature, besides McSaveney's (1978, 227) and Körner's (1983, 96–97) remarks on velocity measurement as well as the comments by Nicoletti and Sorriso-Valvo (1991) demonstrating links between geometry and energy dissipation, little systematic analytical work is available. So it may be worthwhile to recall in short some – admittedly almost trivial – fundamentals. This is done with a side glance at Coulomb's hypothesis which, in spite of not being the unique braking mechanism, at the end of the present chapter appears as particularly important.

The *centripetal acceleration* in a curve normally is expressed by

$$\frac{v^2}{R} = a = R\Omega^2 \tag{5.32}$$

be it as a function of velocity v and curve radius R or as one of R and angular velocity $v/R = \Omega$. This equation can readily be applied for the centre of gravity of a coherent mass and for disintegrated masses in which R is sufficiently large as compared with the lateral extension of the mass (so that errors due to the use of an averaged radius may be tolerated). As soon as a narrower radius has to be considered, various uncertainties become unavoidable, especially in prediction.

First of all, it is not clear a priori how far, within the mass, a transfer of momentum takes place in longitudinal and/or transversal direction. The respective amounts do not entirely depend on the geometry of the track and velocity: also, possible *anisotropies* in the cohesive capacity of the (mainly disintegrated) mass may play an important role, in particular if disintegration at the considered point is only partial. Assume, for instance, two such semi-disintegrated masses, both essentially consisting of rather large clasts, easily passing through curves, but confined to move with their longest axis in the direction of displacement. Now let these clasts in case a (Fig. 5.22) be wide enough to be forced into a motion one be one, in case b let them be slender

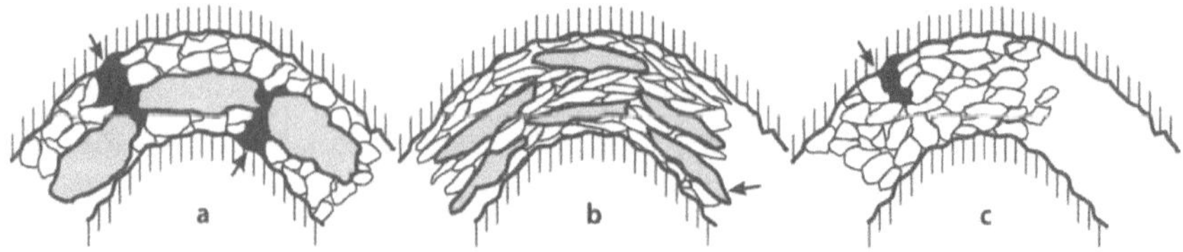

Fig. 5.22. Mass in horizontal curve (motion from left to right), various configurations of determining features. **a** Dominance of large blocks with high lateral friction forcing surrounding debris into motion with equal angular velocity and thus opening cracks at begin and end of curve (stained areae marked by *arrows*). **b** Dominance of slender, slippery blocks allowing debris to maintain velocity so that inside overtakes outside (*arrow*). **c** Mix of **a** and **b** under more current conditions with outside compressed to block by centrifugal force, thus opening crack at beginning of curve (stained area marked by *arrow*); inside able to partly maintain velocity, thus overtaking outside and, at distal end, partly changing side (sketch by Erismann)

so that several such "needles" or "columns" can travel side by side. In both cases let the large clasts be surrounded by smaller particles.

When mass a enters a curve, the large clasts force the smaller particles in between (and, by interlocking and frictional effects, most of the finer-grained material) to take part in their merry-go-round motion. This means that the resulting *angular velocity Ω is equal* for the major part of the mass. Thus centripetal acceleration is best expressed by the right side of Eq. 5.32, and it becomes immediately evident that it is proportional to the local radius. So the surface of the disintegrated parts of the mass becomes part of an – obviously concave – rotary second-order paraboloid. In such instances the assumption of a straight cross-section (as normally assumed) is nothing but a first approximation in which the error can be determined, at least in well-defined cases.

In case b things may be totally different, provided that relative frictional losses between the slender clasts are low (despite their lateral compression by centrifugal forces). To make things clear, negligible friction is assumed as a consciously exaggerated working hypothesis. Under such conditions, the large clasts – and with them practically the entire mass – *will maintain their velocity*, and the expression on the left side of Eq. 5.26 becomes appropriate. Contrarily to case a, equal v means a centripetal acceleration inversely proportional to the local radius, and instead of concave the surface of the mass, in assuming a logarithmic cross-section, becomes convex. In addition, inside clasts overtake outside clasts.

Fortunately, tentative calculations of realistic examples did not yield dramatic speedometric discrepancies between the two cases. In addition, it is not excluded that in a mass elements of *both cases may act jointly* in a mixed form: In the outside portions of a disintegrated mass centrifugal forces, in compressing the particles laterally, increase the tendency to behave like a coherent block, thus favouring case a. The innermost particles, on the contrary, as subjected to a far lower centrifugal compression, have a better chance to maintain their pre-curve velocity, in particular if a low thickness of the mass excludes high vertical compression. In such instances case b is not excluded: the inner portion overtakes the rest of a mass and, once having done so, changes to the outside of the curve (Fig. 5.22c). In a meandering valley such changes of position may occur repeatedly in both directions, thus creating a somewhat jumbled accumulation of clasts at the very distal end of the mass.

All in all, the speedometric aspect of concave and convex surfaces in curves turns out to be less disconcerting than might have been suspected at first sight. The considerations required to clear the situation, however, have given the possibility to have a cursory glance at the fascinating *complexity of a disintegrated mass* in motion, even if looked at in a massively simplifying manner.

*

There is, unfortunately, another aspect which makes questionable the speedometric approach to horizontal curves, and this aspect is far more problematic than the above-discussed one. As announced under Heading 2.7, it has to do with the dynamic behaviour of an essentially disintegrated mass, able to move laterally, in principle like a *giant pendulum*.

The core of the problem consists in the fact that a pendulum, when subjected to a force deviating its position of equilibrium, does not immediately assume the respective new position. According to the determining parameters (in first instance amount and duration of the exciting force, natural frequency, and damping effects) the pendulum may give *dramatically different answers*. And this is true for one single cycle as well as for a sequence of many oscillations. This fact is of particular importance in case of a mass moving through a meandering valley: seldom a succession of curves, when correlated with the velocity of the mass, may be expected to occur in an array entailing a constant frequency of excitation.

Figure 5.23 shows impressive examples of what can happen to an undamped pendulum excited by nothing but a succession of *not more than two short force pulses* (functions represented as rectangles for the sake of simplicity). From a, b, c it can be observed that the amplitude of the answer is less than proportional to the exciting momentum (expressed, in a plot against time, by the area of a rectangle). In parentheses it may be remarked that such proportionality is obtained for a so-called ballistic pendulum (in which the duration of excitation lies far below that of a period in natural oscillation). Example c, besides the mentioned connection with a and b, also can be seen as the start of a new series c, d, etc. in which two equal exciting pulses are separated by an increasing space of time. c in this view is the zero-space starting point

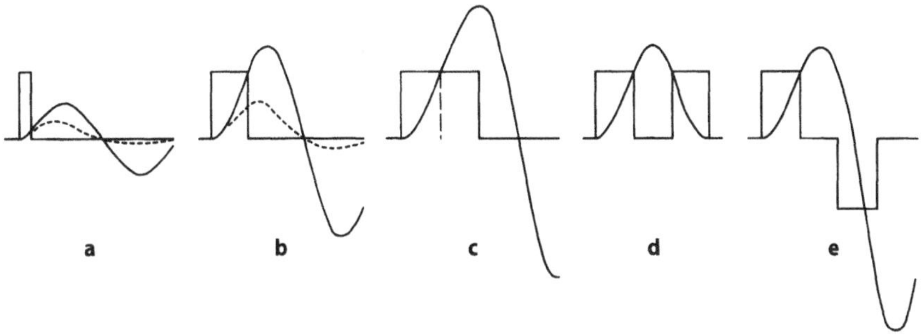

Fig. 5.23. Answers of undamped pendulum (sine waves, for **a, b** also strongly damped, s. *dotted curves*) to various short excitations (rectangular functions). **a, b, c** Increase of answering amplitude grows with duration of short excitation, but less than in proportion thereto. **c, d** Answers to two equal signals separated by growing time lag show that from boosting to annihilation the difference in exciting functions is minute. **d, e** Inversion of annihilating function means resonance (sketch by Erismann)

(s. broken line in c). The transition from c to d is particularly impressive: a second identical exciting signal, following the first after a well-defined delay, not only can fail to increase the effect of its forerunner, it even can completely annihilate it (a phenomenon technically used in certain noise abatement systems in which a servo-controlled loudspeaker "kills" the acoustic emissions of a source of noise). Finally, example e shows a typical case of resonance in which a negative signal shifted by half a period doubles the effect of the first signal. A positive one shifted by a complete period and thus continuing c, d, etc. would do as well.

It is well understood that in reality the differences usually are less dramatic than those shown in Fig. 5.23. On the one hand, in the figure the most striking (though certainly not impossible) cases are picked out; on the other hand, undamped oscillations are presented while in reality a certain *damping effect* must be expected. To give an idea of the attenuation entailed by strong damping, in Fig. 5.23 (a, b) dotted curves show arbitrarily chosen effectively damped oscillations (attenuation by a factor 1:16 in a period).

Provided that friction is mainly Coulombian (an assumption certainly justified as a first approximation), the trivial relation

$$\mu_y = \mu \sin \operatorname{atan} \frac{v_y}{v_x} \tag{5.33}$$

expresses the apparent coefficient of friction μ_y in transversal direction as a function of the actual coefficient of friction μ, the transversal (i.e. pendulous) velocity v_y of the mass, and its longitudinal velocity v_x. The trigonometric expression, perhaps somewhat disconcerting at first sight, serves to extract the transversal resistance from the oblique motion occurring against the resistance proportional to μ (Fig. 5.24a). It will be observed, by the way, that μ_y, contrarily to μ, is not constant but a function of v_y. Nevertheless a simple, easy-to-apply method for the *quantitative determination of damping* is given.

Use of this easiness was, for instance, made in the calculations mentioned under Heading 2.7 in connection with the motion of the *Huascarán* mass in and after Shacsha Narrows. The method might as well have been applied to Pandemonium Creek; tentative calculations have, however, shown that corrections in the particular case would have been comparatively small. The basic differential equation used for Shacsha Narrows was

$$\frac{dv_y}{dt} = \frac{v^2}{R} + \left(\frac{dz}{dx} - \mu\right) g \sin \operatorname{atan} \frac{v_y}{v_x} \tag{5.34}$$

where t is time, x the longitudinal, and z the vertical displacement. The terms on the right side of the equation require some comments. As R is the local radius of the track, the first term expresses the effect of centripetal acceleration. The second term, proportional to the local (descending) slope dz/dx, is the driving agent within the system, and the third, proportional to μ, is its braking counterpart. The proportionality of these two terms to gravitational acceleration g needs no explanation, and the trigonometric factor already has been commented here-above. Attention: as dz/dx and μ are used without trigonometric corrections, these terms are valid in the presented form only for not too steep a descent.

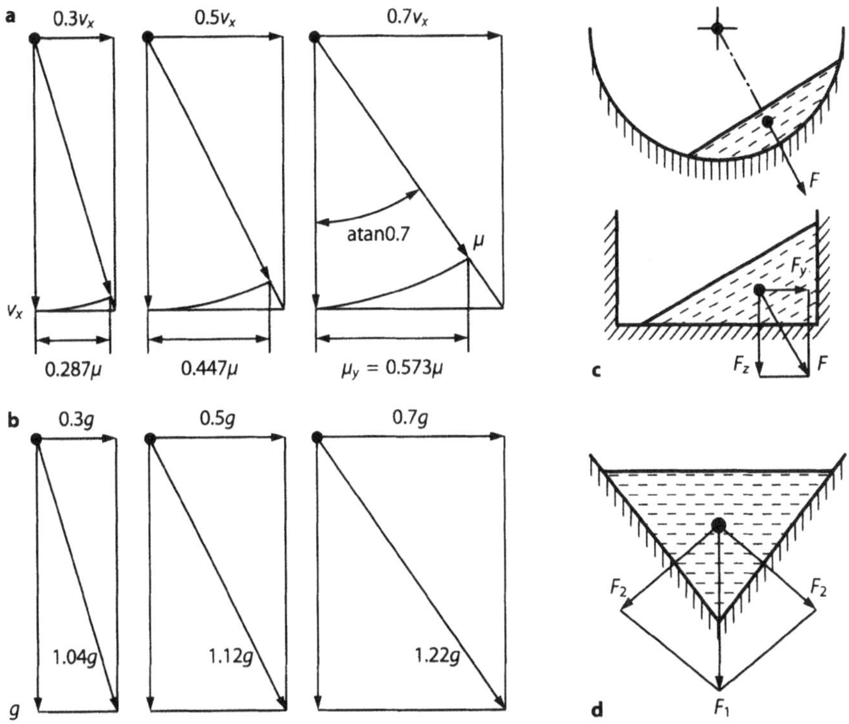

Fig. 5.24. Trigonometric effects in curved and straight motion under Coulombian friction. **a** Resistance against transversal component v_y of velocity, expressed as fraction of total coefficient of friction μ (longitudinal component is v_x); friction is opposed to resulting velocity; attention: for small quotients v_y / v_x trigonometric expression $\sin(\mathrm{atan}(v_y / v_x))$ can be replaced by v_y / v_x. **b** Horizontal centripetal acceleration (expressed on top as fraction of coefficient μ) increases total acceleration (s. *diagonal arrows*) and accordingly frictional force. **c** Top: in track of circular cross-section, most of force F resulting from weight and centrifugal force acts almost at right angles to ground so that seldom corrections are required. **c** Bottom: in track of rectangular cross-section, each component (gravitational and centrifugal) of force F has to be accounted for at full scale. **d** Wedging effect in straight track; weight F is split into components F_1 and F_2, thus increasing friction (sketch by Erismann)

In the present context the reader's attention should be drawn to *four essential points* which, in spite of being implicitly comprised in the above equations, might be overlooked. (1) The expression $\sin[\mathrm{atan}(v_y / v_x)]$ is required only if v_y / v_x is substantial (say, at least 0.3, s. Fig. 5.24a); for small values it can be replaced by v_y / v_x, thus making the equation linear in v_y. In any case, (2) the coefficient μ_y is smaller than μ (for low v_y / v_x even much smaller) so that damping may be lower than expected at first sight. All in all, (3) damping is nearly proportional to the lateral velocity in spite of friction being Coulombian. The proper introduction of the topographic parameters (4) requires a step-by-step input to Eq. 5.34.

The application in the case of Huascarán confirmed the presumption that, to be reliable, speedometry based on superelevations requires a careful *analysis of resonance effects* to avoid unexpected (negative as well as positive) errors.

Once the question of trigonometric effects having been raised, another aspect of horizontal curves should not be omitted. The total *force transmitted to the ground*, by virtue of including a centrifugal component, is increased. For strictly horizontal direction of said component the increase is shown in Fig. 5.24b. It is negligible for centripetal accelerations small in comparison with that of gravitation. In such cases also the resulting increase of lateral frictional resistance (if Coulombian or similar) can be neglected.

If the *increase of total acceleration* is large enough to be accounted for, the geometry of force transmission to the ground has to be considered. For example, in a valley of circular cross-section, adding a centrifugal force to gravitation does not change the geometric conditions of force transmission (Fig. 5.24c, top). If, however, said section is essentially rectangular, the centrifugal force acts at right angles on one side-wall and has to be accounted for at full scale (Fig, 5.24c, bottom).

This phenomenon is obviously nothing but a special case of a rule valid for all kinds of tracks, no matter whether curved or not. In fact, any *wedge-like cross-section* increases the forces acting onto the ground and therewith also Coulombian or similar friction. Figure 5.25 shows an impressive demonstration in a simple experiment. The highest amount of force increase is obtained if the force (which may be single or resulting from more than one component) lies in the plane of symmetry between the

Fig. 5.25. Experiment showing wedging effect upon single pebble in inclined plastic box. **a** Pebble on plane surface. **b** Pebble in 90° inside edge between two planes. For obvious reasons slopes look steeper than they are; ultimate angles before release were 20.2 and 28.6°. Tangents thereof yield quotient 1 : 1.48, corresponding fairly well to theoretical figure 1 : 1.41 if primitive set-up of experiment is considered (photos by Erismann)

two involved walls and cuts their line of intersection at right angles, as for instance the weight F_1 of a mass between two symmetrical slopes shown in Fig. 5.24d. Then force amplification obeys the rule

$$2F_2 = \frac{F_1}{\cos \beta} \tag{5.35}$$

where F_2 is one of the two output forces and β the lateral slope angle.

If acting under perfectly favourable conditions, this *wedging mechanism* can compress a mass (both in coherent and in disintegrated state) with such a vigour that motion either cannot start or, if already started, can be "strangled" to a full stop (it should be borne in mind that $\beta = 45°$ means additional 41%, and $\beta = 60°$ additional 100% friction and that the longitudinal motion through the wedge reduces transversal friction practically to nought). This is, at least theoretically, an effect antagonising that of channelling as discussed in Sect. 5.1. The postulated "perfectly favourable conditions", however, cannot be considered as a frequent case: most events, even if passing through a narrow track with steep lateral walls (s. the sections of Shacsha Narrows in Fig. 2.54), remain far from such dramatic conditions so that, as a rule, the influence of channelling is not effectively compensated. Yet in exceptional cases it may be worthwhile to look at wedging in more detail.

*

After this excursion to a more general field, the discussion of curved tracks should be closed by some short comments about "*vertical curves*" i.e. curves in an essentially vertical plane. For a simple reason these curves are far easier to quantify than their horizontal counterparts: the centripetal acceleration, calculated according to Eq. 5.32, usually can be superposed to gravitational acceleration without substantial corrections. Errors thus committed seldom are large enough to be taken into account in a calculation in which an acceptable accuracy in estimating the most important parameters (for instance the coefficient of friction) is the real problem. As a matter of fact, the highest accelerations in vertical curves often occur in points where displacement is approximately horizontal.

This is particularly true for transitions from descent to run-up, one of the most frequent phenomena in rockslides (s. especially Sect. 2.2, 2.4, 2.5). And in such instances centrifugal *forces may be very substantial.* Take the example of Val Pola (Sect. 2.5) where, according to the author's calculations, the centre of gravity of the mass entered a vertical curve of, roughly speaking, 600 m radius at a velocity of about 75 m s^{-1}. Thus Eq. 5.32 yields a centripetal acceleration in the range of 0.96 g so that the pressure upon the ground – and therewith the Coulombian friction – practically was doubled! But also without run-up, for instance in the transition from the steep wall of Huascarán to Glacier 511 (Sect. 2.7, 6.2), impressive accelerations are observed. The combination of $R \approx 500$ m and $v \approx 97$ m s^{-1} results, at the entrance of the transition, in about 1.92 g, twice as much as calculated for Val Pola! It is a curious fact that the occurrence of such excessive compressions apparently has not been commented as sensational in the literature (not to speak of a systematic consideration in velocity calculations…).

The effect of *force reduction* in case of the curve's centre being located below the track is, perhaps, even more dramatic. In fact, a reduction by approximately one half needs, at a quite frequent velocity of, say, 50 m s^{-1}, no less than a radius of 500 m. And Coulomb's rule may be correct or not, there can be no doubt about the fact that friction fades out when compression approaches zero. Nevertheless only the spectacular phenomenon of taking off is usually reported while considerations about reduced friction as a consequence of centrifugal force excel by rarity.

Obviously it will take some time until the *importance of vertical curves* will count like other established facts in the minds of those who deal with rockslide dynamics.

Perhaps a simple mathematical trick may help to overcome the reluctance against the *quantitative treatment of vertical curves.* Imagine a transition from a steeper to a softer section of a slope, with a short length L, a small radius R, and a moderate angular difference $\Delta\beta$. Now consider Eq. 5.32 and assume an extremely low radius. Then centripetal acceleration becomes large enough to allow neglect of g over the short time of transition. So the process of braking can be treated with (μa) as unique deceleration. The respective differential equation can be written in the form

$$\frac{dv}{v} \approx -\frac{\mu v\, dt}{R} = -\frac{\mu\, dL}{R} \tag{5.36}$$

where v is velocity and t time. Integration yields

$$v_1 = v_0\, e^{(-\mu\Delta\beta)} \tag{5.37}$$

where v_0 and v_1 are the velocities before and after transition, while $\Delta\beta = L/R$ is obvious.

The trick consists in the fact that in Eq. 5.37 *no influence of size* appears. In other words: the entire braking process can, by reducing R and L to infinitesimal minuteness (and without influencing the result), be contracted to the point of intersection P between the two sections of track (Fig. 5.26). There remains only the error due to the dissimilarity of lengths between the angular path $P_0 \longrightarrow P \longrightarrow P_1$ and the circular path $P_0 \longrightarrow P_1$. And this dissimilarity, expressed by $\text{arc}(\Delta\beta/2)/\tan(\Delta\beta/2) - 1$, refers to a small part of a track and is small for moderate angles (1.0% for 20°, 2.3% for 30°, 4.3% for 40°). Its influence is confined to forces of non-centrifugal nature.

It might be added that, if the ground's hardness is lower than that of the moving clasts, the *coefficient of friction* μ in Eq. 5.37 is distinctly higher than under less dramatic conditions: a considerable amount of external energy is dissipated by "ploughing" which assumes a pulsating character as a consequence of repeated internal collisions.

A practical application of the method will be presented in Sect. 6.2.

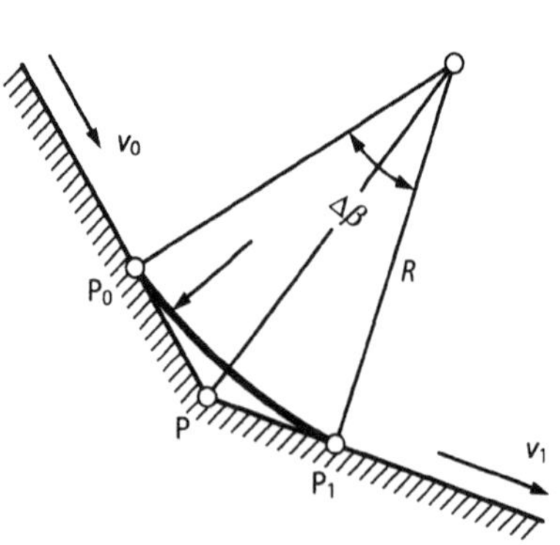

Fig. 5.26. Velocity reduction $v_0 \longrightarrow v_1$ by centrifugal effect upon Coulombian friction in transition from steep to smooth track section. As radius R is eliminated in Eq. 5.37, braking effect can be concentrated from curve $P_0 \longrightarrow P_1$ to point of intersection P. Remaining error is due entirely to non-centrifugal forces along the minute difference of length between curve $P_0 \longrightarrow P_1$ and angular path $P_0 \longrightarrow P \longrightarrow P_1$ (sketch by Erismann)

In the discussion of collisions between bodies the scope of Sect. 4.2 forbade the mention of certain implications which may play an important role in the course of a rockslide or a rockfall. In first instance such implications are found in the "no-man's land" between the core of an event and its environment. It is not useless to consider more in detail these *marginal regions*, not directly threatened by total destruction but dangerous enough owing to the sporadic presence of clasts, be they isolated or in small clusters. As will be shown hereafter, the tendency to spread out in this manner is inherent to any disintegrated mass.

In the comments to Fig. 4.5 it was omitted on purpose to draw the reader's attention to the further development after the last collision of the series. Now it will be observed that, by the end of the 551 milliseconds considered, the most proximal particle is the slowest of all. So, if the serial collisions will remain the only acting mechanism, this particle will for ever and ever remain behind, continuously increasing its distance from the others and thus no longer taking part in the process of equalising velocities described under Heading 4.2. And what will happen at the distal end of the group? The velocity of the remaining particles increases from the leading to the trailing end. So, irrespective of previous internal collisions, sooner or later (during the two following seconds) the most distal particle will be hit from behind and accelerated to a higher velocity than the rest. Therewith its fate will be sealed in perfect symmetry to that of the most proximal: it will for ever and ever hasten away, beyond the reach of the others. And the process thus described will not be terminated: the second particles at the distal and proximal end will now undergo the same kind of expulsion, though, owing to the reduced internal kinetic energy (Sect. 4.2), in a milder form. The generalised conclusion to be drawn from this simple mental experiment is obvious: collisions between the particles of a disintegrated mass *equalise velocities at the expense of spreading*, beginning by the outermost particles.

Now in reality this process is far more complicated, though in principle analogous. Above all, the tendency to spread, in spite of being ubiquitous both in longitudinal and transversal direction, may be counteracted, mainly by lateral confinement. Even in vertical (or quasi-vertical) sense such a tendency does exist, though entirely annihilated by gravitation which forces back expulsed clasts even without perceptible loss of their contribution to internal kinetic energy (this loss takes place in the collisions after falling back). Anyhow, *longitudinal "kicking"* yields the most spectacular, and probably also the most dangerous mechanisms of this kind.

Once again the reader's attention has to be drawn to Sect. 4.2 where Eq. 4.6, 4.7, 4.8 express the basic rules of semi-elastic collisions. Now assume that a relatively small particle m_1 is hit by a larger one, m_2. Assume further that m_1 in the moment of collision either is at rest (e.g. as belonging to the debris of an earlier event) or moves from the opposite side to meet m_2 (e.g. falling back after having spent its kinetic energy in a run-up). Let the masses be linked by $m_1 = c_m m_2$ and the respective pre-collision velocities by $u_1 = -c_u u_2$ (where c_m and c_u are dimensionless factors). Finally assume that no energetic relevance must be attributed to secondary mechanisms (e.g. rotation of the involved masses). In such instances the *post-collision velocity* v_1 of m_1, by adjustment of Eq. 4.8, can be expressed in the dimensionless form

$$\frac{v_1}{u_2} = \frac{1 - c_m c_u + J + J c_u}{1 + c_m} \tag{5.38}$$

Some of the results obtained by variation of the parameters determining this equation perhaps will be regarded as surprising.

As m_1 was assumed to be the smaller of the involved particles, the range $0 < c_m < 1$ is considered. For c_u the falling-back scenario after a run-up hardly will yield more than $u_2 / 2$ so that in principle $0 < c_u < 0.5$ would suffice; but for the sake of transparency (s. hereafter) the more complete range $0 < c_u < 1$ is adopted. For J a single value is taken into account, namely 0.7 as already used for Fig. 4.5. The *quantitative results* within said ranges are shown in Fig. 5.27.

Two *particular features* are worth some special comments. On the one hand, a small clast, when kicked forth by a larger one, can reach a velocity substantially higher than that of the last-mentioned. If starting at rest, "amplification factors" between 1.5 and 1.7 are possible. If boosted by an initial reverse motion of the small particle, factors may exceed 2.0. On the other hand, if c_m equals J, the terms containing c_u on the right side of Eq. 5.38 cancel each other out, and the remaining terms yield unity. In other words: $c_m = J$ purely and simply means $v_1 = u_2$, no matter what amounts the involved parameters may assume apart from this equality (s. horizontal line on level 1.0 in Fig. 5.27).

While this second feature is little more than a mathematically interesting physical coincidence, the first has the character of a *serious threat*. As mentioned here-above, reach increases, roughly speaking, proportionally to the square of velocity. And this is valid both in horizontal and in ascending motion. A certain moderating effect lies in the fact that the mechanism is confined to the smaller particles of an event which,

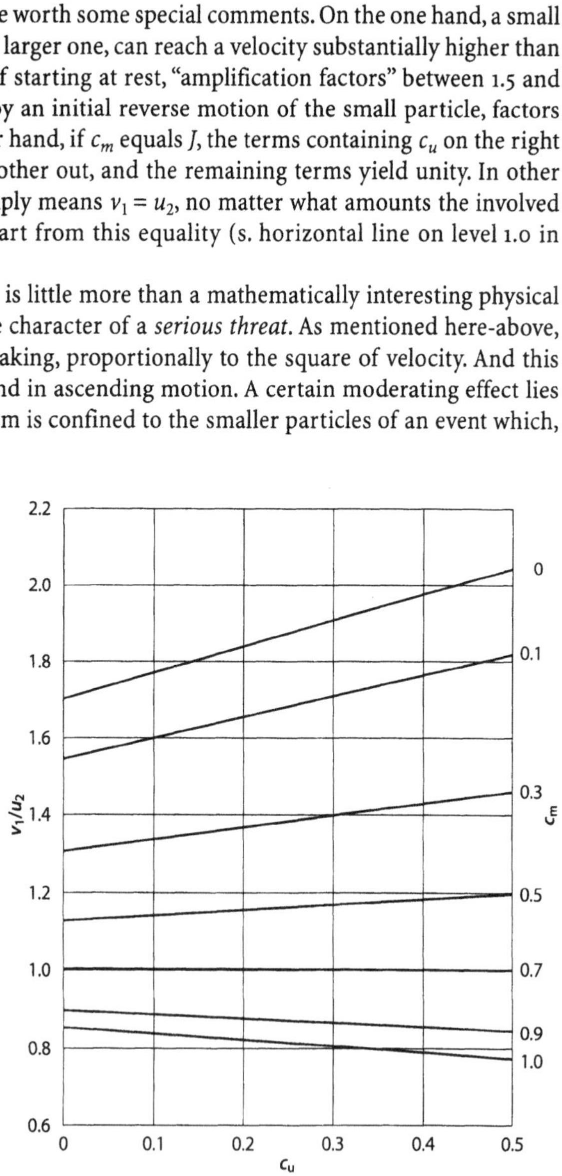

Fig. 5.27. Kicking of small mass m_1 by larger m_2 in standardised form. Parameters: $c_m = m_1 / m_2$; $c_u = -u_1 / u_2$: quotient of pre-collisional velocities, m_1 moving backward, m_2 forward; v_1/u_2: quotient of velocity amplification, post-collisional velocity v_1 of m_1 referred to pre-collisional velocity u_2 of m_2. Dangerous range is concentrated on relatively small c_m and high c_u. Mark horizontal line for $c_m = 0.7$ which is equal to the assumed velocity restitution coefficient of the material (sketch by Erismann)

being more or less solitary, are arrested faster than if moving within a large mass. Nevertheless, a clast of one cubic metre, kicked by one ten times more voluminous and speeding at nearly twice the velocity of the larger mass, is a formidable battering ram. And, although not any spattered zone necessarily must be due to kicking, there certainly exist many cases in which the probability of this mechanism is high, especially if in the tongue of the main debris very large boulders are found. An example will illustrate the situation.

At first sight the blocks found by Eisbacher (1977, 238–240) on a shelf about 600 m above the foot of the valley in the event of *Avalanche Lake* might give the idea of having been accelerated to the estimated velocity of more than 100 m s^{-1} by a mechanism of kicking. An argument in favour of such a hypothesis would lie in the low Fahrböschung (overall slope) of $\tan 8° = 0.140$. On the other hand, considering a remarkably careful study by Evans et al. (1994), a rather continuous pushing by the following core of the mass appears more probable than the brutal mechanism of kicking: despite the small portion of only 2.5% of the total mass (which was 0.2 km^3) having reached the shelf, number and size of the clasts on top of it fit badly into a scenario of having been kicked. There are, however, few blocks of the mass which were found almost another 90 m higher than the rest of the shelf lobe (pp. 755, 762). They are assumed to have arrived at their present position by large bounces. It cannot be excluded that this singularity (that reduces the Fahrböschung below $\tan 6.5° = 0.132$) is due to being kicked in the ascent to the shelf where small-scale fallbacks would not be improbable. The required velocity at the crest of the shelf, something around 115 m s^{-1} according to a tentative calculation using Eq. 5.11, is far from prohibitive in the case of a smaller clast falling back and being kicked by an ascending larger one.

Anyhow, kicking, both in distal and lateral direction (as well as oblique in certain cases), in spite of not representing the main danger of a rockslide or a rockfall, should not be considered as a comparatively harmless fallout. It is a *dangerous mechanism* of its own right and merits careful investigation, especially in connection with prediction.

*

In Sect. 2.2 (refer especially to Fig. 2.10) the *transport capacity of a rapid water jet*, pressed out by the weight of a rockslide moving over water-saturated gravel of a valley fill, was demonstrated. Here this capacity as well as its inherent sorting power will be discussed quantitatively using a simplified model.

In principle the *differential equation* of an initially immobile single spherical particle of given density and given drag coefficient is considered in a vertical jet of water moving at given velocity under turbulent conditions. The minimal parameters adopted for velocity and particle size (20 m s^{-1} and 0.1 m respectively) yield Reynolds numbers far within the turbulent range. And the velocities considered are by no means exaggerated as 20 m s^{-1} correspond to a rockslide thickness of not even 9 m, and 60 m s^{-1} to one of 80 m (not to speak of possible additional dynamic effects in the moment of hitting the valley floor). The other simplifications (shape of particles and densities of both liquid and particles) entail a certain scatter and also may shift the results systematically by several percent (plus or minus). Yet they cannot seriously influence the general trends displayed by the results. A certain blurring due to unavoidable collisions between particles has, in any case, to be expected.

Assuming a drag coefficient $c = 0.5$, velocities w and v for water and particles respectively, and a water-to-rock quotient of densities $\partial_w / \partial_r = 0.4$, the acceleration

$$a = \frac{3}{4} c \frac{\partial_w}{\partial_r} \frac{(w - v)^2}{D} - g \frac{\partial_r - \partial_w}{\partial_r} \tag{5.39}$$

numerically is reduced to $a = 0.15(w - v)^2 / D - 5.886$. As usual, g is gravitational acceleration. The twofold *integration of this equation* over time yields the gain in altitude z that a particle accomplishes in time t (Fig. 5.28). Water velocities and particle diameters serve as parameters.

Although Fig. 5.28 is almost self-explanatory, a short *comment* may be useful. A definite, tough not dramatic tendency to sort the particles coarse on bottom is clearly perceptible. This tendency is certainly somewhat reduced by the above-mentioned blurring effect of collisions. Sorting intensity increases with increasing water velocity. This is particularly well perceptible when including into the comparison the sinking travels shown at larger scale at the bottom of the figure. A remarkable fact, to make an end: a water jet moving at no more than 20 m s^{-1} is able to hold in suspense a boulder of more than 6 m in diameter!

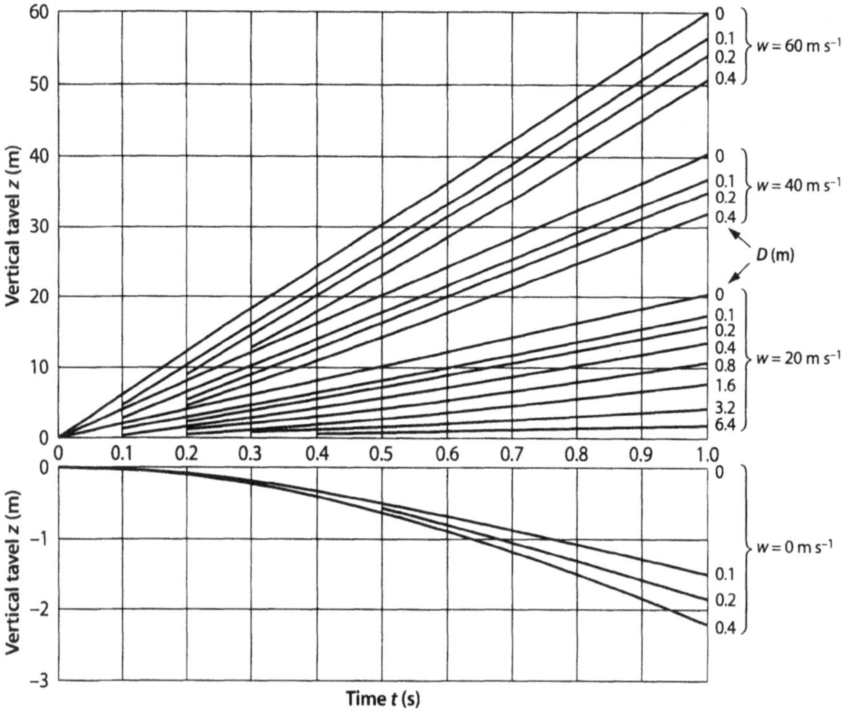

Fig. 5.28. Vertical transport of approximately spherical particles by water jet of given velocity. D: diameter; t: time; w: velocity of jet; z: vertical travel. Attention: scales used for top and bottom portions of figure are different. Mark almost lag-free motion at $D = 0.1$ m and levitation at $D > 6$ m. Refer to Fig. 2.10 (sketch by Erismann)

Chapter 6

From Analysis to Prediction

In examining things present, we have data from which to reason with regard to what has been; and, from what has actually been, we have data for concluding with regard to that which is to happen hereafter.

James Hutton

6.1 Some Fundamentals

So far, this book essentially bears an analytical character. It deals with mechanisms which, when considered individually, readily can be used as a more or less quantitative raw material for analytical statements. Their use in forecasting, however, besides the necessity to get at the working parameters, depends on the possibility of inserting them into a *coherent line of thought* intended to synthesise an event as a whole. Such a logical backbone is required from the moment when a mass is released somewhere high in the mountains to its final standstill (and possibly that of other masses mobilised by it) in the depth of a valley or a plain.

Certainly the most appropriate natural link in this context is *velocity*, by definition the omnipresent parameter of motion which, in the course of an event, can drop to zero only in case of a momentary singularity (e.g. in starting to fall back after having spent the entire kinetic energy in running up). Anyhow, the importance of velocity as a central parameter of mass displacement is such that the consequences of an event practically can be estimated if velocity at any point of a track is sufficiently well known.

And as, in the complex multitude of phenomena entailed by a mass of rocks rushing downhill, a calculation of velocity a priori cannot be envisaged owing to lacking knowledge of parameters, the necessary empirical basis for prediction consists in establishing start-to-stop velocity schedules of pre-existing events and in deducing therefrom – as explicitly as possible – the required parameters. In other words: *no prediction without previous analysis* is possible. As a consequence, Sect. 6.2 entirely will be devoted to the processes used to determine velocity.

The respective situations of *experts in analysis and in prediction* are markedly different. Analytical work per se is not confined in time, libraries and data banks are available on a world-wide scale, and high performance computers can be used to solve complex numerical work in reasonably short time. Contrarily thereto, an expert facing the possibility of an approaching catastrophe is, in addition to the stress of being responsible for hundreds or thousands of human lives, in a dramatically worse position. The (1) available time is restricted and normally uncertain; work may occur (2) in rough terrain, far from other means of communication than those carried along. So the indispensable special outfit may be described as a (3) more or less reliable map, and a (4) portable calculator or computer allowing to treat problems quantitatively. Of course, the expert's situation is greatly improved if he has at his disposal (5) the equipment required for on-line, bi-directional information exchange with a well-equipped scientific institution. It should, perhaps, be noticed in this context that all calculations required in setting up this book were, in spite of the immediate presence of more powerful equipment, carried out on a simple 350-byte programmable calculator.

The most severe restrictions certainly are found in Item 1 where the lack of time is mentioned in combination with an almost unbearable stress, further increased by the uncertainty about the actually available space of time. Hence highly sophisticated and differentiated algorithms – as might be expected in connection with certain sections of Chap. 5 – normally must be replaced by relatively *simple, easy-to-apply methods* in which a complex succession of physical mechanisms is more or less accurately reduced to a black box model containing only few key parameters (Sect. 6.3; Ashby 1961). Essentially, the necessity to simplify is not a problem of calculation; it is far more a question of how to acquire, in due time, the parameters needed for prediction.

So it will be one of the main scopes of the present chapter to show that, despite the partly Spartan postulates formulated here-above, the results obtained in a rapidly prepared process of calculation – after careful inspection of the site and using comparatively primitive equipment – must by no means be less reliable than those known from the literature, no matter how generously the latter had been endowed in terms of working time and mathematical refinement. The nostrum that will pave the way for such an issue has no mysterious features, it is, on the contrary, most trivial as it simply consists in spotting and *eliminating the main sources of errors* and in selecting methods which minimise the total amount of time to get reasonable accuracy. In the review of errors required for the purpose it will, by the way, be found that at least one of the major errors is "unforced" (in the sense of tennis jargon) inasmuch as it is not an integrating constituent of a method, but the result of erroneous application. It also will emerge that an increased degree of sophistication does not automatically warrant improved accuracy.

Of course, a long discussion might take place about the somewhat sibylline term of "reasonable accuracy", especially with respect to velocity. Such a discussion would, however, bear a rather theoretical character as it is needless to say that the requirements in accuracy cannot be higher than dictated by the knowledge of determining parameters and the simplifications subsisting even in case of a most sophisticated method. So, while in analysis the best possible accuracy unconditionally has to be aimed at, in prediction the *philosophy of risk management* has to be applied, for instance by calculating the most probable and the worst possible cases and by taking further decisions after due consideration of both.

Notwithstanding the circumstances of a particular event, the expert has to bear in mind the fact that he risks nothing but his personal reputation of infallibility (or, at the worst, the sympathy of an interested party, as might have occurred in the case of Vaiont, s. Sect. 2.6) if he predicts an exaggerated reach, but that *human lives are at stake* in the opposite case.

6.2
How to Determine Velocity?

In the primeval times of quantitative work on rockslides and rockfalls the estimation of velocity post eventum essentially was based on the determination of the *time elapsing between two positions* of a mass. In the famous Elm slide (1881), for instance, a boy had had the presence of mind to observe both the beginning and the end of motion although in the meantime he was running for his life. From the distance he had covered the time of motion could approximately be deduced. So Heim (1932, 93) obtained an average of about 50 m s^{-1}. A similar situation is critically commented under Heading 2.7 for the Huascarán event.

In Heim's considerations about Elm some inconsistencies cannot be dissimulated. A maximal velocity exceeding 100 m s^{-1}, more than twice as high as the estimated average, fits badly with the 83.5 m s^{-1} calculated later (p. 150) and with the idea of an almost immediate transition from full speed to dead stop where "*... nobody... could observe a slowing down of motion...*" (p. 93, translation by Erismann). These details are mentioned as forerunners of a general trend that developed later. Otherwise they would have passed uncommented, eclipsed by Heim's fundamental contribution to the quantitative treatment of rockslides. As a matter of fact, in spite of not explicitly aiming at prediction, he was aware of the high value inherent in an integral *start-to-stop knowledge of velocity*. So more than half a century after the event of Elm, at the age of 83 and somewhat uncertain of his recollection in physical matters, he addressed himself to his friend Müller-Bernet, a physicist. The respective section of Heim's book (1932, 143–152), after the beautiful sentences used as a motto for the first chapter of the present work, reports the wording of Müller-Bernet's letter of August 17, 1932. This letter was no less than the first strictly scientific approach to velocity in the investigation of downhill mass displacement.

The method has been described more than once in the literature. So hereafter the shortest possible form will be presented. Consider a vertical section along the path of the centre of gravity (CoG) of a mass in its displacement from release to stop (Fig. 6.1). Draw a straight line E (the "*energy line*") from the CoG of the ready-to-move mass to that of the mass in its final position of rest. Obviously the slope $\tan\beta_e$ of the energy line represents the average of the slope on which the CoG has moved. Now assume a constant Coulombian coefficient of friction $\mu_e = \tan\beta_e$. This simply means that, if the CoG of the mass were moving along the energy line instead of its real path, there would be a permanent, velocity-independent equilibrium between driving acceleration and frictional deceleration. In other words, once motion somehow having been initiated, the mass would gently accomplish its travel at constant velocity, entirely transforming the driving energy of descent into heat at the surfaces subjected to friction. In reality the CoG passes at a vertical distance Δz below the energy line. And the potential energy thus released can be converted into no other form than kinetic. So, according to Eq. 2.1, at each point of the track a velocity $v = \sqrt{(2g\Delta z)}$ results if g is gravitational acceleration.

Some *comments* may, despite their partly almost trivial character, be useful to avoid misunderstandings.

1. The vector of velocity assumes, of course, the direction given by the motion of the CoG at the considered moment. In other words, this velocity (except for horizontal motion) is not identical with its horizontal component.

2. It would be an error to believe (as might be concluded from certain publications) that the velocity determined according to the energy line method is valid only for a mass in coherent motion. In reality a physically conclusive claim to validity can be laid for this method as long as one condition is fulfilled: that the summed up energy-dissipating effects acting upon the elements of the mass result in a constant quotient μ_e of total frictional resistance over total gravitational force exerted at right angles upon the ground. In such instances the assumption of a spreading process between start and stop, as shown in Fig. 6.1, represents no violation of physical consistence.
3. In view of spreading and other mechanisms entailing an elongation of a disintegrated mass, it is obvious that its distal end may move slightly faster, the proximal end slightly slower than the CoG which represents the mass as a whole.
4. The last sentence needs further specification as it is reasonable to apply the method to the longitudinal section through the CoG. Lateral portions may move somewhat slower, especially if size (in principle thickness) effects are expected. In such cases a certain correction would be required to establish the velocity of the CoG of the entire mass. However, to account for the most dangerous (i.e. central) parts of an event, such corrections will be abstained from. In other words: the CoG of the central longitudinal section will be treated as if it were the CoG of the entire mass.

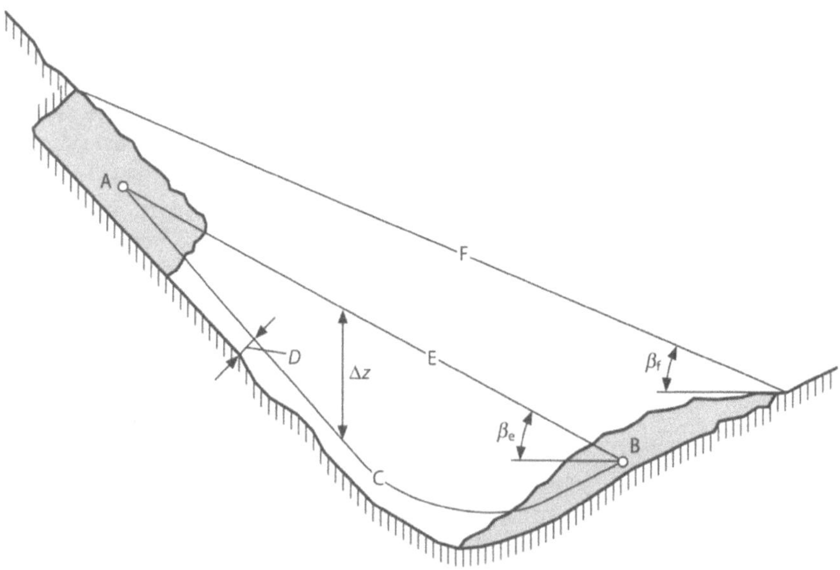

Fig. 6.1. Velocity measurement by energy line method. *A*, *B:* Centre of gravity before and after displacement; *C:* path of centre of gravity (note distance *D* from ground, generally reduced due to fading thickness of disintegrated mass, but locally increased in vertical curvature); *E:* energy line; Δz: elevation of energy line above path *C*, proportional to square of local velocity for constant coefficient of friction μ; *F:* overall slope (Fahrböschung), approximately parallel to and often misused as energy line (result: exaggerated velocity); β_e: slope angle of energy line; β_f: slope angle of Fahrböschung (sketch by Erismann)

The *weak point of the method* lies in one of its strong points, the simplicity. When looking at the various effects presented in Chap. 5, the assumption of a constant coefficient of friction appears questionable. Yet the fact cannot be denied that at least the above-defined averaged coefficient of friction μ_e is determined correctly if the positions of the CoG at both ends are known. And it has to be accepted that, by virtue of this fact, at least the average velocity cannot be dramatically wrong. According to Sect. 5.2, the same applies also to the maximal velocity. These are sufficiently strong arguments to refrain from immediately cancelling the energy line method from the list of candidates on which prediction might be based. Further arguments will turn up hereafter.

It was a strange irony of fate that, of all people, the creators of the method, Heim and Müller-Bernet, were first to falsify it by *wrong application*. As a matter of fact, in his letter, after a correct representation of the principle (pp. 144–145), Müller-Bernet recommends that the energy line be drawn between the extreme limits of the mass, i.e. top of scar and most distal debris. From a remark it becomes obvious that he had in mind a forecasting system based, both in the previous analysis and in prediction proper, on the total extension of the involved area. He forgot, however, that in the determination of velocity only the path of the CoG counts. So in his longitudinal section of the Elm slide the energy line is located at least 200 m higher than it should. Figure 6.2, a slightly simplified copy of Müller-Bernet's respective sketch (p. 149), shows, by the way, a second error of minor importance (and even slightly compensating the first one): instead of the path of the CoG, the surface of the debris was used to determine velocity. Anyhow, a maximal difference in elevations of 356 m was obtained, resulting in a velocity of 83.6 m s^{-1}. In the figure a more probable energy line, somewhat steeper than the one assumed by Müller-Bernet, and the approximate path of the CoG are represented, yielding Δz in the range of some 180 m and a maximum velocity below 60 m s^{-1}.

Certainly, the error thus committed might be considered as a rather amusing faux pas of one who had effectively contributed to the scientific treatment of mass movements – if its consequences were not found in an impressing majority of massive overestimations of velocities. Whatever the psychological background may have been, it

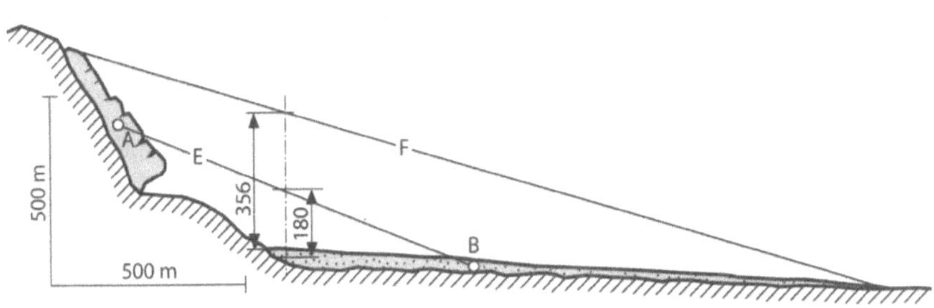

Fig. 6.2. Müller-Bernet's "original sin". Instead of straight line *E* from initial to final position of centre of gravity (*A*, *B*), connection *F* from crown of scar to most distal deposits is used as energy line. Velocity-relevant differential elevation is exaggerated by 356/180 m, in terms of velocity 83.6/59.4 m s^{-1}. This error has been adopted by many later authors, not excluding Körner (simplified and generalised copy from Heim 1932, 149; note slightly differing horizontal and vertical scales)

is nothing but a fact that the once-committed *error became current practice* so that even nowadays no velocity figure can be accepted without knowing how it was obtained. More than that: the virus so effectively propagated over the scientific world that other models of calculation caught the disease of using total instead of CoG elevations. The methods of Francis and Baker (1977; s. Sect. 2.5) and Körner (1976, 1977, 1983; s. Sect. 2.7), both referred to hereafter, have by no means demonstrated immunity.

Dramatic are the cases where the CoG passes through a rapid transition between two different slopes, in particular if a *descent is followed by a run-up* so that leading and trailing portions of the mass simultaneously are high above the foot of the crossed valley, thus lifting the CoG by many tens (or even hundreds) of metres. Impressive examples have been presented in detail for the fast events of Pandemonium Creek (Sect. 2.2) and Val Pola (Sect. 2.5), and Avalanche Lake was shortly mentioned (Sect. 5.7).

Few are the studies in which a *correct CoG treatment* in speedometry is applied or at least proposed as a possible improvement. At least for the three above-mentioned model events as well as for Köfels (Sect. 2.4) such publications do exist. Consider the following examples: for Pandemonium Creek and Avalanche Lake: Evans et al. (1989, 442, in a marginal remark; 1994, 765, with sections showing CoG paths); for Val Pola: Crosta (1991, 104) and Kienholz et al. (1993, 305–309); for Köfels: Erismann et al. (1977, 91–94, unfortunately without stressing the importance of the point). In a general manner, Straub (1997, 419) pleads for a correct use of CoG paths. A list of articles in which Heim's and Müller-Bernet's error has been adopted would turn out to be substantially longer…

*

Once the *model of Francis and Baker*, published in an article of little more than a page (1977), has been mentioned, it is, perhaps, worthwhile to give a brief description of its physical background. First of all, however, it should be pointed out that the Eq. 2.1 and 2.2, used in connection with Val Pola, contain the essence of the idea not only in a notation coherent with that of the present book, but also in a more condensed form than presented in the original. As already insinuated in the context of Val Pola, the method is based on two assumptions, namely that (11) the effectivenesses of energy conversion from potential to kinetic form and vice versa be equal and that (12) said conversion comprises all relevant sources of energy dissipation.

In *comparison with the constant μ method* as proposed by Heim and Müller-Bernet, it is obvious that both yield equal results if the slope angles of descent and ascent are equal and if the transition between the two phases is negligible. Francis and Baker, aware of the last-mentioned fact, explicitly remarked that results would be modified by a horizontal distance between the two phases considered. Still it must be added that absolutely negligible transitions do not exist and that, on the contrary, their contribution to energy dissipation is overproportional (s. considerations concerning vertical curves with centre above ground in Sect. 5.7). Now in the constant μ model the transition counts at least on a basis of equal rights with other parts of the track. So the model of Francis and Baker, being confined to special topographic conditions and offering no advantage with respect to accuracy, certainly cannot be used as the universal means for determining velocity aimed at in this section.

Its incidental use is nevertheless justified in the rare occasions in which an almost *instantaneous first estimate* is required for an event consisting of descent and run-up with a short (though well-rounded) transition in between. The whole work required is (21) estimating the elevations of the CoG before the descent, at its lowest position above the foot of the valley, and at the highest position of run-up; (22) forming Δz_1 and Δz_2 by subtraction; and (23) calculating $\Delta z = \sqrt(\Delta z_1 \Delta z_2)$ and $v = \sqrt(2g\Delta z)$ in accordance with Eq. 2.2 and 2.1 respectively. But even in such instances the gain in time is relatively small as the main work is spent in determining the approximate positions of the CoG. And this work is required for any physically plausible model.

*

After this intermezzo a far more important approach has to be considered. Starting with a review of the literature available at the time, Körner (1976, 1977, 1983, 96–105) developed the method that bears his name. In first instance his endeavours were influenced by *Voellmy's work about avalanches* (1955). On the assumption that the behaviour of moving snow is similar to that of liquids, Voellmy had postulated (1955, 214–216) a similarity also for the coefficients of resistance. Speaking in the terminology of the present book, he assumed that the basic differential equation of displacement corresponded to Eq. 5.20 and that the coefficient of the quadratic term, c_v, was inversely proportional to the thickness of an avalanche. Thus he combined the dynamically important introduction of a quadratic term with the quantitative expression of a size effect – a remarkable performance.

It is not intended to discuss here the validity of Voellmy's hypothesis for snow avalanches. Only two subjects are interesting for this book: the fact that Körner, unsatisfied with Scheidegger's (1973) size-dependent prediction curve as well as Scheller's (1970), Mellor's (1968), and Salm's (1966) comments on velocity-dependent resistance, made use of *Voellmy's model also for rockslides*, including the mentioned role of thickness as a determining parameter. So for resistance a model with two terms resulted, one of which was constant, the other quadratically proportional to velocity. As can be concluded from the remarks made in Chap. 5 (mainly in Sect. 5.4), this concept, congruent with the mentioned Eq. 5.20, is promising in principle. In fact, rare are the occasions in which resistances other than velocity-independent or dependent on the square of velocity were claimed as being important.

Of course, to *work with two parameters* is necessarily more complicated than with a single one – constant, to top it all. In Müller-Bernet's model, by drawing the line from CoG to CoG (i.e. from departure to arrival) and measuring its inclination, the parameter μ is unequivocally determined. Here, on the contrary, an infinite multitude of parameter combinations (μ, c_v) exists which can bring the mass from its initial to its final position (when μ is increased, c_v decreases accordingly). Somehow additional information must be obtained to find the right pair of said parameters. Now each of the combinations yields its individual total time of displacement differing from all the others. Unfortunately, however, the total time is practically never known with sufficient accuracy (Sect. 2.5, 2.7; here-above). So Körner decided (31) to calculate an event – i.e. to solve Eq. 5.20 with the sequence of local slope angles β as inputs – several times, each time assuming another total time of displacement. The next step (32) consisted in finding particular information about locally observed velocities (mainly from bounces or run-ups). Finally, (33) by inserting these local results into the plot of velocity versus distance, the combination best suited had to be found (Körner 1976, 245, 1983, 97).

The further *discussion of Körner's model* has to be divided into two parts, one referring to the execution, the other to the inherent qualities of the method. The easier task, dealing with execution, will be treated first.

It is an almost tragic misfortune that Körner (who really could have known better), after having developed a remarkably clever method, immediately walked into the trap that Müller-Bernet so unconsciously, though perfectly camouflaged by Heim's great name, had set up for himself and others. All relevant longitudinal sections (Körner 1976, 244, 1977, 102, 1983, 74) infallibly show *erroneous energy lines* covering the total length of the event, from top of scar to most distal deposits. So the errors criticised in Sect. 2.7 with respect to the local determination of velocity, despite a certain importance of their own, assume the role of spoilsports in a more important context.

While errors thus committed are easily corrected (as far as the required positions of the CoG can be determined in a satisfactory way), the inherent problems of the model necessitate a more detailed discussion connected in first instance with the coefficients μ and c_v in Eq. 5.20. The trouble lies mainly in the fact that these coefficients (like μ in Müller-Bernet's model) are used as if they were constant during the entire travel. Körner knew that this is not the case and expressed (1976, 235), in connection with Salm's (1966) considerations about higher orders of approximation, the expectation that – despite inconsistencies in the physical interpretation of other terms than constant and velocity-quadratic – more sophisticated formulations would become necessary to account for *en route variations of coefficients*. This opinion cannot pass uncommented. In Körner's model a higher order of approximation means a larger number of points in which the calculated velocity can be made to comply with values otherwise obtained. This would, however, occur at the cost of forcing the differential equation into the Procrustean bed of fulfilling, besides the wanted velocity compliances, the conditions of continuity inherent to the used algorithm. Yet the conditions of a track by no means warrant such continuity.

Consider *Pandemonium Creek* as an example (Sect. 2.2). On the glacier, resistance certainly was very low, perhaps even subjected to a reversal of velocity-dependence owing to self-lubrication (Sect. 5.2, mechanism R; Sect. 5.5); also in the following descent a relatively low resistance was given by lateral confinement; high energy dissipation resulted from centrifugal compression in crossing the valley; running up and back were additional sources of losses (it was with a good reason that Evans et al. 1989, 441, introduced a "hydraulic jump" to account for the two last-mentioned portions); finally the "bobsleigh run" probably offered the possibility of pressurised water as a lubricant in combination with efficient lateral confinement. All these portions of track followed each other more or less immediately so that hardly a length exceeding that of the moving mass can be envisaged for transition. From such evidence the question arises how far at all Körner's twin parameter model or higher order algorithms can improve velocity analysis.

A preliminary answer is found in Sect. 5.2 where, in Fig. 5.6 and 5.7, various mechanisms are compared and, under Item 21, relatively low differences of maximal velocities are diagnosed. It certainly will be observed that mechanism N stands for a straight energy line and mechanism Q for one due to velocity-quadratic resistance. A simple mental experiment will complement this rather general statement by a somewhat more specific *comparison of energy lines* associated with the two said mechanisms (Fig. 6.3). Consider the purely quadratic mechanism Q in Fig. 5.7, running out on the soft slope of diagram 5.7c (to avoid the theoretically infinite run-out on horizontal ground). As done in Sect. 5.2, the resistance due to transition from descent to run-up is neglected.

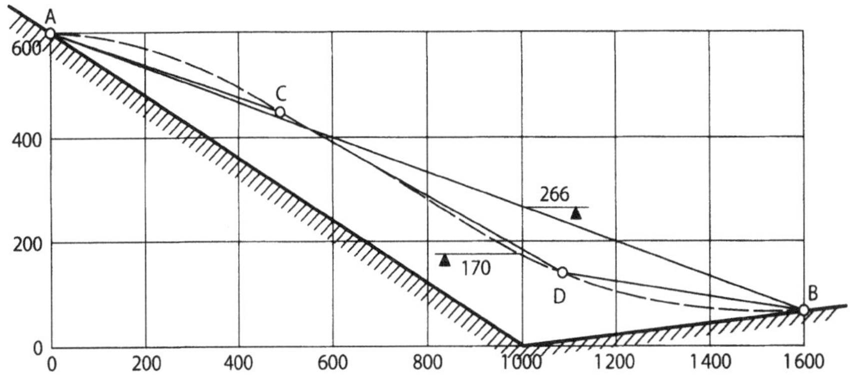

Fig. 6.3. Limits of Körner's model illustrated by energy lines (all dimensions are in metres). *A, B:* Centre of gravity before and after displacement (thickness of mass is assumed to be negligible); *straight line A → B:* Körner's energy line for $c_v = 0$ in Eq. 5.20, identical to Müller-Bernet's model; *curved line A → B:* Körner's energy line for $\mu = 0$, identical with curve *Q* in Fig. 5.7a and 5.7c; *C, D:* additional known points (at least one required to set up Körner's function); *zigzagging sequence of straight lines A → C → D → B:* partial energy lines for $c_v = 0$. Note rapid approximation of straight and curved energy lines when additional points are known (differential elevations of 266 and 170 m mean velocities differing by 14.4 m s^{-1}). Attention: for abruptly varying tribological conditions of track, superiority of continuous line is questionable (sketch by Erismann)

In such instances, the line for quadratic dependence suggests a maximum velocity of 57.8 m s^{-1}, the straight energy line (as postulated by Müller-Bernet) one of 72.2 m s^{-1}, i.e. 14.4 m s^{-1} more – a substantial difference. As there is no way to decide which one of these results is nearer to reality, two conclusions can be drawn: the maximal velocity is 65.0 ±7.2 m s^{-1}, and the mechanism may be anything between pure Coulombian friction and velocity-quadratic resistance, i.e. between the two limits of Körner's model as given by one of the parameters μ and c_v in Eq. 5.20 equalling zero.

Now assume that two reliable *velocity measurements are added*, one at a horizontal distance of 500 m from the point of release, the other at 500 m before the point of standstill. Let these velocities be in perfect coincidence with those given by the energy line for quadratic resistance. Obviously there can be no objection against drawing a tripartite polygonal energy line from the point of release A to the point of standstill B via both additional points C and D. The figure shows that in such a case not only the suggested maximal velocity is almost identical for the compared methods (the difference being about 2 m s^{-1}): also the largest local discrepancy is halved to 7.7 m s^{-1}. In discussing such results, a supporter of quadratic resistance might stress the coherence of "his" energy line while an advocate of Coulomb's rule, in view of the variety of local circumstances, would describe as suspect precisely this quality. Neither would be entirely wrong, and neither entirely right. As a matter of fact, except for particularly steady conditions, the knowledge of velocity at four points of a track (two in motion and two at rest), in spite of overdetermining Körner's energy line, turns out to be too weak a basis for a complete start-to-stop description of the mechanism(s) of displacement.

To exclude misunderstandings, it should be remarked that a purely quadratic mechanism never has been envisaged by Körner, and that it has been introduced here only to demonstrate the width of the field covered by his method and ranging between $c_v = 0$ for the purely linear and $\mu = 0$ for the purely quadratic form of Eq. 5.20. The main conclusion emanating from the presented line of thought is that said field narrows down rapidly with any increase of the information beyond the "three-point" minimum required to set up Körner's model. In other words: at least at the four-point level of information as shown in Fig. 6.3 the practical value of a *polygonal energy line is competitive* with a twin-parameter line. Thereby it has to be borne in mind that, as far as physical correctness is considered, Körner's solution is clearly superior; yet, as everybody knows, the colour of a cat has little importance as long as it is efficient in catching mice.

So far the problem explicitly was discussed on the basis of *reliable speedometric information*. This is obviously in contradiction with the critical remarks made in connection with the Huascarán event (Sect. 2.7). In view of possible errors of this kind, a further argument might be raised in favour of a polygonal energy line: a wrong measurement falsifies nothing but the immediately adjacent straight portions while in a twin-parameter model the entire length of the energy line somehow reflects the error.

*

The reader might *ask for a non plus ultra*. Such ideally would consist in the possibility to apply Körner's method to reasonably chosen, mutually independent portions of the track. But where is an event giving away its secrets so generously that sufficient information would be available for the purpose? Pandemonium Creek (Sect. 2.2) may raise some hope in the "bobsleigh run", but even the apparently simple measurement by superelevation turns out to be more tricky than expected at first sight (Sect. 5.7). So, at least for the time being, speedometry remains a difficult job in analysis...

This does certainly not mean that speedometry on the simple basis of polygonal energy lines is hopeless in case of well-investigated events. The contrary is demonstrated hereafter for the particular *example of Huascarán* (Sect. 2.7). Figure 6.4 shows the longitudinal section of the – physically dominant – Matacoto lobe. The initial and final positions of CoG (C, H) are admittedly estimated with moderate accuracy. It is, however, easy to see that the resulting energy lines would not suffer drastic changes if a displacement within a reasonable range would take place: the general character in any case would remain intact and the parameters would be corrected by several percent. The extraordinary length of the track entails a small scale so that the problem of establishing the CoG path can be replaced by bearing in mind that this path lies a few tenths of a millimetre above the line representing the ground. The straight line from C to H is, of course, the classic energy line according to Heim and Müller-Bernet (1932). Its slope of 0.266 is slightly lower than what would – in anticipation of Sect. 6.3 – be expected as Fahrböschung for an average event of 0.05 km^3, namely 0.284. So the first impression might suggest little more than a near-to-normal event. But the reality is quite different.

An attempt to *represent the reality of Huascarán* is shown by the polygonal energy line in the figure. And this attempt is worth a couple of special comments. As the mass from its start contained ice and, in addition, moved partly on Glacier 511, partly between lateral confinements, the "normal" coefficient of friction (mainly used between C and E) was assumed to be 0.25, i.e. somewhat lower than Fahrböschung. D is a "jump" used to account, according to Eq. 5.37, for the transition from the steep wall to Gla-

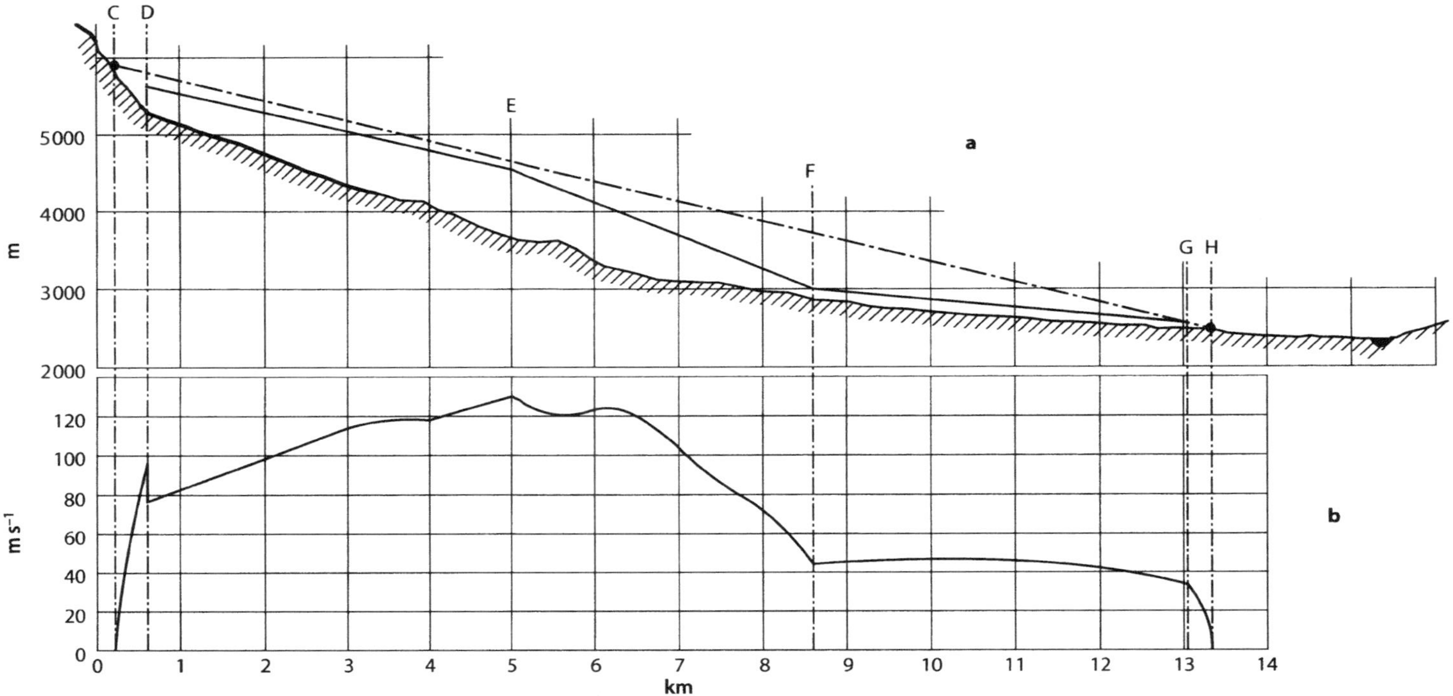

Fig. 6.4. Energy lines in longitudinal section of Huascarán (**a**) and velocity (**b**) resulting from polygonal line. *C:* km 0.2, release position of centre of gravity (CoG); *D:* km 0.65, velocity "jump" due to transition from steep wall to Glacier 511; *E:* km 5.0, massive increase of friction in wide uneven area; *F:* km 8.6: Quebrada Shacsha with pressurised water lubrication; *G:* increase of friction by spreading, *H:* final position of CoG. Small scale makes mass thickness almost imperceptible. Topographic details in Fig. 2.48–2.50

cier 511. The energy loss was calculated on the basis of a local μ of 0.4 and a difference in slope angles of 33°, resulting in a loss of 478 – 302 = 176 m, in terms of velocity almost 20 m s^{-1}. The dramatic increase of resistance in the section E → F is topographically justified by the violent ups and downs, by the fact that a great part of the mass followed Quebrada Armapampa – a far longer distance with a sharp angle, and by the necessary collisions of arms when reuniting after having taken separate tracks. The finally used value of 0.423 was calculated from the necessity to obtain, near km 10, the restricted velocity postulated in Sect. 2.7. Eventually, in descending Quebrada Shacsha, great changes of velocity could not occur, otherwise either motion would have been arrested or Matacoto would have been devastated. So μ was assumed to have been 0.09, approximately equal to the slope – in view of the high probability of pressurised water lubrication a plausible value. A similar value is, by the way, probable for the "bobsleigh run" of Pandemonium Creek (Sect. 2.2).

The example shows that a reasonable velocity analysis is not excluded, even if nothing but a sequence of constant-resistance sections is assumed. Nevertheless, it cannot be denied that, at the exorbitant velocities of Huascarán, a non-negligible improvement could be obtained by taking into account directly (and not via adjustment of μ) the resistance component proportional to the square of velocity: especially the maximum of velocity (in Fig. 6.4 about 133 m s^{-1}) would be reduced, and a somewhat lower resistance between E and F would suffice as a brake.

*

Anyhow, the reliability of *prediction would enormously profit* from a more accurate start-to-stop knowledge of velocities which, with moderately sophisticated mathematical means, could be converted into a corresponding knowledge of the coefficients controlling the downhill ride of rock masses. This aspect, including possible approaches to feasible methods, will be the central theme of the two following sections.

6.3 The Size Effect – a Useful Tool

As in rockfalls, owing to lower interaction between the moving clasts, the size of the total mass is of secondary importance for the distance travelled, the following reflections refer *mainly to rockslides.*

In Heim's book (1932, 121–130) seven rockslides of different sizes are listed and described in detail. For each event the list (depicted in Fig. 6.5a) shows the volume V and the angle of Fahrböschung or overall slope (or β_f in the notation of the present book). In the context of speedometry by the energy line method this angle was presented in Fig. 6.1. To recall the definition: β_f is the angle between the horizontal and the line connecting the crown of the head scarp with the most distal debris along the midstream path of the mass. The tangent of this angle, i.e. the quotient of the vertical distance between said points, divided by their horizontal distance (the latter again counted along the path of the mass), will play an important role in the following considerations. And, wherever opportune in view of a repeated use, the *abridged notation* $f = \tan\beta_f$ will stand for Fahrböschung or overall slope.

With his list, Heim had the intention to illustrate that, as he and others had observed, large events generally travel more economically than smaller ones. With good reasons he considered this effect as important. So, from the relatively small number of rockslides reasonably well investigated at the time, he made an appropriate selection showing this dependence, the *size effect*, in a persuasive manner.

Of course, this – probably first – attempt to establish an essential *quantitative correlation between parameters* of rockslides makes a somewhat primitive impression when looked at almost seven decades later. Neither the number of considered events nor the knowledge of the involved parameters can, at the present time, be judged as satisfactory. Nowadays, owing to extensive lists established in the meantime, a far richer selection of examples is possible (Scheidegger 1973, Girsperger 1974, 74; Abele 1974, 45; Hsü 1975, 133; Cruden 1976, 9; Lucchtta 1978, 1604; Tianchi 1983, 476–477; Rouse 1984, 500–501; Nicoletti and Sorriso-Valvo 1991, 1366–1367; Costa 1991, 22–23. Attention: some events bear other names than used here; take notice of: Cima di Dosdé → Corno di Dosdé; Monte Zandilla → Val Pola; Scima da Saoseo → Cima di Saoseo; Siders → Sierre; Tamins → Säsagit; Yungay → Huascarán). Later, in connection with the discussion of interpolations by regression techniques – probably not just familiar to Heim – it will be demonstrated that the set of figures he chose, though in itself allowing for a reasonably well-fitting interpolation, turns out to be less optimal as soon as further events are accounted for.

Unfortunately, from more extended lists it also becomes evident that the real *amount of scatter* is distinctly higher than Heim probably had expected. In fact, he considered it opportune to explain (by a more intense lateral spreading, p. 123) the higher Fahrböschung of the Frank slide (H4 in Fig. 6.5a) as compared with the straight connection from that of Elm (H3) to that of Goldau (H5). Now the respective deviation is no more than 0.035 in terms of f, an almost negligible amount in a set of figures in which the overall slopes of equal-sized events may differ by almost ten times as much. An anticipating look at the "starred sky" presented in the subsequent Fig. 6.5–6.6 certainly will make obsolete further comments.

In spite of being blurred by scatter and thus expressing rather the chaotic superposition of many different effects than the stringent physical consequences of a single

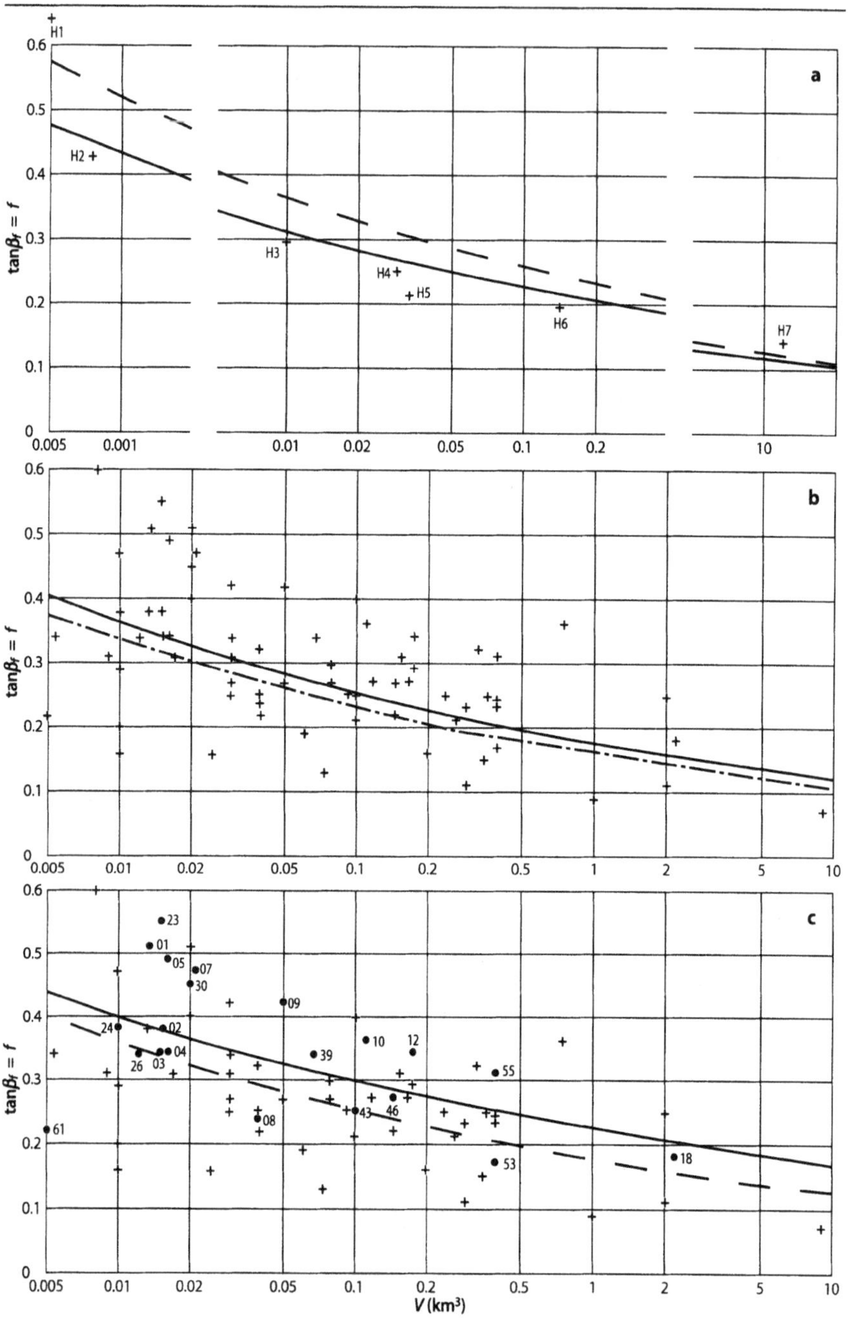

Fig. 6.5. Scheidegger functions averaging Fahrböschung f versus volume V. **a** Heim's seven events (*plain line;* data: 1932, 121), compared with standard reference (*broken*). *H1:* Airolo, *H2:* Monbiel; *H3:* Elm; *H4:* Frank; *H5:* Goldau; *H6:* Kandertal; *H7:* Flims. **b** Author's standard reference from 69 events (*plain;* Table 6.1). Note coincidence with original Scheidegger function (*dots and dashes*). **c** 21 events = crystalline part of **b** (*circles; plain line*). Note large scatter and moderate difference with standard reference (*broken*) (sketch by Erismann)

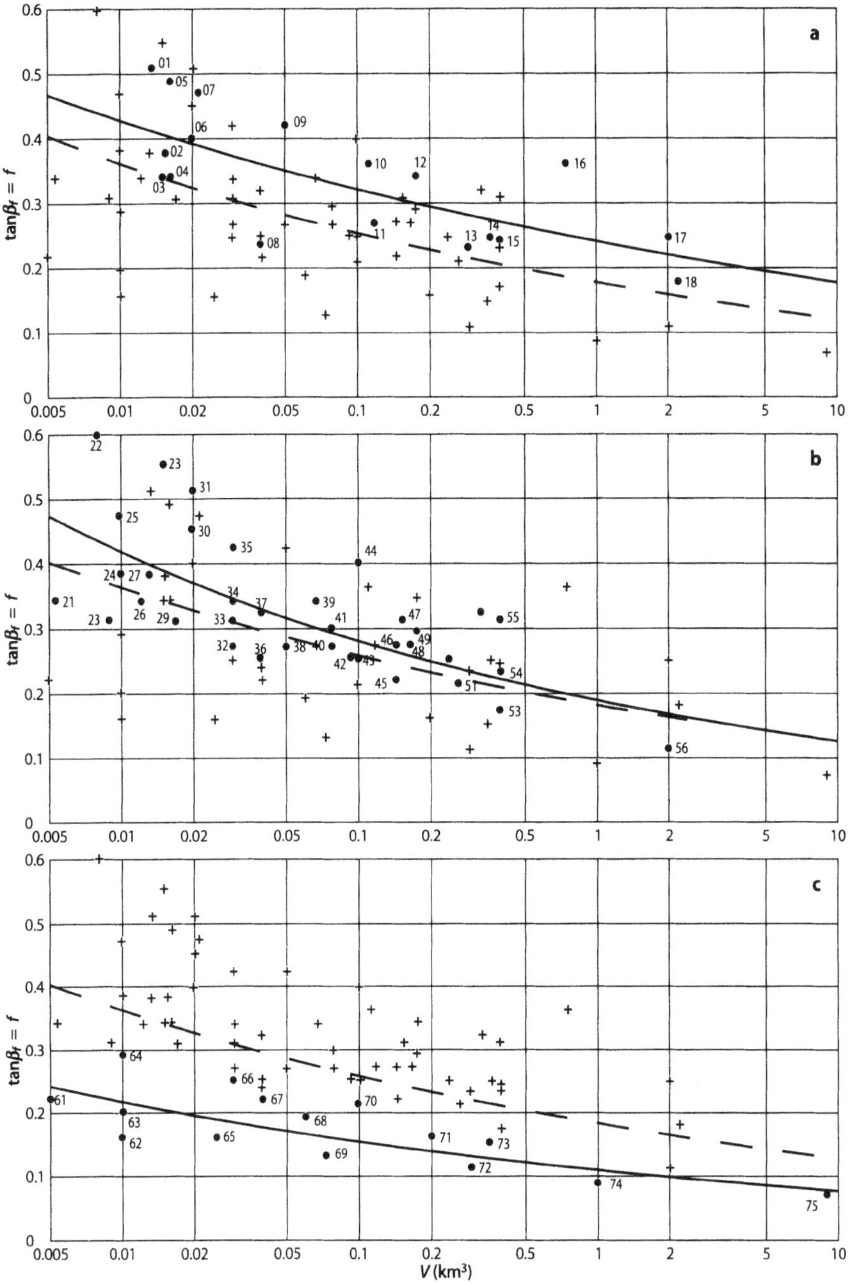

Fig. 6.6. Scheidegger functions obtained by selecting, from the 69 events of the standard reference (Fig. 6.5b), those characterised by special geomorphological features. For data s. Table 6.1. **a** 18 events of V type (collision of mass with solid slope in V-shaped valley). **b** 36 events of M type (neither L nor V). **c** 15 events of L type (tracks often confined; lubrication by ice and/or pressurised water). Compare functions (*plain lines*) with standard reference (*broken lines*) (sketch by Erismann)

Table 6.1. Data of 69 well-investigated rockslides. *i:* Reference number; *Ref. A:* reference number in Abele (1974, 171–198); *M:* material (carbonate if not specified; *CR:* crystalline); V_i: volume of mass in km^3; f_i: actual Fahrböschung (overall slope); f_c: overall slope calculated by regression within group; f_i/f_c: quotient indicating compliance between f_i and f_c; italics: *two largest and two least figures in group*

i	Ref. A	Country	Event	M	V_i	f_i	f_c	f_i/f_c
Group V (Fig. 6.6a): Mass arrested by massive obstacle – high energy dissipation								
01	3207	Switzerland	Ludiano	CR	0.014	0.51	0.410	*1.244*
02	3302	Switzerland	Disentis	CR	0.015	0.38	0.406	0.935
03	5403	Austria	Kals	CR	0.015	0.34	0.406	0.837
04	3203	Switzerland	Brione	CR	0.016	0.34	0.403	0.843
05	1406	France	Grand Clapier	CR	0.016	0.49	0.403	1.215
06	5501	Austria	Mordbichl		0.020	0.40	0.392	1.020
07	1409	France	St. André	CR	0.021	0.47	0.390	1.206
08	6403	Italy	Monte Avi	CR	0.040	0.24	0.360	*0.668*
09	7502	Italy	Pontives	CR	0.050	0.42	0.350	1.201
10	1410	France	Col de la Madelaine	CR	0.115	0.36	0.315	1.143
11	2111	Switzerland	Oeschinensee		0.120	0.27	0.313	0.862
12	3501	Switzerland	Lago di Poschiavo	CR	0.180	0.34	0.298	1.141
13	7603	Italy	Vaiont/Longarone		0.300	0.23	0.279	*0.823*
14	5504	Austria	Dobratsch A. Schütt E		0.360	0.25	0.273	0.915
15	7309	Italy	Lago di Molveno		0.400	0.24	0.270	0.890
16	2409	Switzerland	Glärnisch-Goppen		0.730	0.36	0.250	*1.440*
17	2302	Switzerland	Engelberg		2.000	0.25	0.220	1.134
18	5206	Austria	Köfels	CR	2.200	0.18	0.218	0.826
Group M (Fig. 6.6b): Events neither of V nor of L type – medium energy dissipation								
21	7306	Italy	Tucketthütte		0.0055	0.34	0.461	0.737
22	7307	Italy	Val Brenta Alta		0.008	0.60	0.432	1.387
23	7305	Italy	Vallesinella		0.009	0.31	0.424	*0.732*
24	7520	Italy	Fedaia	CR	0.010	0.38	0.416	0.913
25	4102	Germany	Melköde		0.010	0.47	0.416	1.129
26	6501	Italy	Lago di Autrona	CR	0.012	0.34	0.403	0.843
27	4306	Germany	Hintersee		0.013	0.38	0.398	0.955
28	3206	Switzerland	Biasca	CR	0.015	0.55	0.388	*1.417*
29	7303	Italy	Monte Corno		0.017	0.31	0.380	0.816
30	3505	Italy	Cima di Dosdé	CR	0.020	0.45	0.369	1.218
31	7522	Italy	Lago di Alleghe		0.020	0.51	0.369	1.381
32	5505	Austria	Dobratsch J. Schütt W		0.030	0.27	0.344	0.784
33	7312	Italy	Torbole		0.030	0.31	0.344	0.900
34	4114	Austria	Haiming		0.030	0.34	0.344	0.987
35	2405	Switzerland	Haslensee		0.030	0.42	0.344	1.219
36	2108	Switzerland	Kleines Rinderhorn		0.040	0.25	0.328	0.763

Table 6.1. *Continued*

i	Ref. A	Country	Event	M	V_i	f_i	f_c	f_i/f_c
Group M (Fig. 6.6b): Events neither of V nor of L type – medium energy dissipation								
37	2502	Switzerland	Voralpsee		0.040	0.32	0.328	0.976
38	4301	Germany	Marquartstein		0.050	0.27	0.315	0.856
39	1405	France	Lac Lauvitel	CR	0.068	0.34	0.299	1.136
40	4303	Austria	Lofer		0.080	0.27	0.291	0.928
41	4201	Austria	Pletzachkogel		0.080	0.30	0.291	1.031
42	7601	Italy	Fadalto		0.095	0.25	0.282	0.885
43	5408	Austria	Mallnitz	CR	0.100	0.25	0.280	0.893
44	4106	Austria	Am Saum		0.100	0.40	0.280	*1.429*
45	1701	France	Abîmes des Myans		0.150	0.22	0.261	0.843
46	3504	Switzerland	Cima di Saoseo	CR	0.150	0.27	0.261	1.034
47	2403	Switzerland	Obersee		0.160	0.31	0.258	1.201
48	5503	Austria	Dobratsch A. Sch. W		0.170	0.27	0.256	1.057
49	7201	Italy	Bormio		0.180	0.29	0.253	1.146
50	7302	Italy	Lago di Tovel		0.250	0.25	0.239	1.045
51	2407	Switzerland	Dejenstock		0.280	0.21	0.235	0.895
52	7310	Italy	Marocche del Val Sarca		0.340	0.32	0.227	1.411
53	3406	Switzerland	Totalp	CR	0.400	0.17	0.221	0.771
54	7304	Italy	Monte Spinale		0.400	0.23	0.221	1.043
55	3401	Switzerland	Parpan-Lenzerheide	CR	0.400	0.31	0.221	1.406
56	2113	Switzerland	Sierre		2.000	0.11	0.167	*0.658*
Group L (Fig. 6.6c): Effective lubrication, confinement – low energy dissipation								
61		Canada	Pandemonium Creek	CR	0.005	0.22	0.253	0.869
62		Canada	Damocles		0.010	0.16	0.225	*0.710*
63		Canada	Twin E		0.010	0.20	0.225	0.887
64	2410	Switzerland	Elm		0.010	0.29	0.225	1.287
65	7311	Italy	San Giovanni		0.025	0.16	0.193	0.827
66		Canada	Frank		0.030	0.25	0.188	*1.333*
67	2401	Switzerland	Goldau		0.040	0.22	0.179	1.231
68	5208	Austria	Obernberg		0.061	0.19	0.167	1.141
69		Canada	Nozzle		0.075	0.13	0.161	*0.808*
70	7401	Italy	Lavini di Marco		0.100	0.21	0.153	*1.370*
71	4113	Austria	Tschirgant		0.200	0.16	0.137	1.172
72	4401	Austria	Almtal		0.300	0.11	0.128	0.862
73		Canada	Rockslide Pass		0.350	0.15	0.124	1.206
74	4105	Austria	Fernpass		1.000	0.09	0.104	0.863
75	2411	Switzerland	Flims		9.000	0.07	0.072	0.970

causal connection, *Fahrböschung is a useful parameter*, distinguished by important advantages. In first instance, it often can be determined even if most of the debris has been carried away by erosion. In the frequent cases of rockslides having crossed valleys, for example, the bulk of the mass, after having been originally deposited near the valley's bottom, may subsequently have been removed by a river and/or by a glacier. In contrast thereto, both the head scarp and the foremost debris, owing to their more elevated positions on both sides of the valley, may have been preserved in their original position. Furthermore, the overall slope, as already mentioned under Heading 6.2, as a rule, does not dramatically differ from the slope of the energy line, i.e. one of the most important parameters of a rockslide. So it appears reasonable to take the overall slope into consideration as a second-best substitute and to look how far one can get in applying it.

Relatively easy determination, the main advantage of Fahrböschung, is not confined to analysis post eventum; it also may facilitate *prediction*. In many cases the presumptive location of a head scarp is marked somehow (for instance by a growing fissure). This means that one of the two required points is at hand, and if a plausible value can be guessed for the overall slope, the horizontal distance – and therewith the reach – can be estimated.

Analysis and prediction on the basis of the positions assumed by the *centre of gravity* would, in principle, be preferable. But what can be done if these positions are unknown or difficult to estimate?

*

It is not intended to repeat here the details previously discussed in connection with the *physical background* of the size effect, be it by direct influence upon resistance or, indirectly, by track-making. Refer in first instance to Sect. 5.1, 5.3, 5.4, and 5.5. Most of these effects are positive (larger masses reach farther), but also negative examples exist, e.g. for air lubrication (Sect. 2.3) and "roller bearings" (Sect. 5.3). Anyhow, a review of the mechanisms involved clearly shows that the theoretical deduction of a generally valid dependence between resistance and size (as, for instance, the "area-to-volume" rule, s. Sect. 5.4) would be extremely complex (not to say impossible) owing to the great variety of effects, some of which, as for instance self-lubrication by frictional heat, are highly non-linear. This complexity is further increased by the fact that, as a rule, large events move faster than smaller ones so that the velocity-dependent component of resistance, within its normally moderate role, counteracts the general trend of improved economy connected with the size effect.

In view of such difficulties it makes sense to consider the *size effect as a black box* (as defined by Ashby 1961). Accordingly a rockslide is – at least in the initial phase of an investigation – treated as a system known only by its phenomenological (i.e. externally conceivable) data and not analysed internally by reduction of mechanisms to generally accepted physical laws. Not in the least this means an invitation to forget everything that has been presented when considering the details of motion. On the contrary, it will soon become evident that an unconditional black-box approach is unable to yield useful results, in particular if prediction is aimed at. So, already at the present time, and all the more in the future, improvements only can be obtained by taking into account more differentiated knowledge.

In any case, the first step in making use of the considerable number of known overall slopes consists in choosing an appropriate *interpolating algorithm* allowing for an easy conversion of a large number of $f_i(V_i)$ data pairs into a continuous function $f_c(V)$ able to act as a tool for prediction. The index i is the ordinal number assigned to each of the known events. It is obvious that the quality of such conversions can be expressed by the mean square root of their deviations from the actual values. It is equally obvious that a certain simplicity of the resulting function is required: it shall monotonously decrease for increasing V, aperiodically approaching a very low value or zero (singularities $f_c = \infty$ for $V = 0$ and $f_c = 0$ for $V = \infty$ are acceptable as long as their consequences remain harmless in the considered range of volumes). A monotonous course, by the way, excludes the use of theoretically possible functions (e.g. Taylor series) yielding zero mean square roots irrespective of the number of considered data pairs: the resulting curves would consist of ups and downs and hardly could be set up to avoid crossings with the abscissa axis (not to speak of cases with more than one f_i value being given for a single V value).

*

A function well complying with said requirements was proposed by Scheidegger (1973) who based his argument on the *regression by a straight line* (i.e. a linear function), certainly the most-used method to convert a cluster of points in a plane into the straight line minimising the mean square root of deviations. Now a straight line obviously would not be convenient in the usual representation of $f_i(V_i)$ clusters where f is on linear and V on logarithmic scale: crossing the abscissa axis necessarily would occur. Scheidegger's trick to overcome the problem was simple and particularly useful. He applied the method in a representation of both variables on a logarithmic scale and obtained more than the asymptotic approximation to zero: as minimising of deviations takes place on the logarithmic scale, the quotients f_i/f_c are minimised (instead of the differences $f_i - f_c$). In view of the large range of overall slopes this is very reasonable. Imagine a deviation of ±0.05, once occurring with $f = 0.50$ and once with $f = 0.10$. Deviations labelled "10%" and "50%" certainly are more comprehensive than the statement "equal deviations". And in case of extremely small f there is no risk of obtaining negative values after subtraction of expected deviations...

Hereafter this elegant method will be denominated as "Scheidegger's method" and the resulting function as "*Scheidegger function*".

As mentioned above, the application of Scheidegger's method in the field is remarkably easy. In addition, the following *equations* are simpler than might be assumed at first sight: once the sums of Eq. 6.2 having been formed, the rest is obtained almost immediately. The basic relation is

$$f_c = 10^{C_1 + C_2 \mathrm{Log} V} \tag{6.1}$$

wherein the coefficients are obtained as follows:

$$\begin{aligned}
C_1 &= \frac{\Sigma_1 \Sigma_2 - \Sigma_3 \Sigma_4}{\Sigma_1^2 - n\Sigma_3} \\
C_2 &= \frac{\Sigma_1 \Sigma_4 - n\Sigma_2}{\Sigma_1^2 - n\Sigma_3} \\
\Sigma_1 &= \Sigma \mathrm{Log} V_i \\
\Sigma_2 &= \Sigma(\mathrm{Log} V_i \; \mathrm{Log} f_i) \\
\Sigma_3 &= \Sigma \mathrm{Log}^2 V_i \\
\Sigma_4 &= \Sigma \mathrm{Log} f_i
\end{aligned} \tag{6.2}$$

The number of considered events is n; so each sum contains n values of the respective parameter(s). With due reference to the problem of scatter, Scheidegger proposed Eq. 6.1 as a *basis for prediction*. It soon will be discussed in more detail. But first a more general consideration will be presented in short.

It is obvious that besides Scheidegger's method there do exist *other possibilities of interpolation*. In planning this book, some thereof were, by way of trial, adapted to the purpose and tested. In particular one was considered as interesting owing to its being given by three values: maximal and minimal overall slopes (for $V = 0$ and $V = \infty$, respectively) and maximal differential quotient df / dV. In the normal representation (f on linear, V on logarithmic scale) this function would have looked like the right-hand half of a Gaussian distribution curve – an attractive profile (signalling, by the way, a close mathematical relation). The study was, however, dropped in view of practical difficulties: the three parameters are linked in a highly non-linear manner so that minimising deviations requires a somewhat awkward parameter variation.

*

One of the fundamental problems of Scheidegger's method is shared by all methods based on nothing but a large number of data pairs obtained from field evidence. As being purely empirical, the resulting function depends on the individual data of the events employed for its calculation, thus being subjected to the risk of a non-representative choice and other sources of errors. It is, therefore, worthwhile to have a look at the *strategy of selecting events*. Above all, two general principles should be respected. On the one hand, to avoid too strong an influence by individual peculiarities, the number of considered events should be as high as possible; on the other hand, the criteria for selection and data acquisition should be as uniform as possible (s. the items enumerated hereafter). These principles are contradictory to a certain extent as the quality of "having been determined in different manners" easily may turn out to be synonymous with "excluding the possibility of being applied in common". As the following choice of examples will show, there are more sources of such discrepancies than might be guessed at first sight.

1. The definition of Fahrböschung, though simple and clear enough for current use, sometimes still allows for different interpretations. What, for instance, about a mass having been arrested in a rather compact (though not necessarily coherent) state, by exception of two or three boulders which, for whatever reasons, managed to overtake the rest and to get substantially farther (Sect. 5.7)?
2. Besides the obvious difficulties of estimation ante eventum, the mere definition of volume (which in itself contains the uncertainty of being a substitute for the physically more relevant mass) may be considered as valid for the moment of release or for that of being deposited (with the possibly of substantial interstitial spaces).
3. In various prehistoric events neither volume nor Fahrböschung can be more than roughly estimated owing to changes having occurred in the meantime. Thus the field is open for different well-founded hypotheses.
4. There may be different plausible opinions about the point where motion is arrested. For example, a momentary stop (or quasi-stop) may occur at the highest point of a run-up. If it is followed by an extended displacement along a valley as in the case of Pandemonium Creek (Sect. 2.2), nobody will disagree with the idea of a regular continuation of motion. In a case like Val Pola (Sect. 2.5) where the subse-

quent motion (especially in the catastrophic northern lobe) was almost completely opposed to the first descent, one is rather tempted to speak of a run-back and to consider the main motion as finished by the quasi-stop after the run-up. In fact, this view is adopted in the evaluations following hereafter (in full awareness of possible good arguments for a different opinion).

5. The most difficult question perhaps lies in striving for a representative distribution of mechanisms and topographical features in the samples selected for evaluation. As reliable data are available only for a restricted number of events, there is only the hope that, more or less automatically, a larger number of samples will fulfil the condition better than a smaller one.
6. Finally also the sword hanging over any scientific notion is present as new findings can change the situation. Abele's and Heim's figures for Flims differ in both volume and Fahrböschung, not to speak of the question of the Domleschg tongue (Sect. 2.2); hereafter, according to Abele's modified concept, this tongue is attributed to Flims.

In such instances it is well understood that the quantitative results presented hereafter are far from being considered as the philosopher's stone valid for eternity. Yet it can be claimed that a certain *care in selecting the events* was used. For his book about Alpine rockslides (1974), Abele had critically and most conscientiously studied all available publications, and he had visited all important events. So it was nothing but natural to base the mathematically evaluated list on the information presented in said book (pp. 45 and 172–197), thus obtaining, in addition to the required care, a good uniformity of criteria in selection and data acquisition. Several events were crossed off the list owing to uncertainties which might have had a substantial influence upon the results. Few data were modified on the basis of later studies (including those of the authors). And, as an important complement, a number of well-documented American events were added (the respective publications are mentioned under Headings 2.2 and 2.7). A complete list will be presented somewhat later.

*

A first Scheidegger function is used in Fig. 6.5a to demonstrate that the seven *events presented in Heim's book* – by exception of the "jump" between Airolo and Monbiel (H1 and H2 in Fig. 6.5) – fit quite well into an interpolating function. In the same figure a second curve is represented as a broken line. It is a Scheidegger function used as standard reference and commented in the next paragraph. The relative position of the two curves shows that, on the one hand, Heim had displayed a remarkable skill in selecting near-to-average events, but that, on the other hand, a more generally accepted interpolation based on nothing but this narrow basis would probably yield too low Fahrböschung values.

Figure 6.5b shows the "starred sky" of 69 rockslides resulting from the above-mentioned careful application and complementation of Abele's work and listed in Table 6.1. The plain line is the above-mentioned Scheidegger function obtained therefrom and used hereafter – for lack of anything better – as a *standard reference*. It will be observed that the curve passes slightly below the centre of gravity of the cluster of events. This impression is not an optical illusion: it is due to the mentioned minimisation, on the logarithmic scale, of the deviations between points and curve. The line of dots and dashes in this figure is the function originally proposed by Scheidegger. The constant values in Eq. 6.1 are $C_1 = 0.7447$, $C_2 = -0.1539$ for the author's and $C_1 = 0.6242$, $C_2 = -0.1567$

for Scheidegger's original curve (volume is expressed in km^3). Within the entire represented range the differences between both are about 10% of the respective values of *f*, showing that the used events in both cases were selected in a similar (though not identical) manner – a satisfactory result in a field containing so large a degree of scatter. This is all the more true as a closer comparison of Table 6.1 with Scheidegger's table shows that only 15 events appear in both, and only two thereof perfectly coincide in both volume and Fahrböschung.

Although this is without any doubt an encouraging result, the mentioned starred sky spreads wide enough to wipe out too optimistic expectations. In fact, the largest deviations between reference curve and individual events in Fig. 6.5 yield, roughly speaking, quotients between 1/2.8 and 1.9/1, and – more important than such consequences of isolated exceptional circumstances – also the average deviations are recognised as substantial before any calculated results are at hand (which will be presented hereafter). In such instances reliable *prediction is problematic*, even if based on the best possible Scheidegger function (and nothing else). So the search for more accuracy is stringent, and this means accounting for additional information and evaluating it at a somewhat higher level of sophistication.

*

In 1991 Nicoletti and Sorriso-Valvo proposed what they called the "geomorphic control of shape and mobility" of rockslides. The essential idea was to describe quantitative correlations between the geometry of the deposits and Fahrböschung, considering the amount of internal deformation as a dominating indicator of energy dissipation. For the distribution of debris post eventum, *three basic shapes* were discerned, denominated by the self explanatory terms "elongated hourglass", "nearly oval", and "deformed T". They were respectively correlated with low, moderate, and high energy dissipation.

At the time Nicoletti was in close contact with Abele whose book (1974) he was translating into Italian (1990–1994). So it is nothing but normal that their ideas of principle became congruent in various respects. However, as reported mainly under Headings 2.2 and 5.5, Abele introduced approximately at the same time the additional notion of pressurised water lubrication which easily fits into the general concept of Nicoletti and Sorriso-Valvo, though shifting the emphasis from the purely geometric to a more general topographic aspect in which the condition of the surface acquires a dominating position. This view is adopted hereafter. Accordingly the idea of a *tripartite subdivision* of event types is slightly modified. In addition, the mathematical treatment follows Scheidegger in minimising deviations on the logarithmic scale of Fahrböschung. Besides the types described in Items 11, 12, 13, intermediate configurations necessarily must occur so that a certain additional uncertainty cannot be avoided.

11. Type "V" ("V-shaped valley") is characterised by an extended, solid obstacle able to arrest motion, e.g. the steep slope of a V-shaped valley. Energy dissipation takes place by disintegration and/or deformation. T-shaped deposits are not necessarily a criterion if disintegration is intense.
12. Type "M" ("moderate") means the absence, on the one hand, of massive obstacles, on the other hand of effective lubrication, lateral confinements, or other mechanisms substantially reducing resistance. Deformation mainly is accomplished by spreading.

13. Type "L" ("long run-out") is defined by the presence of effective mechanisms reducing, over a substantial portion of the track, resistance against motion. In first instance lateral confinement and lubrication (pressurised water, melting ice or rock) are envisaged.

The information for establishing the functions, arrayed in accordance with the three types, is presented in Table 6.1, and the results are shown in Fig. 6.6 (Fig. 6.5c will be discussed subsequently; it has been inserted here for reasons of optimal arrangement). Though by no means sensational, these *results are encouraging.*

Before adopting definitely this concept, however, a side-glance was thrown at the possibility of other than the above-mentioned features being relevant for Fahrböschung. In particular the question appeared interesting how far an influence of the material might go. Thus a separate Scheidegger function was established for the 21 events in crystalline rock belonging to the used selection (labelled "CR" in Table 6.1). The result (Fig. 6.5c) shows a curve differing from the all-events reference by distinctly less than +0.05 over the entire considered range. This issue could have been expected as only one of the L-type events (Pandemonium Creek) belongs to the group. The question whether this small difference might suffice to treat crystalline events separately, is left open., especially in view of the lack of large masses.

To obtain a reasonably well-founded judgement about the predictive power of Scheidegger functions and subdivision in groups, a systematic *quality evaluation* was carried out. Quotients f_i/f_c were formed with the four reference functions described under Items 21–24, and their logarithms were summed up over the three groups V, M, L, as well as over all 69 involved events. From each of the 16 sums thus obtained the average quotient was extracted according to the formula

$$\left.\frac{f_i}{f_c}\right|_{\mathrm{av}} = 10^{\sqrt{\frac{\Sigma(\mathrm{Log} f_i - \mathrm{Log} f_c)^2}{n}}} \tag{6.3}$$

where n was, of course, the number of samples in the respective context.

The *reference functions* used are defined as follows:

21. A constant value, namely the arithmetic average of all involved f_i values, is considered as the simplest possible basis of reference.
22. Forming groups is a first refinement. So the arithmetic average of the f_i values within a group becomes the reference for the group.
23. Alternatively the single Scheidegger function taken from all considered f_i values can act as reference function.
24. Eventually Items 22 and 23 can be applied simultaneously, thus yielding for each group its individual Scheidegger reference function.

The resulting averaged quotients are directly applicable as indicators of compliance – and therewith as *yardsticks for the predictive power* of the considered degrees of sophistication. They are listed in Table 6.2 and can be commented as follows.

31. The use of twin figures for each result is a consequence of the mentioned minimisation of divergencies on a logarithmic scale (if the logarithms were shown, a sim-

Table 6.2. Averaged quotients of actual Fahrböschung over reference basis. As quotients may lie below or above unity, each result is expressed by two reciprocal figures (e.g. 1.355 = 1 / 0.738). Synopsis of improvements obtained by subdivision of 69 events into groups *V, M, L*, by optimised Scheidegger functions, and by combination of both. Improvements are visible by comparison of lines. Last line shows that by combining subdivision into groups and Scheidegger functions useful (though not unconditionally reliable) results can be expected

Reference basis	Averaged quotients			
	V type	M type	L type	69 events
Arithmetic average of 69 *f* values	1.355 0.738	1.380 0.724	2.008 0.498	1.536 0.651
Arithmetic average of f values per group	1.337 0.748	1.382 0.724	1.462 0.684	1.389 0.720
Scheidegger function taken from 69 events	1.348 0.742	1.265 0.791	1.713 0.584	1.404 0.712
Scheidegger functions taken from events of the group	1.215 0.823	1.237 0.809	1.230 0.813	1.230 0.813

ple "±" would suffice, but there is no sign for "multiplied or divided by"). Obviously the pairs are mutually reciprocal.

32. Also the somewhat puzzling fact that for group M the arithmetic average taken from all 69 events is – though very slightly – lower than that taken from the group alone, has to do with the log scale: the arithmetic average, obtained by a linear algorithm, differs a tiny bit from the logarithmic one.
33. When looking at the results in a lump (i.e. concentrating on the column at right), the improvements by averaging in groups and by using Scheidegger functions are, roughly speaking, equal. Both reduce the predictive uncertainty by about a quarter. The combination of both measures approximately halves this uncertainty.
34. Within the groups the improvements are distributed quite unevenly. Group L, initially showing a catastrophic result, profits, when measures are applied, almost exclusively if compared with the other two groups. Eventually (s. bottom line of table) all groups acquire approximately equal levels of uncertainty.
35. When combining both measures, 50 of the 69 events (i.e. 72.5%) remain within the limits given by the pair of figures presented for each group.

*

The results thus obtained can be *summarised* in a few sentences. Both a reasonable subdivision in groups and the interpolation by Scheidegger functions are useful means to improve the quality of prediction, separately, however, only to a moderate degree. Applied in common, they allow to reduce uncertainty by about 50%. So a level is obtained that can, with the necessary caution, be used for prediction. As formulated above, the situation is encouraging, though certainly not sensationally favourable. The prospect to be confronted in 27.5% of cases with a travel missing the expected value by at least –20 or (on the critical side) +26.3%, is not a very pleasant one for a consulting expert.

So *further improvements* remain a challenging goal.

6.4 Improvements – Possibilities and Limitations

In principle, the level of quantitative knowledge in a particular field can be raised in two different manners. On the one hand, hidden information can be brought to light, for instance by a new mathematical approach making obvious hitherto unknown connections between known data. On the other hand, the acquisition of new data may enhance knowledge, be it on the basis of known or of new methods of interpretation. In practical work, especially if new data require a large amount of investigations in the field or in the laboratory, the first-mentioned manner often turns out to be easier. So the following considerations will be started by the question of what can be done *using available quantitative information.* This principle shall be respected as long as improvements on this basis are considered as feasible. Certainly the goal remains unchanged: it is the improvement of predictive power.

In such instances, the main basis for prediction is the knowledge of data taken from a sufficient number of events, allowing reasonably reliable interpolation. As shown in Sect. 6.3, data in ample numbers are readily available only in connection with overall slope (Fahrböschung, $\tan\beta_f$ or f). So prediction in a practical case consists, roughly speaking, of (1) estimating the volume of the mass, (2) attributing the event, after analysis of its character, to one of the groups V, M, L (Sect. 6.3), or to an intermediate state, (3) using Fig. 6.6 to find the approximate value of f, and (4) introducing this value into a longitudinal section of the expected path, thus obtaining a first estimate of the probable reach. This procedure is, of course, nothing but a rough basis, and it is well understood that various additional considerations are required to account for a particular case.

There is no doubt that, besides the difficulty of estimating volume (which, however, is a rather tolerant parameter: an error by a factor 2 means about ±10% in terms of f), the most problematic operation is Item 2. In fact, to find out the character of a slide ante eventum may be an awkward business. One of the possible solutions is to look out for a known event which, after due consideration of the tribologically relevant circumstances, can be considered as "similar". The trouble thereby is, of course, that not only surface quality, gross geometry, possible mechanisms of energy dissipation and energy saving (lubrication) should coincide sufficiently well, but also size. In other words, if a single, easy-to-obtain, size-independent *indicator of motional economy* could be defined, there might be some hope for an essential improvement.

*

Although originally not conceived for the quantitative prediction of rockslides, Hsü's "*excessive travel distance*" (ETD) parameter (1975, 133, 138–139) might, at first sight, be suspected to be a candidate. As it was created with the intention to replace Fahrböschung by a more useful parameter, it is legitimate and, perhaps, worthwhile to discuss it somewhat more in detail. ETD (Fig. 6.7) is defined as the difference between the overall horizontal length of the considered event and a virtual reference event of equal overall height, characterised by $\beta_f = 32°$. As its main advantage, the new parameter claimed an approximately linear increase in function of the logarithm of volume – as compared with the curved function resulting in the normal plot of f against the same logarithm. For more than one reason it is not easy to follow this argumentation.

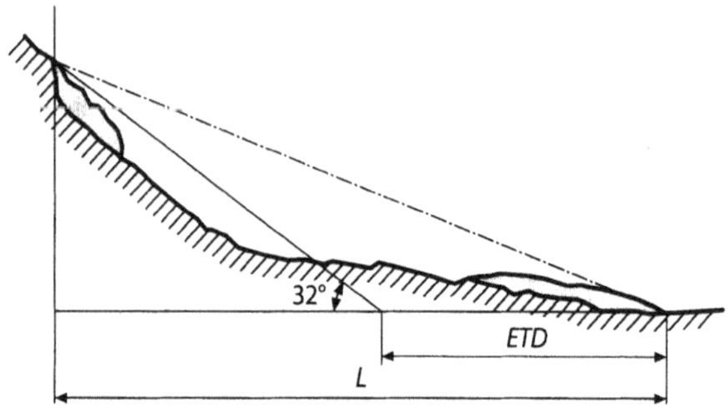

Fig. 6.7. Definition of excessive travel distance (*ETD*) according to Hsü (1975). *ETD* is the part of the overall travel *L* which exceeds that of an event of equal height with 32° as angle of Fahrböschung (sketch by Erismann)

11. If a linear function were considered as particularly important in a plot of f against volume, it would suffice to represent both variables on logarithmic scales as Scheidegger had done in 1973 (obviously his study was unknown to Hsü – and others, as can be concluded from a study by Tianchi 1983). On the other hand, when representing ETD, the wide range of values compresses most of the points into a very small portion of the diagram. A plot on logarithmic scale is impossible as negative values are not excluded (point H1 in Fig. 6.5a, the slide of Airolo, is an example).
12. If ETD had any particularly essential physical meaning, the introduction of an arbitrarily chosen (not a physically significant) standard would have depreciated it. An overall slope line, on the contrary, though not exactly, is at least approximately parallel to the physically important energy line (Sect. 6.2).
13. No information is hidden in ETD which cannot be extracted from f. In fact, the two parameters are convertible into each other by the formula $\text{ETD} = L\,(1 - 1.600 f)$. L is the overall length, and $1.600 = \text{ctan}\,32°$.

When taking into account these facts, it becomes clear that *ETD cannot be of any substantial help* in finding an approach to a more differentiated prediction. Other solutions must be considered to make use of similarities between known and expected events. It will be shown hereafter that there are still possibilities concealed in the concept of Fahrböschung.

*

Thus, returning to information hidden in Fahrböschung, the further line of thought is almost trivial: could not the existence of a Scheidegger function, based on a large number of well-selected events, be used to bridge the gap between similar events of different size? To find an answer, it has to be recalled that a Scheidegger function is nothing but an optimised average of the quantitative relation between f and the size of the considered mass (expressed by its volume V). So the assumption is at least plausible (though admittedly not compelling) that for events of a similar character the particular functional character of $f(\text{Log}\,V)$ is not dramatically different from that of the average. Hence, after appropriate adjustment of Eq. 6.1, "*prediction by similarity*" may be considered as a valid candidate for a somewhat more sophisticated method to estimate Fahrböschung of a prospective event. It can be imagined along the following simple line of operations.

21. Find a known event *i* resembling as well as possible the impending one, *k*.
22. Adjust the Fahrböschung f_i of *i* so as to obtain that of k, f_k, using the equation

$$f_k = f_i 10^{0.1539(\mathrm{Log}V_i - \mathrm{Log}V_k)} \quad (6.4)$$

where V_i and V_k are the respective volumes.
23. Use f_k as described in Item 4 here-above.

The importance (and, hopefully, also the improving capacity) of this second approach consists in using mainly a single, but similar, event as a basis instead of an average. Averaging is, so to say, reduced to a tool helping to *bridge the gap of volumes* between *i* and *k*.

Once the principle being made clear, further refinements are easily found. First of all, it appears useful to adjust all events listed in Sect. 6.3 to an arbitrary standard volume. It is well understood that the choice of this volume is unconditionally arbitrary as the final results obtained with any standard value (be it 0.001 or 10.0 km^3) are perfectly congruent. Obviously the simplest possible form of Eq. 6.4 is obtained if the standard volume is set at $V_k = 1\ km^3$. And as this is a size within the range considered (though far above its average), there can be no serious objection against making this choice, thus yielding $\mathrm{Log}V_k = 0$ and streamlining the equation to

$$f_s = f_i 10^{0.1539\,\mathrm{Log}V_i} \quad (6.5)$$

wherein f_s is the "*standard Fahrböschung*" (standard overall slope), i.e. the approximate Fahrböschung of an event similar to *i*, but having the standard volume of 1 km^3 (Table 6.3). The inverse operation,

$$f_i = f_s 10^{-0.1539\,\mathrm{Log}V_i} \quad (6.6)$$

obviously shows the simple way from a known standard Fahrböschung to an unknown one, as a basis of prediction.

The *plausibility of prediction by similarity* is, to a certain extent, backed up by the fact that events spontaneously recognised as similar are found in close proximity to each other. A showhorse: the f_s values of Almtal, Fernpass (southern tongue), and Pandemonium Creek (Fig. 2.4, 2.6) lie between 0.090 and 0.102 in spite of volumes covering a span of 200:1. Certain irregularities also can easily be discerned. Glärnisch-Guppen, for instance, owes its extremely high standard overall slope of 0.343 to a particular geometry: the bulk of the mass crossed Linth Valley and was arrested in running up; yet a small appendix performed a lateral run-out along the very floor of the valley and placed the farthest distal tongue at an excessively low level, thus increasing Fahrböschung, at least in the sense of its accepted definition. For Sierre, attributed to group M and appearing amidst representatives of L, a questionable classification is by no means excluded: most of its travel was along the wide Valais Valley, probably on a water-saturated fill which, even without lateral confinement, could justify classification L at least as well as M (s. Abele 1974, 118).

All in all, the table shows that L seems to be very well defined while the *differentiation between M and V* is less distinct and probably should be reviewed critically. A look

Table 6.3. Standard values f_s of overall slope, obtained by adjusting the actual values to the standard volume of 1 km^3 using Eq. 6.5. List is arrayed according to the values of f_s. Thus similarities are directly perceivable, and prediction is simplified by use of Eq. 6.6. Ref VML: reference numbers of Table 6.1, showing group by position (*V:* left, *M:* middle, *L:* right)

f_s	Ref V	Ref M	Ref L	Event	f_s	Ref V	Ref M	Ref L	Event	f_s	Ref V	Ref M	Ref L	Event
0.079			62	Damocles	0.166		29		Monte Corno	0.208	15			L. di Molveno
0.087			69	Nozzle	0.170		38		Marquartstein	0.214	14			Dobratsch AE
0.090			74	Fernpass S	0.172		26		L. di Autrona	0.219	06			Mordbichl
0.091			65	San Giovanni	0.173		51		Dejenstock	0.223		49		Bormio
0.091			72	Almtal	0.174		42		Fadalto	0.225		39		Lac Lauvitel
0.098			75	Flims	0.175		43		Mallnitz	0.231		25		Melköde
0.098			63	Twin Slide	0.178	03			Kals	0.234		47		Obersee
0.102			61	Pandem. Creek	0.180	04			Brione	0.245		35		Haslensee
0.122		56		Sierre	0.181		33		Torbole	0.246		30		Cima di Dosdé
0.124			68	Obernberg	0.183		40		Lofer	0.258	10			Col d.l. Madelaine
0.125			71	Tschirgant	0.187		24		Fedaia	0.259	07			St. André
0.128			73	Rockslide Pass	0.191	13			Vaiont	o.259	05			Grand Clapier
0.134			67	Goldau	0.195	11			Oeschinensee	0.261	12			L. di Poschiavo
0.143			64	Elm	0.195		37		Voralpsee	0.264	01			Ludiano
0.146			66	Frank	0.195		27		Hintersee	0.265	09			Pontives
0.146	08			Monte Avi	0.198		34		Haiming	0.269		55		Parpan-Lenzerh.
0.147			70	Lavini S. Marco	0.199	02			Disentis	0.271		52		Marocche del V.S.
0.148		53		Totalp	0.200		54		Monte Spinale	0.278	17			Engelberg
0.150		23		Vallesinella	0.202		46		Cima di Saoseo	0.279		31		Lago di Alleghe
0.152		36		Kl. Rinderhorn	0.202		50		Lago di Tovel	0.281		44		Am Saum
0.153		21		Tucketthütte	0.203	18			Köfels	0.285		22		Val Brenta Alta
0.157		32		Dobratsch JW	0.203		41		Pletzachkogel	0.288		28		Biasca
0.164		45		Ab. des Myans	0.206		48		Dobratsch AW	0.343	16			Glärnisch-G.

at Table 6.4 adds to this impression. At natural scale, columns V and M show only small differences in average and scatter; in column L scatter is distinctly lower. At standard scale, a slight differentiation of doubtful significance between the averages of V and M is observed, and L excels by a better reduction of scatter against that at original scale.

Spontaneous recognition of similarity as postulated for Almtal (Abele 1997, 13–15), Fernpass (Abele 1991), and Pandemonium Creek (Evans et al. 1987) raises the question of a more general approach to *similarity criteria*. The tribological essence in the particular case lies in the fact that in the three mentioned events a steep descent of moderate length, a more or less distinct run-up/run-back, and a more or less acute change of direction are succeeded by a dominating portion of track (about 80% of the overall length for Almtal and Fernpass, 55% for Pandemonium Creek), following

Table 6.4. Comparison of average and scatter values of Fahrböschung at natural and standard scale (for data s. Tables 6.1 and 6.3, respectively), calculated for all 69 events and for groups *V*, *M*, *L*. Reduction of scatter (~1:2) by transition from natural to standard scale is more than size effect (expressed by reduction of average, ~2:3). Note similar results for *V* and *M*, contrasted by distinctly different values for *L*. For further comments s. text

Fahrböschung *f*		Group V 18 events	Group M 36 events	Group L 15 events	All 69 events
Natural scale	Average	0.3372	0.3219	0.1740	0.2938
	Scatter Δf_n	0.0941	0.1007	0.0582	0.1154
Standard scale	Average	0.2289	0.2026	0.1122	0.1898
	Scatter Δf_s	0.0463	0.0435	0.0230	0.0590
	Quotient $\Delta f_s / \Delta f_n$	0.4920	0.4320	0.3952	0.5113

the course of a valley with high probability for effective compressed water lubrication (Sect. 5.5) and, at least partly, excellent lateral confinement (Sect. 5.2). The handicap of relative shortness in the case of Pandemonium Creek is – as far as not reflected by about 10% additional standard Fahrböschung – counterbalanced by particularly favourable conditions in the initial descent (naked rock, glacier ice, lateral confinement) and the geometry of running up and back. This configuration allowed for the kinetic energy still available at the beginning of the "bobsleigh run".

In a more general approach, *three basic aspects* have to be observed in establishing similarity, namely

31. the tribologically relevant character of the main sections composing the track, i.e. effects increasing reach (absence of obstacles, smooth surface, lubrication, lateral confinement, etc.) or reducing it (obstacles, unevenness, narrow passages or curves, topography favouring lateral spreading, etc.);
32. the sequential order of said sections;
33. the percental distribution of said sections within the total length.

As in this stage of work the respective effects are quantified in the lump and not individually, *no clearing between sections* is acceptable and similarity can be assumed only if, in a rough approximation, all three aspects coincide. The residual uncertainty postulated by speaking of a "rough approximation" may be reduced by considering known similarities (e.g. the mentioned one between Almtal, Fernpass, and Pandemonium Creek).

Anyhow, application of standard Fahrböschung to prediction by similarity obviously has *strong and weak points*. On the strong side there is, besides the simplicity, the possibility to "consult" more than one known event in case of uncertain similarity. And, of course, the error due to the adjustment of volume can be minimised if a similar event is available with not too different a volume. On the weak side, the problem mainly consists in a correct judgement about the physical relevance of similarity indicators, especially in case of prehistoric events. It may, for example, be difficult to make sure whether a considered part of a track at the time of an event was formed by the surface of a glacier or not, and the correct succession of incidents may be difficult to reconstruct (Patzelt and Poscher 1993; Abele 1997b, 5–12). But even in less enigmatic cases the attribution of particular features to an event will not always be an easy task. Anyhow, the possibility of a second manner of estimating overall slope certainly is more than only a means of cross-checking the first.

So far, the rule formulated under Heading 6.1 with respect to the *simplicity of outfit* for prediction in the field may be considered as rather strictly observed, provided that it does not count as a violation to take along several descriptions of possibly similar events. Essentially the same view can be taken for the next steps in the pursuit of possible improvements. Although the required preparatory work on the side of evaluation will be substantially increased, application will remain simple.

*

In applying the method of prediction by similarity, individual features of a number of events necessarily have to be considered. So a natural continuation may consist in aiming at a *more differentiated knowledge* than that expressed by nothing but Fahrböschung. It might, for instance, be worthwhile to find out the influence of particular effects upon the overall slope. It might, to quote a practical example, be interesting to know the differential effect upon f obtained by exchanging naked rock against glacier ice in a given portion of track.

The final issue could, for instance, consist in the attribution, on the standard volume scale, of a certain particular "coefficient of friction" μ_k to each partial horizontal length Δx_k of a track characterised by a perceptible tribologically relevant feature. The index k would be an ordinal number used for each of n considered track sections, counted from top scar to distal deposits (if a feature is observed more than once in a track, the respective sections might be added up in one single Δx_k). The quotation marks used for the coefficient are intended to signal that the coefficient stands for the averaged sum of all linear and non-linear resistances, whatever their physical nature may be (as Shreve did in 1966 in connection with air lubrication, s. Sect. 2.3). If the described attribution is feasible, and if the most important systematic errors can be kept under control, it should be possible to build up the overall slope of an event by vectorial addition of n partial lines with slopes μ_k and lengths Δx_k (Fig. 6.8 which,

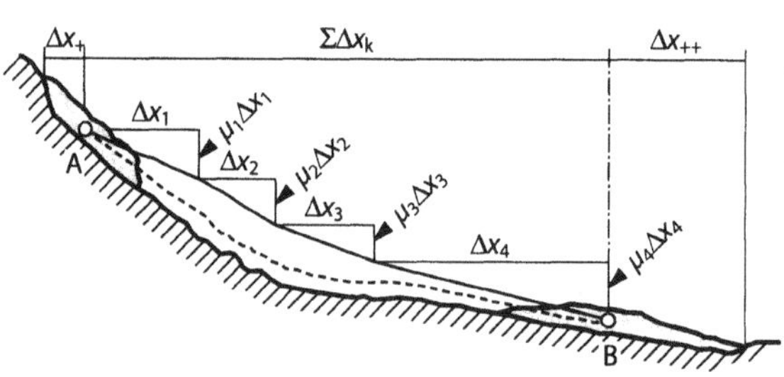

Fig. 6.8. Prediction by local coefficients, working principle. Coefficients $\mu_1 \ldots \mu_4$, averaged from many known events (s. text), are attributed to horizontal components of track Sect. $\Delta x_1 \ldots \Delta x_4$. Travel is set up section by section from initial (A) to final (B) position of centre of gravity (its path is *dotted line*). Polygonal connection is energy line. Note additional lengths Δx_+ and Δx_{++} due to initial and final extension of mass (Δx_{++} important for risk estimation). Note analogy to speedometric considerations used in Fig. 6.4. Attention: straight line from top scar to most distal deposits would yield Fahrböschung. Thus set-up would be more straight-forward, but accuracy lower than in presented configuration (sketch by Erismann)

anticipating further considerations following hereafter, is based on a CoG instead of an overall geometry). The result would be a "*prediction by local coefficients*".

The overall slope of a single event would be expressed by

$$f\Sigma\Delta x = \mu_1\Delta x_1 + \mu_2\Delta x_2 + \ldots + \mu_k\Delta x_k + \ldots + \mu_n\Delta x_n \qquad (6.7)$$

and, after division of both sides by the total length $\Sigma\Delta x$, definition $\Delta_{ik} = \Delta x_{ik} / \Sigma\Delta x$ (i.e. substitution of absolute dimensions by relative ones), and introduction of q known events with ordinal numbers i, the dimensionless set of q linear equations

$$\begin{aligned}
f_1 &= \mu_1\Delta_{11} + \mu_2\Delta_{12} + \ldots + \mu_k\Delta_{1k} + \ldots + \mu_n\Delta_{1n}\\
f_2 &= \mu_1\Delta_{21} + \mu_2\Delta_{22} + \ldots + \mu_k\Delta_{2k} + \ldots + \mu_n\Delta_{2n}\\
f_3 &= \ldots\\
f_i &= \mu_1\Delta_{i1} + \mu_2\Delta_{i2} + \ldots + \mu_k\Delta_{ik} + \ldots + \mu_n\Delta_{in}\\
f_{i+1} &= \ldots\\
f_q &= \mu_1\Delta_{q1} + \mu_2\Delta_{q2} + \ldots + \mu_k\Delta_{qk} + \ldots + \mu_n\Delta_{qn}
\end{aligned} \qquad (6.8)$$

with the n unknown values μ_k results. By application of error analysis, Eq. 6.8 can be transformed into *Gaussian normal equations* (in the given situation preferable to Tshebysheff's method), i.e. a set of n linear equations with n unknown values. To simplify things, hereafter the case of $n = 3$ is presented:

$$\begin{aligned}
&\mu_1\Sigma_i(\Delta_{i1}^2) + \mu_2\Sigma_i(\Delta_{i1}\Delta_{i2}) + \mu_3\Sigma_i(\Delta_{i1}\Delta_{i3}) - \Sigma(\beta_i\Delta_{i1}) = 0\\
&\mu_1\Sigma_i(\Delta_{i1}\Delta_{i2}) + \mu_2\Sigma_i(\Delta_{i2}^2) + \mu_3\Sigma_i(\Delta_{i2}\Delta_{i3}) - \Sigma(\beta_i\Delta_{i2}) = 0\\
&\mu_1\Sigma_i(\Delta_{i1}\Delta_{i3}) + \mu_2\Sigma_i(\Delta_{i2}\Delta_{i3}) + \mu_3\Sigma_i(\Delta_{i3}^2) - \Sigma(\beta_i\Delta_{i3}) = 0
\end{aligned} \qquad (6.9)$$

whereby the index of Σ_i signals sums calculated over all values of i. The equations, easy to solve even with simple means, are, once more, optimised in the sense of reducing to a minimum the sum of the squared differences between the real values of f_i and those obtained by using the μ_k values resulting from Eq. 6.9 to reconstruct Eq. 6.8.

All this sounds very promising. Before recommending the method for practical application, however, various *sources of errors* (more substantial than those committed in determining the points of transition between sections) should be considered in some detail. Besides considering the questions which can be treated a priori, several simple cases have been tentatively calculated with data taken from well-investigated events so that the practical experience thus acquired will be of some use.

41. Less than the entire track is passed by the entire mass (see, in Fig. 6.8, extensions Δ_+ and Δ_{++}). Consider the proximal and distal ends of Vaiont (Fig. 2.36c, d) where a mass of almost 1500 m length was displaced by hardly more than 500 m. Obviously the weight of the track's features is not necessarily equal at every point.
42. Velocity-dependent resistance may influence results. Only in case of strictly dominating Coulombian friction do equal features mean equal resistances at various locations within the track.

43. Even in purely Coulombian friction the local degree of disintegration can influence the resistance in a given section of track (Sect. 5.1).
44. Tentative calculations have shown that results obtained from few events are highly sensitive to relatively small variations of inputs (absurd results, like negative μ values, are not excluded). Thus large numbers q (in other words: $q \gg n$) are stringent. So data acquisition may become precarious.

The verdict is clear: in the presented form and on the basis of the knowledge available at present, prediction by local coefficients, in spite of its seducing aspects, cannot be regarded as a ready-for-use practical solution. Still its prospects justify, perhaps, a review of possibilities which might *give it a more realistic character.*

In first instance, a line of thought similar to that applied in the context of Scheidegger functions might be useful. In fact, *subdividing events into groups* automatically brings together rockslides with certain tribologically relevant similarities. Of course, in the most proximal and distal portions the uncertainties of Item 41 will remain; of course, velocity-dependent resistance (Item 42) and local degree of disintegration (Item 43) still will partially defy the assumption of topographic features as unique criteria of resistance. Yet an improved coincidence of decisive combinations (criteria applied, parameters involved) in long-term, multi-event evaluation on the one hand, in short-term, single-event prediction on the other hand cannot be disputed if the features are selected to be sufficiently uniform in tribological matters. Still the ironical remark might be made with a good reason that error symmetry between evaluation and prediction sometimes might entail a correct result...

A fundamental question, however, begins to emerge with increasing clarity. Can the local variation of coefficients of friction be made sufficiently effective if based on no other data than the overall dimensions of an event, combined with tribologically relevant features of the track? As this critical argument will reach beyond the bounds of the considerations so far made, it will be treated hereafter separately.

*

The question at the end of the preceding paragraph has already once been answered in the related context of speedometry (Sect. 6.2). The overall geometry, expressed by Fahrböschung, as being limited in its physical content, should be replaced by the more meaningful geometry based on the *path of the centre of gravity* (CoG) of the considered mass. And the task of the following considerations is to disclose possibilities and limitations of predictions resulting from this fundamentally superior approach.

As mentioned under Heading 6.2, Item 4, it is normally reasonable to use, instead of the entire mass, its midstream longitudinal section for velocity determination. The arguments used are valid for a CoG geometry with equal rights as for an overall geometry. The advantage consists in reducing the three-dimensional problem to a two-dimensional one at the expense of a moderate loss (and in some cases even a moderate gain) in accuracy. In Sect. 6.2 the reader's attention also was drawn to sporadically feasible en-route measurements of velocity as a means to introduce polygonal energy lines, and it was shown that such energy lines may even be competitive with twin-parameter calculations using a constant and a velocity-square-dependent member. So here it remains to look out for *more systematic methods* of determining CoG positions (before and after displacement) and connections between topographic features and resistance against motion.

Almost in all cases of near-to-present rockslides longitudinal sections, both for the situations before and after displacement, are available (or easy to establish) so that it is not very problematic to find the respective *CoG locations* with acceptable accuracy. The impossibility to reconstruct certain details in the top region of Huascarán is an exception (which, owing to the extension of the track, does not exclude a fairly reasonable estimation). For prehistoric events the situation can vary from perfectly clear to hopeless, and the handicap of a substantially reduced number of appropriate events has to be admitted.

As a general rule, the chance for a useful *estimate of CoG locations* is not too bad in those cases where a scar is perceptible and substantial parts of the deposits have remained in place. In such instances reconstruction is, roughly speaking, confined to the original outline of the surface. Thereby it may be of great help if the approximate volume is known (which is required anyway as a basis for making use of the size effect). In the area of release, extreme eccentricity of CoG is not a very probable phenomenon, and the same applies also to masses arrested by bulky obstacles (generally steep slopes). The following remark might be useful in this context. Compare a rectangular longitudinal section of a mass with a trapezoidal one of equal area and length, having a ratio of 2 : 1 between the thickness at the proximal and the distal end. The longitudinal difference of CoG positions then will be only 5.6% of its length. Contrarily, deposits extended over a considerable length require the measurement of their depth at several points to obtain a sufficiently reliable result.

A positive aspect has to be stressed in this not very encouraging situation. Even moderately accurate CoG positions probably will yield definitely narrower scatter bands than known from overall slopes. To understand this prognosis it is useful to recall the particular sources of *scatter in long run-outs*. Various mechanisms, rated as marginal in the context of general motion, gain importance in the special circumstances of low thickness. Hereafter they are recalled to the reader's mind. Small variations of the slope angles of bed and distal talus (Fig. 5.4) result in large variations of length; rolling, as soon as the dominance of the overburden fades out, becomes possible both as near-to-surface free motion and as roller bearing (Sect. 5.3); the "energetically bankrupt" rebounding – on slopes far smoother than required to overcome friction – remains possible as long as there is sufficient (external and/or internal, s. Sect. 4.2) kinetic energy to excite small bounds (Sect. 5.3); the mechanism of "playing volleyball" (i.e. dynamic lubrication) is no longer strangulated by a low ratio between lifting and lifted masses (Sect. 5.5); kicking in the distal region, occasionally also by large boulders overtaking the rest (Sect. 4.2), contributes to stretch the tongue (Sect. 5.7); in certain cases, a substantial portion of water in the debris (not specially treated in this book, s. Sect. 2.7) entails additional mobility. All these phenomena are involved in a highly complex general interaction, and their particular contributions to scatter add up in a manner difficult to treat quantitatively. It is nothing but natural that a substantial total scatter necessarily results. When the CoG is considered instead of the most distal elements, the internal effects are, to a certain extent, ruled out, and it is not unrealistic to rate this gain higher than the drawback due to a somewhat imprecise CoG determination.

Anyhow, a stringent necessity for the future consists in *re-writing Tables 6.1 and 6.3* in terms of energy line slopes instead of Fahrböschung. As far as possible, geomorphologically perceptible and physically relevant features of the tracks should be

Table 6.5. Features to be accounted for in re-writing Tables 6.1 and 6.3, visualised by example. Δx: Length of section, referred to total length (sum is 1.0000); state $C \rightarrow D$: disintegration within section; state D: disintegrated

Δx	State	Slope	Track	Remarks
0.0303	C→D	1.00	Naked rock	Disintegration near end of section
0.1591	D	0.33	Glacier ice	Temperature unknown (summer)
0.1477	D	0.45	Naked rock	Slight lateral confinement
0.0341	D	0.00	Rock, valley fill	Vertical curve, $R = 190$
0.0511	D	0.36	Rock+humus	Run-up, half-curve on slope
0.0701	D	−0.39	Rock+humus	Half-curve, run-back impeding run-up
0.3636	D	0.13	Valley fill+H_2O	Lateral confinement, 6 curves, $235 < R < 470$
0.1440	D	0.07	Valley fill	Spreading zone, H_2O uncertain, lake at end

added according to Table 6.5 in order to make possible an improved description of similarities and/or a more consistent subdivision into groups.

It is well understood that the required work is very extended, both in the study of existing literature and in the field. And it also has to be admitted that the way from known events to an expected one is made more complicated by the necessity to conclude from topography (as far as discernible) to positions of CoG and vice versa. But there can be no doubt that such efforts will not be made in vain.

*

What follows, is an abridged repetition of reflections developed, with overall slope in mind, in this and the preceding section. Each of the items describing a *method of prediction* might be introduced by the words: "*Provided that a sufficient number of events can be made available*", and the consecution of items is given by the condition that from item to item the number of required events remains constant or is increasing. It is well understood that in this context the term "required" includes the meaning of "*sufficiently well investigated to allow for successful application*".

51. Establishment, in the same way as for Fahrböschung (and using all available events), of a "CoG Scheidegger function". Prediction based on expected volume. The average of many predictions will be better than that based on Fahrböschung. Whether any advantage can be obtained in comparison with more sophisticated Fahrböschung-based predictions, is an open question.
52. Prediction by similarity based on CoG Scheidegger function. If number and variety of used events are sufficient, better average results than with any Fahrböschung-based method may be expected.
53. Use of CoG Scheidegger function to establish a standard-volume, CoG-based table of energy line slopes β_e (in perfect analogy to Table 6.3); formulation and solution of Gaussian normal equations. Prediction by local "coefficients of friction". Advantage: coefficients can, after adjustment to volume, readily be used for any new event. At the time being, accuracy is promising, though difficult to estimate.

54. Subdivision of events in groups with well-defined characteristics (analogous or identical to V, M, L groups of Sect. 6.2) and establishment of a CoG Scheidegger function per group. Prediction based on determination of group and expected volume. Comparison with Item 52 as concerning accuracy is difficult.
55. Combination of Items 51, 52, and 54 by subdivision of events in groups, establishment of the respective CoG Scheidegger functions, and prediction by similarity. If number and characteristics of used events are satisfactory, better accuracy than with any other method presented so far may be expected (including the above items).
56. Combination of Items 53 and 54 by establishment, per group, of a standard-volume, CoG-based table of energy line slopes, including formulation and solution of Gaussian normal equations. Prediction by local coefficients. Advantage: coefficients can readily be used, after adjustment to its volume, for any new event. At the time being, accuracy is promising though difficult to estimate.

It is evident that by no means can one of the proposed methods be recommended by itself as "the best of all possible solutions". On the contrary, it seems reasonable to use, say, two and thus to obtain a means of mutual cross-checking. And it is well understood that the resulting zigzagging energy lines, as yielding continuous information about velocity, can be checked against other methods of velocimetry (s. Sect. 2.7, 6.2).

A general drawback, inherent to any CoG geometry in comparison with methods based on Fahrböschung, lies in the necessity to estimate the *additional reach of debris* beyond the final position of the CoG (Δx_{++} in Fig. 6.8). In other words: as soon as disintegration acquires a certain degree, the thickness of the mass starts fading, and a larger distance than that at release has to be added to the travel accomplished by the CoG. It is evident that in all Fahrböschung-based methods this extra is included. A toll has to be paid for any envisaged gain in accuracy. Where Δx_{++} is small as compared with the total length of track (not to speak of masses remaining in an essentially coherent state), the estimation may be a rough approximation. In cases of extreme run-out length the problem can become serious. To decide how far studies on granular motion can be helpful, is a challenge for future investigations (s. second-last paragraph of this section).

If the method of local coefficients would turn out to give reasonable results, it certainly would be a seducing challenge to attack the problem of a direct approach to the parameters μ and c_v as postulated in Eq. 5.20 and thus to quantify the *amount of nonlinearity* reigning in the mechanisms of rockslides. From the present study it appears probable that, for the practical work of reach prediction, this additional knowledge would not yield dramatic results. Still it would be an important step forward on the scientific side if Körner's daring dream of 1976/1977 would come true on a more reliable basis.

The difficulties of such an endeavour, however, should not be underestimated. As far as they lie *on the mathematical side*, they are not linked to the process of synthesising an energy line. In fact, the differential Eq. 5.20 has, for the velocity v at the end of a track section length $\Delta s = \Delta x / \cos\beta$, the elementary solution

$$v = \sqrt{v_\infty^2 + (v_0^2 - v_\infty^2)e^{-2c_v\Delta s}} \qquad (6.10)$$

where $v_\infty = \sqrt{[g(\sin\beta - \mu\cos\beta) / c_v]}$ is the velocity of equilibrium reached after an infinite travel under constant external conditions, v_0 the velocity at the beginning of the section, g the gravitational acceleration, and μ and c_v the coefficients of Eq. 5.20. As a more general alternative, this equation also can be integrated in quasi-infinitesimal steps, and then more than two members (one constant, one proportional to the square of velocity) can be taken into account (if at all required). More troublesome is the problem of getting at the coefficients μ and c_v which can no longer be determined by a set of linear equations. Probably parameter variation would be the recommended approach, and the requirements for computer capacity accordingly increased.

The real difficulty, however, resides in a physical question. The velocity-dependent coefficient c_v *cannot be considered as constant.* Like the micro-rugosity of a pipe passed by a liquid, the macro-rugosity of the track may entail, by number and intensity, more or less decelerating collisions with elements of the moving mass. And thereby the degree of disintegration plays a most important role (Sect. 5.1). So it is not an easy task to set up a system of equations, fed with granulometric and rugosimetric data (as far as known) and yielding a set of locally assorted coefficients after solution.

This is true in spite of the fact that various *studies treating granular motion*, at least implicitly, express the opinion that "granular flow" can be formulated relatively easily by equations similar to those used in flow dynamics. Some of these studies (as well as the resulting models, experiments, and simulations), no matter whether based on Bagnold's work (1954, 1956) or not, display a high scientific standard and also may be of great use under certain circumstances – unfortunately rather in the field of moving bulk goods than that of rocks. To fulfil the conditions stipulated in the preceding sections of this book, most of these studies lack sufficient consideration of some facts which are essential for the displacement of a rockslide. Here the main examples: the role of a large overburden; the variety not only of particle sizes, but also of their shapes; the tendency of the particles to assume a position with the shortest axis at right angles to the ground; the importance of lateral confinements; and, last but not least, the size effect. Some of these publications already have been mentioned in other sections. In Sect. 5.1 refer to: Savage (1979), Savage and Hutter (1989), Nohguchi et al. (1989), Bak and Chen (1989), Wiesenfeld et al. (1989), Hutter (1991), Hutter and Koch (1991); in Sect. 5.3–5.6: Dent (1985), Campbell (1989). It is not intended here to enumerate further references, as a well-commented (though in certain respects perhaps somewhat optimistic) review is given by Straub (1997).

Anyhow, there is a wide field open for perhaps rewarding (and certainly laborious) investigations, a wide field of half-acquired knowledge which James Hutton (1795, II, 105) might have commented by his beautiful argument about the principle of trial and failure: "*If man must learn to reason,… he must reason erroneously before he reasons right…*"

Chapter 7

Secondary Effects

In descending, an immense mass had filled a verst of the gorge's length, thus damming up Terek River. Guards ... observed that the outflow rapidly was fading, and in a quarter of an hour it was completely quiet and exhausted. It took the Terek no less than two hours to force its way through the slide. Then, indeed, it became terrifying!

Alexander S. Pushkin
(after having crossed the Caucasus; original in Russian)

7.1 Flood Waves

As an agent in the mechanisms governing the dynamics of descending rocky masses, water has been mentioned more than once in this book. And the roles it had to play in various contexts reach from the "anti-lubricant" postulated by Terzaghi (1960, 91) to Abele's highly efficient pressurised lubricant (1997b, 2), both backed up by extensive field evidence. The role of *water as the very destruction-bearing mass*, however, though stressed for the catastrophes of Val Pola (Sect. 2.5) and Vaiont (Sect. 2.6), has, so far, not been discussed with respect to its physical background.

It may be considered as an open question how far *such a discussion should take place* in this book. On the one hand, occurrences are, of course, frequent, and there are physical relations between fluid and rockslide dynamics; on the other hand, substantial parts of a comprehensive excursion into this field would mean little more than retelling things which have existed in the literature for a considerable time, and (as stipulated in Sect. 2.5 and 2.6) often the usual equations cannot be readily applied to obtain reach, especially in run-ups. So it appears judicious to confine more detailed comments to the particular aspect of long run-outs and high run-ups and to refer to existing studies for the rest. This is all the more plausible as far reach often is connected with heavy damages and as the mechanisms behind it, at least to the author's knowledge, hitherto not have been treated with due care in the literature.

*

Among the various publications dealing with *field evidence* of waves generated by masses plunging into a large volume of water (a lake or the sea), two of the most impressive have been chosen for a short comment. The first case occurred in 1958 in Lituya Bay, Alaska. Giant waves having run up as high as about 530 m (!) were reported by Miller (1960). His study contains (p. 60) a table of similar events in Japan, Norway, and the U.S.A., which, however, did not exceed a maximum run-up height of 80 m. This contrast, obviously not a consequence of erroneous observation, focuses attention on the particular topographic situation in which a high kinetic energy must have been concentrated in a comparatively low volume of water. The second example concerns the eruption of Mount St. Helens in 1980. In their elaborate description of the large rockslide entailed by the partial decapitation of the mountain, Voight et al. (1983, 249, 255–258) mention the wave generated in Spirit Lake. It ran up 260 m and was followed by a lifting of the lake's level by 60 m. This tour de force was accomplished by the easternmost lobe of the slide which may have represented a mass of somewhat between 0.5 and 1 km^3 (it is difficult to estimate the portion that, after having contributed to the generation of the wave, turned to the left when encountering the resistance of the opposite slope); the lake, with a surface of about 8.5 km^2, must have had a comparable volume. This makes the case interesting for the present investigation as, owing to the difference in density, the mass of rock very probably exceeded that of water by more than a few percent.

On the *theoretical side*, Noda (1970) developed a practical model of a slide plunging into a lake, with ready-to-use equations allowing, for instance, to calculate, for a given water depth, the wave amplitude resulting from a slide of given velocity (or vice versa). A more recent study, backed up by laboratory tests and presenting the key information in a particularly compact form, was published by Vischer (1986).

It was with a good reason that Costa (1991, 26), in his thoughtful comments on Val Pola (Sect. 2.5), pointed out that application of Noda's equations would result in a violation of one of the basic assumptions made. In fact, the model is set up for everyday use, and as waves most frequently are a consequence of relatively small incidents (local rockfalls etc., as presented e.g. by Huber 1982, 566–570, in a historical review of Swiss events), it is nothing but natural to assume the *water-to-rock ratio* of masses as being far higher than unity. Now in the case of Val Pola the northern lobe of the slide, at least 0.01 km^3, hit a water volume that was about 200 times smaller. For Vaiont a water-to-rock mass ratio of 1/20 already has been mentioned (Sect. 2.6), and even if only the western lobe is considered, there still remains approximately a power of ten. In both cases the velocity of the rock mass was relatively low: in Val Pola the water was hit in a run-back, and for Vaiont even a top speed of 30 m s^{-1} is excluded. As observed here-above, things are not so clear for Mount St. Helens: very probably the mass of rock was distinctly larger than that of water, and the velocity of the rock mass was higher than in the two Italian events. Anyhow, in two of the three cases the run-up heights on the water side were remarkably high if considered in their relation to the velocity of the rock mass. So it appears worthwhile to think about mechanisms determining the displacement of a limited mass of water by a larger mass of moving rock. Thereby due attention has to be paid to the inherent mobility of water.

*

In a somewhat idealised form the mechanism of *sweeping away the water* rather than plunging into it (as characterised in Sect. 2.5) is represented in Fig. 7.1a–c. Three phases can be discerned.

1. The more or less wedge-shaped frontal slope of the moving rock mass exerts – be it with or without friction – a lifting and accelerating force upon the water (Fig. 7.1a). This force decreases with the fading out of differential velocity between the two masses. Simultaneously the water loses thickness by longitudinal and/or lateral spreading. The phase ends in a momentary standstill of the water with respect to the slope (Fig. 7.1b).
2. Under the influence of gravitation the water starts re-descending. Thereby it assumes a velocity more and more exceeding that of the rock mass. Spreading goes on. The maximum of velocity is attained more or less simultaneously with passing the distal edge of the rock mass (Fig. 7.1c).
3. After having left the rock mass behind, the water moves forth on the account of its own kinetic energy. The height of a possible run-up of its centre of gravity (CoG) is, roughly speaking, proportional to the square of the velocity at the foot of the run-up.

To obtain an *easy quantitative approach* to the essential points of the mechanism, secondary effects should, in a first consideration, be left aside. Thus let (11) the ground

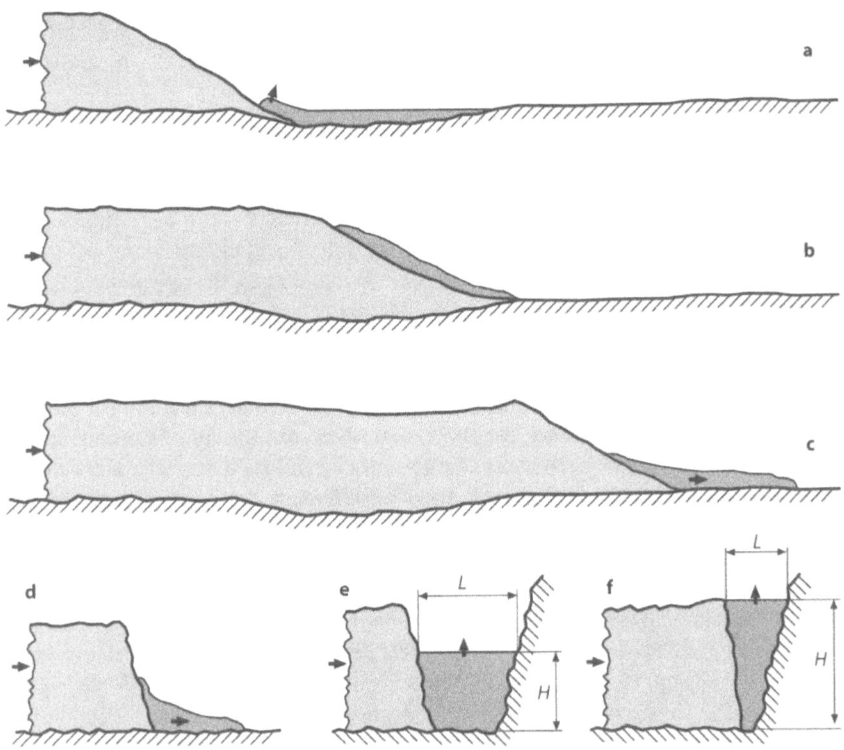

Fig. 7.1. Mechanisms of effective energy transfer from large rock volume to smaller water volume. *Arrows:* Directions of motion. **a, b, c** displacement by sweeping away with "soft" restitution requiring considerable relative travels. **a** Approaching rock mass starts vertical and horizontal displacement of water; **b** water in most elevated position, momentarily immobile with respect to rock mass; **c** water, moving at top speed, on the verge of leaving behind rock mass. **d** Steep front of rock mass, unable to lift water effectively, hardly can accelerate "bow wave" to distinctly higher velocity than its own. **e, f** Forced expulsion of water from gully formed by immobile and advancing rock walls: velocity of water is approximately proportional to H/L (refer to text); note that sketches **e** and **f** may be snapshots of one and the same process (sketch by Erismann)

be horizontal. Let (12) the water-to-rock mass ratio be low enough to base calculation on a constant velocity of the rock mass. Let (13) the resistances against relative displacement for the pairs rock/ground and water/rock (for the second-mentioned including the first – possibly collisional – contact) be negligible while the mechanism is working. Let (14) the rock mass be practically impermeable. And eventually, to simplify things, let (15) the audacious observer travel with the rock mass, thus considering this mass as immobile.

In such instances the water is observed as being shot (at the unperceived velocity of the rock mass) against the frontal slope of the rock mass, then running up the slope until its kinetic energy is spent, finally running back and leaving the rock mass. Except for the spreading of the water and the loss of CoG elevation resulting therefrom, the entire process is – by virtue of Items 13 and 14 – perfectly symmetrical, and no energy is dissipated. So the velocities of the water in contacting the rock slope and in leaving it differ only by their directions. Looked at with the eyes of an immobile observer, this means that the water returns from the slope at *twice the velocity of the rock mass.* In other words: the water acquires a kinetic energy (and a running-up capacity) exceeding by four times that of a water mass moving at equal step with the rock mass. This rough quantification promises an explanation for excessive run-ups.

Before acceptance of such a conclusion it appears, however, advisable to scrutinise Items 11–15 for *energetic leakages.* Immediately Items 11 and 15 can be dropped as, on the one hand, the initially horizontal surface of the water forces the considered part of the track into a near-to-horizontal direction and, on the other hand, the viewpoint of the observer is physically irrelevant – it was chosen for purely didactic reasons. The central parameter of the mechanism is the water-to-rock ratio of Item 12, so far assumed to be infinitesimally low. Its increase necessarily modifies the energy transfer from rock to water.

Here it is opportune to recall the collisions presented in Sect. 4.2 and (in the context of kicking) Sect. 5.7. In fact, the above-described consecution of running up and back on the slope of the rock mass energetically resembles a perfectly elastic collision of a large solid body with a small one. The difference, besides the particular features of water, mainly consists in the duration of interaction which is not infinitesimally short. The term of "*soft restitution*" might be used to mark the difference with respect to restitution by collision. In all other respects there is a perfect analogy so that the equations used for collisions are, mutatis mutandis, applicable. For a parabolic slope profile the process evolves in direct analogy to the loading and unloading of a soft coil spring. Photos a and b of Fig. 7.2 show, in a primitive toy railway experiment, the effect of such a spring that grants a high degree of energy restitution and therewith a long run-out of a small body "softly kicked" by a larger one. Photos c and d of the figure, anticipating Item 21, represent basically the same configuration, the spring, however, being replaced by an energy-dissipating crumple zone of aluminium foil. For further comments see the legend.

Motional resistance and permeability of the rock mass are closely interconnected so that Items 13 and 14 will be treated in common. As a matter of fact, according to the permeability of the rock mass, different scenarios with different hydrodynamic resistances can be imagined.

21. If the interstitial voids in the rock mass are sufficiently accessible and large to swallow the entire water, the return, if at all taking place, may be reduced to feebly seeping rivulets. Energy is mainly dissipated by the water being pressed through numerous narrow, badly shaped passages, and the water assumes more or less accurately the velocity of the rock mass (Fig. 7.2c, d).

Fig. 7.2. Experimental demonstration of "soft restitution" between large driving and small driven ▶ mass as discussed for rockslide sweeping away small water volume. **a, c** Initial situations with heavy vehicle ready to descend and light vehicle ready to be hit. **a** Light vehicle bears soft coil spring as energy accumulator. **c** Heavy vehicle bears slightly pre-crumpled energy absorber of aluminium foil. **b, d** Final situations after displacement. **b** High energy transfer between vehicles by loading/unloading of spring, low dissipation (travels after collision: heavy vehicle 4 rail lengths, light vehicle almost 11 rail lengths). **d** High energy dissipation by complete crushing of absorber, low transfer (after collision equal travel of both vehicles: 5.5 rail lengths). Light-to-heavy mass ratio in configuration **a, b** 0.18; in configuration **c, d** 0.16 (photos by Erismann)

a
b
c
d

22. The other extreme is a practically impermeable front where the water cannot lose much of its volume. This is the case of more or less coherent rock masses. As in such instances the surface normally is rather smooth, excellent restitution may be expected as far as the shape of the front allows for an appropriate lifting of the water: a very steep or overhanging profile forces the water into a bow-wave shape: run-up is low, and the ability of restitution is poor (Fig. 7.1d).
23. The third possible configuration, a moderately loose rock mass, is particularly interesting by not excluding, under certain circumstances, a remarkably high coefficient of restitution. Necessarily, on the uneven surface of such a mass, an energy-dissipating turbulent boundary layer starts to develop. If the "inlets" of the mass, though unable to swallow a considerable portion of the water, can do so with most of this layer, the turbulence in the unswallowed water is effectively reduced. The mechanism is in perfect analogy to the technique of sucking off the boundary layer at the critical portions of an airfoil, a solution known to aircraft engineers for more than 60 years and achieving excellent resistance coefficients in the wind tunnel, but hitherto hopeless in practice owing to the problems of effective low-cost, low-energy sucking arrangements. In the rockslide the toll paid is nothing but a certain loss of water which is driven into the mass essentially by the weight of its liquid overburden.

Obviously energy losses in this set of mechanisms can vary within a wide span so that for its coefficient of restitution the range $0 < J < 1$ has to be considered. Similar are the conditions for the water-to-rock mass ratio $c_m = m_w/m_r$ which, by definition, is restricted to $0 < c_m \ll 1$ (it is needless to say that m_w and m_r are the respective masses of water and rock). Then in Eq. 5.36, owing to the absence of an initial velocity of the water, all terms containing the initial ratio of velocities $c_u = u_w / u_r$ (where u_w and u_r are the initial velocities of the two masses) can be cancelled, and the equation assumes the simplified form

$$\frac{v_w}{u_r} = \frac{1+J}{1+c_m} \tag{7.1}$$

where v_w is the *resulting velocity of the water.* Two trivial consequences are worth mention: whenever $c_m > 0$ is true, the rock mass loses velocity in the mechanism; and for $J = 0$ both masses finally move at equal velocities. This is the case of irreversibly swallowed water, as mentioned in Item 21 and visualised in Fig. 7.2d.

As a reasonable estimation of c_m in many cases is not excluded, the essential difficulty in *prediction* normally lies in determining J. If a rather coherent rock mass can be expected, the shape of its front perhaps can be assumed more or less exactly. But for a disintegrated mass a previous differentiation between the descriptions of Items 21 and 23 is an extremely problematic task. For $J = 0$ the resulting range is $0.5 < v_w / u_r < 1.0$ and for $J = 1$ it is $1.0 < v_w / u_r < 2.0$. And, as mentioned here-above, a factor 2 in velocity means a factor 4 in run-up height…

The circumstances of *Val Pola* point to a sweeping away scenario. When hitting the water, the mass of rock definitely was in a disintegrated state. And as the lake had a maximum depth of only 5 metres and lay in the relatively flat area near Morignone (Fig. 2.32), a mechanism similar to that shown in Fig. 7.1a–c can be imagined. In view of a negligible value of the water-to-rock mass ratio c_m, coefficients of restitution J between 0.5 and 0.7 would have yielded water velocities 1.5 to 1.7 times that of the rock. Without this acceleration, the water would have run up only 62 to 48 m instead of the 140 m partly visible in Fig. 2.34.

*

The inherent mobility of water makes possible another mechanism able to accelerate a relatively small mass of water to a high velocity. Imagine an essentially coherent rock mass with a steep distal front approaching a steep slope on the opposite side of a valley containing a sufficient amount of water. Thereby a narrowing water-filled gully is formed between the two rock walls (Fig. 7.1e, f). As a necessary consequence the *water is forcibly expelled* by the rock taking its place. Similar situations certainly did occur in the case of Vaiont (Fig. 2.36).

The velocity which the water can assume depends on the velocity of the rock mass and the geometry. In fact, the mean forced velocity of the water is proportional to the velocity of the rock mass and to the hydraulic ratio of transmission between attack and escape of the water. And this ratio is the quotient of the sectional area of attack divided by the sectional area of escape. Both areas have, of course, to be considered at right angles to the respective displacements. Now assume an approximately plane flow for the central portion of the event (the total width being sufficient to neglect, in this portion, lateral escape of water). Thereby the process is reduced from its originally three-dimensional to a two-dimensional level and the quotient of areas to a quotient of lengths. So the mean *velocity of escape* can be written

$$v_w = v_r \frac{H}{L} \tag{7.2}$$

where v_w and v_r are the respective velocities of water and rock, H the thickness of the area of attack (at right angles to the direction of attack), and L the longitudinal extension of the area of escape (at right angles to the direction of escape).

H and L are shown in Fig. 7.1 in two different configurations. The ratio of transmission of the "hydraulic gear" is somewhat below unity in Fig. 7.1e and above 2 in Fig. 7.1f. Now both sketches may be considered as two snapshots of one and the same process. In other words: the ratio of transmission is increasing while the mechanism is going on. This means that in the final phase, provided that the rock mass is large enough to overcome the rapidly growing pressure of the water, *excessive escape velocities* are not excluded. They may exceed by several times the initial velocity of the rock mass.

Where exactly, in the case of *Vaiont*, the centre of gravity (CoG) of the water at the end of its run-up might be found, is an open question. Yet, as the highest level reached by water was at about 950 m, i.e. 240 m above its initial surface (not its CoG), a conservative estimate for the CoG's run-up would perhaps yield 130 m. For a perfect conversion of kinetic energy of rock into potential energy of water this would mean an initial velocity of 50 m s^{-1} or, if the maximum velocity of the slide was 25 m s^{-1}, twice this value. In addition, certain sources of energy losses, both in conversion from rock to water and in running up, cannot be ignored. Hence it is as good as certain that the overall ratio of rock to water velocity exceeded the level 2.0, i.e. the absolute limit found for the sweeping away mechanism. This is, besides the topographic evidence (Fig. 2.36), a strong argument in favour of forced expulsion as the mechanism having accelerated the water in Vaiont Valley.

It should be repeated, in this context, that the *ratio of transmission* v_w / v_r, as defined in Eq. 7.2, continuously increases during the process of water expulsion. It can, therefore, not be compared directly to the ratio v_w / u_r of Eq. 7.1 where the final mean velocity of water was related to the initial velocity of rock.

*

To all appearances, the two presented mechanisms give sufficient arguments for a physically plausible understanding of the *unexpectedly dramatic effects* observed in the events of Val Pola and Vaiont: water reached farther and ran up higher than could be guessed by attributing it a velocity equal to that of the rockslide which mobilised it.

For both mechanisms fundamental *difficulties of prediction* remain. In case of sweeping away, an exact estimation of the coefficient of restitution can become a knife-edge question between Items 21 and 23. And for forced expulsion, the question of how far the rock mass is decelerated in the moment of maximal water velocity, cannot be predicted with reasonable accuracy. In both cases even a careful estimation contains a substantial uncertainty, especially if the result has to be expressed as a run-up height. In both cases extreme caution is recommended.

7.2 Damming Effects

In the preceding section water has been presented as a medium sometimes active at places so distant from a rockslide that its appearance was unexpected. Val Pola (Sect. 2.5), contrarily to Vaiont (Sect. 2.6), can be considered as a model case in which correct prediction would have been very difficult. In addition to a precise forecast of the rock displacement in running back, it would have required a very special knowledge of rock-to-water energy transfer (Sect. 7.1). Hereafter a possibly even more perfidious quality will be commented: water can bear its destructive capacity over hundreds of kilometres, and such strikes may be *delayed by hours, weeks, or millennia*, thus creating similar situations of threat, uncertainty, or false certainty as often do exist in the time preceding the release of a rockslide.

Nevertheless, the *usual configuration* is almost the simplest that can be imagined. A rockslide (or a large rockfall as in the case of Randa, Noverraz and Bonnard 1991; VAW 1991, 30–32), after descending from one of the lateral slopes of a valley (or along a tributary valley), dams up an existing water course, thus forming a lake. In the key events (Chap. 2) this configuration entailed actual damming in four cases, namely: Köfels (Sect. 2.4) where a lake did exist for several millennia in the basin of Längenfeld, at the south of the deposits (Heuberger 1975, 234–235; Heuberger et al. 1984, 347); Val Pola (Sect. 2.5) with a succession of two dangerous effects, the precursory little lake, cause of the actual death toll, and the main one that threatened for some weeks the densely populated Valtellina Valley, but finally was controlled by well-planned drainage (Costa 1991); Vaiont (Sect. 2.6) where part of the artificial basin was additionally dammed by the slide without generating additional damage (Fig. 2.42); finally Huascarán (Sect. 2.7), particular by the alertness of the inhabitants which, in taking flight from the threatened valley, avoided a possibly even more disastrous catastrophe than that of Yungay (Plafker and Ericksen 1978, 297; Stadelmann 1983, 64).

It is by no means extraordinary that four in six key events – chosen on account of criteria alien to damming – display this phenomenon (and that Pandemonium Creek would have represented a fifth example if the mass had crossed a valley of lesser steepness). In fact, it has to be taken into account that valleys crossed by a descending mass are frequent, and so is also the chance of the material to be ample and resistant enough not to be immediately washed out by a course of water. So damming is necessarily a *frequent phenomenon.*

*

It is needless to say that normally the main danger is not that of dwellings with their inhabitants being engulfed by the lake growing behind the dam: only in exceptional cases the rate of level-raising can be high enough to exclude the possibility of timely escape. The main danger resides in the accumulation of a substantial volume of water, followed by its rapid release due to a failure of the dam and resulting in the *inundation of the valley downstream.*

The description and discussion of the various mechanisms of failure cannot be treated in detail in a book mainly dealing with motional dynamics of rocks. Some general remarks, however, will be presented later. But first a short choice of citations from the *literature of the twentieth century* will be discussed. It will show that the interest mainly remained centred on geomorphological questions while dynamic aspects only sporadically were considered, prevailingly in spartanian shortness.

Russel's (1927) description of several *slide-generated lakes in the North-western Great Basin* (U.S.A.), in the section dealing with extinct lakes, only once gives a somewhat more extended comment concerning a mechanism (pp. 243–244). It runs as follows: "*... draining of Lake Jess apparently took place rapidly without any notable resting stages as but one shore line marks the level of the lake. This suggests that once the Pitt* [River] *succeeded in eroding the steep western side of the slide to the extent of tapping the lake, the further removal of slide material was rapid...*"

Approximately at the same time in the U.S.A. the opportunity was given to go somewhat more into detail owing to the catastrophic *flood of Gros Ventre* (Wyoming) in which, on May 18, 1927, the village of Kelly and several ranches were destroyed and six or seven persons drowned. Alden (1928, 353–358) presented a well-illustrated report about the flood. The collapse of the dam (generated in 1925 by a rockslide) had taken place after almost two years of more or less peaceful seepage balancing the influx to the lake. So a false feeling of security had developed. The flood was preceded by a winter rich in snowfalls, followed by rapid melting and heavy rain. As apparently nobody had witnessed the break-through, Alden undertook to reconstruct "*...what really happened at the dam...*". The essential sentences certainly are worth literal citation: "*Probably seepage increased rapidly as the lake rose and this tended to undermine at the same time as overflow at the lowest point cut the initial channel... Rapid deepening of the trench across the crest must have caused rapid slumping at the sides and the loose material must have at once been swept out of the way, opening a broad outlet toward which the enormous body of water started moving with rapidly increasing velocity...*" Of course, it is an open question how far a combined action of erosion from the top and undermining at the bottom did exist. Anyhow, the description sounds realistic. In fact, Fig. 3.12 shows that immersed spherical blocks of 6 m diameter (corresponding to the maximum of 20 ft. reported by Alden) require a velocity below 4 m s^{-1} to be displaced by sliding (not to speak of rolling!). This is what can be expected at a passage driven by a differential elevation of about 0.8 m – an impressive little illustration of the transporting capacity residing in flowing water.

In his famous book, Heim dedicated a section to slide-generated lakes (1932, 168–174). Four basic types were discerned: (1) damming of a valley by deposits coming from one of its slopes or a lateral tributary valley; (2) damming of a tributary valley by deposits in the main valley; (3) lake in a local depression of the deposits; (4) lake between scar and deposits of a slide with short run-out. For each type, several examples, mainly in the Swiss Alps, are mentioned, in certain cases also described. Among the ten completely or partly extinct lakes mentioned in this context, *Buzza di Biasca* (Ticino, Switzerland) is the most disastrous and the most interesting at once. Covering, in 1512, a vertical span of 1990 m, this rockslide was very high. The lake, formed by the river Brenno, had a volume in the range of 0.2 km^3. It engulfed the villages of Malvaglia and (partly) Semione, but the inhabitants had had the time to leave their dwellings. In 1514, not even two years after the slide, the dam gave way. A catastrophic flood destroyed the large village of Biasca, killed about 600 persons, and caused damages along Ticino River and parts of Lago Maggiore, at a distance of more than 30 km. As in the case of Gros Ventre, only guesses about the phenomena preceding the break-through are made. Sagging, cracks, and violent erosion at the outlet are supposed.

Among the lake-generating events briefly mentioned by Heim is also that of *Disentis* which, in comparison with Buzza di Biasca or Huascarán (s. below), was remarkably harmless. The slide, in 1683, barred the mighty water course of Vorderrhein for three hours. Then the dam was overflown, but its giving way was the gradual formation of a deepening passage so that no catastrophe took place.

It is well understood that in their *anthologies of alpine events*, Montandon (1933, 295–296, 304, 336) and Abele (1974, 121–125, 182–183) not only mentioned Buzza di Biasca and Disentis but also reserved the space required for the presentation of case histories with collapsed dams (Montandon enumerates no less than 25). However, these authors also concentrated their interest mainly on geomorphological aspects.

For the extremely dangerous dam failure of *Huascarán* (Sect. 2.7; Plafker and Ericksen 1978, 297; Stadelmann 1983, 64) some important facts can be found in the literature. Besides rock of various granular sizes, the dam contained large quantities of mud and ice. For a short space of time, mud may have played the role of a sealant, but certainly washing out began quickly and initiated an increasing process of seepage. The high temperature not only resulted in an ample supply of melt-water to Rio Santa: it also could not fail to melt the ice in the dam so that local collapses in the originally coherent mass were programmed. Once this twofold mechanism of destruction having been started, a complete annihilation of any damming effect must have been the work of few minutes. All in all, the massive break-through occurred about half an hour after the event proper.

*

The term of "*dangerous dam*" as used above raises the question of criteria allowing to recognise a dam as – at least potentially – dangerous. Such conditions are contradictory to a certain extent. On the one hand, to be dangerous, a dam should be tight enough to exclude a continuous draining by seepage, and strong enough to withstand the hydrostatic pressure at a substantial water depth. On the other hand, it should bear the germ of rapid destruction, or, more precisely, of a rapidly opening passage large enough for the formation of a surge exceeding by far the normal flow through the valley. These are the – so far unspecified – basic qualities of a potentially dangerous dam.

To establish a first systematic order in the consideration of pre-failure mechanisms it appears useful to make a difference between the effects of *seepage and overflow*. For both, however, the general rule should be borne in mind that the transporting capacity of water is a function of the differential pressure acting upon an obstacle. This pressure may be essentially static (if the water's velocity is non-existent or negligible) or dynamic (if the obstacle is exposed to a flow). Quantification of the static case is trivial (s. hereafter), that of dynamic sliding follows Eq. 3.7 and Fig. 3.12 – with certain corrections if the criteria of validity are unduly violated, i.e. if the body is only partially immersed or if the ground is oblique. If rolling takes place, it is always more economic than sliding. Both in rolling and in sliding the critical velocity of transition from immobility to motion – e.g. v_c in Eq. 3.7 – is proportional to $\sqrt{(V/A)}$, i.e. to the square root of the longitudinal extension for geometrically similar bodies. In other words, smaller particles are easier to move than larger ones – another size effect.

At this point serious differences between seepage and overflow come up. In a subterranean environment any particle may become involved in the complex system of force transmissions resulting from the overthrust load. In such instances the particle's frictional *resistance against the flow of seepage* is dramatically increased. Near the downstream end of a seeping zone this effect fades out owing to decreasing overthrust thickness, and the chance of a particle to be definitely washed out is increased.

A particularly important role in this context is that of particles moving in spite of being (as a rule slightly) loaded: in being pushed ahead, they necessarily entail a redistribution of forces in the support of overthrust. Sooner or later such incremental effects may add up to a local collapse which, in its turn, may contribute to phenomena perceptible at the surface. This is probably what Alden meant when he spoke of "undermining". And as this process of undermining prevails downstream, it entails local steepening of the dam's slope and possibly prepares a bed for an overflow channel.

This is, of course, only one half of the truth about seepage. The other half lies in the *velocity distribution of seeping*. As shown by Eq. 3.5, the force exerted upon a body in a – prevailingly turbulent – flow is proportional to the square of velocity. Now the highest velocity occurs in the narrowest passages. So particles undergo the highest driving force where they almost clog the passages. The extreme case would be total sealing or, in other words, the transition from a dynamic to a static mechanism. In such instances the initial situation of fine loose material within a dam is important. If it is mainly on the side of the lake, it must be pushed from there through the entire thickness of the dam to find outlets on the downstream side. And in each narrowing passage the particles, on the one hand, accelerate their own motion by generating an increased pressure drop, but, on the other hand, they risk being caught, possibly (e.g. by wedging) in a more than transient manner. If, however, such material initially is found rather on the downstream side of the dam, its way is shorter, and its chance to get at an outlet is improved. There is no question that this configuration means a shorter lifetime for dam and lake.

In seepage, the level of a lake's surface is nothing but the determining parameter of the pressure drop through the seeping zone. Contrarily, this level must attain the dam's crest to fulfil a trivial precondition for starting a *surge by overflow*. How long it takes to reach this situation, depends on various parameters: the required volume, the water supply, and its losses (mainly by seepage, as evaporation normally is negligible). As soon as overflow has begun, a balance can be defined between supply and escape of water: the level of the lake must exceed that of the crest sufficiently to warrant an escape (by overflow and seepage, if any) equalling the entire supply. Of course it is, however, an open question whether this state of balance is at all attained. In fact, from the very start of overflow, the water exerts hydrodynamic forces upon the particles in its way. So, once more on the basis of Eq. 3.7, the conditions of balance may change by displacement of particles. The process is not unambiguous: disappearance of a particle locally increases the free section, thus reducing velocity of flow; but it also may lower the mean altitude of said section, thus increasing both the differential elevation and velocity. It is obvious that, in analogy to seepage, the longitudinal gradation of particles is important. And it also is obvious that a particle, once having started motion on the more or less horizontal crest, has a good chance to move in the subsequent descent.

All in all, these short considerations should make plausible three facts. Owing to the multitude and variability of determining parameters, (11) the space of time between formation and failure of a dam may vary within an extremely wide range; the existence of an abundant water supply (12) warrants early overflow but does not necessarily mean a catastrophic discharge (compare the issues of Huascarán and Disentis); and (13), in spite of basic processes allowing expression by relatively simple equations, the formulation of straight-forward algorithms for reliable lifetime *prediction* is not an easy task, especially not in circumstances favouring a false feeling of security as in the cases of Gros Ventre and Buzza di Biasca.

7.3 Various Secondary Effects

Of course, rockslides and rockfalls entail, besides those connected with water, a considerable number of other effects, most of which are of lesser importance as far as immediate damage to human lives and goods is considered. Nevertheless some of them are worth mention in the present context, mainly owing to the interesting mechanisms involved.

*

Various events are reported to have been accompanied by violent *gusts of wind*. As mentioned in connection with Huascarán (Sect. 2.7), boulders (and clusters thereof) could not accomplish their giant bounces without carrying along a more or less extended boundary layer of air. All the more a large, badly streamlined mass – no matter, whether coherent or disintegrated – cannot travel at high velocity without transmitting a minute portion of its kinetic energy to the surrounding air. Thickness and velocity distribution of such layers are not easy to determine. In any case the mass in its immediate vicinity imposes its own velocity on the air, and somewhat farther off eddies are detached so that effects can be observed at distances of several metres or even tens of metres. For Val Pola, to quote the example of a fast event, chopped trees (Crosta 1991, 101, and author's observation) are found at definitely more than a dozen of metres beyond the traces of moving debris. In such instances it is by no means excluded that persons may be knocked down or killed by a gust.

Observations pointing to a *high turbulence* unintentionally were reported by an author who definitely was not an expert in aerodynamics. In his inexhaustible book, Heim (1932, 142–143) mentions, for the case of Elm, a very uneven local distribution of gust effects (as typical in case of eddies). More problematic is his subsequent statement that many people had been lifted by the gust. It should be borne in mind that such a tour de force needs – mutatis mutandis according to Eq. 3.5 – a vertical component of velocity in the range of 80 m s^{-1}, a figure exceeding both the maximal velocity of the mass in its descent (about 60 m s^{-1}, s. Sect. 6.2) and the velocity required to knock down a man. So it is probably not excluded that persons knocked down by the unexpected force of a gust had, in the extreme stress of the situation, a feeling of having been airborne for a moment. And, of course, a sudden fall of this kind, according to the particular circumstances, may represent a serious danger. The hypothesis of a psychological delusion is, to a certain extent, backed up by the fact that many of these persons are reported to have been put down gently. In fact, the idea of a genuine aerial transport crowned by a gentle landing is more than what a person acquainted with aerodynamics can cope with.

*

The second interesting mechanism, connected with recent investigations, is dealing with one of the key events of this book. It has to do with the rather plausible fact that a large disintegrated rockslide mass can *bring to light materials formerly safeguarded* by impenetrable rock dozens or hundreds of metres thick.

Thereby three partial mechanisms require special attention. All are connected to disintegration. (1) Certain chemical and physical processes (reactions, emanations etc.) tend to be substantially accelerated by an increased surface of the particles in a given volume of rock. It should be borne in mind that halving the linear dimensions of particles means doubling their total surface. (2) In the process of spreading, the distance between an active zone (be it in a physical or in a chemical sense) to the surface has a chance to diminish. Even in the case of a mass having travelled in coherent state and being disintegrated by a last dramatic collision, lateral and/or distal portions usually cannot fail to show cropping out material that originally was quite distant from the surface. (3) In a disintegrated mass a liquid or gaseous medium, even if generated at a certain distance from the surface, sooner or later may get there owing to the interstitial voids. In this context the importance of water as a vehicle for originally non-fluid products is obvious. It is, however, clear that effective delivery is fast delivery so that the lifetime of a phenomenon is confined by its fastness.

In such instances the ideal set of preconditions for phenomena of said kind is given in a large (i.e. having enclosed its treasures in a solid safe), coherent (i.e. not having spent much of said treasures before descending), young (i.e. not having had the time to spend much after descending) rockslide, drastically disintegrated (i.e. having opened the safe) in its descent and allowed to spread in running out (i.e. generously spending).

Passing in review the key events, no ideal candidate can be identified. Three are large, but Blackhawk is old, Köfels did not accomplish much spreading in running out, and Vaiont was not disintegrated at all. Yet Köfels has been cut by the Maurach Gorge ("Maurach-Schlucht" in Fig. 2.18). In addition, certain spreading zones are found at both ends of the gorge.

As early as 1940 Krüse, in a study dealing with the radioactivity of springs in Tyrol, showed that in the rockslide area substantially higher values were observed than elsewhere. This potentially interesting statement passed almost uncommented in the shadow of the discussion about the cause and nature of the event (s. Sect. 2.4). In the early nineties, however, the question found its way to the press and assumed a rather charged role in local politics (times had changed, and radioactivity of a spring no longer was considered as a selling point...). In fact, the village Umhausen (Fig. 2.18), partly standing on the deposits of the slide, turned out to be exposed to an alarmingly strong radiation (Purtscheller et al. 1995). As could be expected, the noble gas *radon* (generated by fission of uranium via radium) was the cause – a perhaps surprising demonstration of the fact that a rockslide can continue to be harmful over many thousands of years. It might be added that simple mitigating measures do exist for buildings contaminated by radon. Usually it suffices to equip the cellar (i.e. the badly ventilated room next to the ground) with a massive concrete floor of a thickness dictated by the unshielded radon flux.

The connection with the slide was not only backed up by the topographic distribution. A special expedition to the "Himalayan twin" in Langtang yielded – mutatis mutandis – similar results.

*

More than that: the rockslide of Köfels, generous (though sometimes malicious) in presenting results of particular interest, added to this physically generated material one of chemical origin, brought to light by a similar mechanism. Near the entrance of Maurach Gorge (again at a location that might be expected) no less than 16 *sulphurous springs* were found, and it could be shown that they came from pyrite in the small-grained debris. Water had effectively taken part in the process of bringing to the surface the active material.

This brand-new information and the permission to use it here is owed to Professor H. Mostler, Director of the Institute of Geology and Paleontology at Innsbruck University (oral communication obtained through the help of Professor H. Heuberger, Salzburg).

*

Of course, various further secondary effects could be enumerated, especially if human reactions (as presented mainly in Sect. 2.5, 2.6, 3.3) are not excluded. However, their connection with the general theme of this book probably would be found questionable or even non-existent. And an unmistakable signal in this respect was given by the last-mentioned example: when the discussion of a phenomenon, essentially considered as a byword for danger and destruction, gets at a product potentially usable as a means of healing, a boundary is attained and the time is ripe to close game, set, and match...

Review of Highlights

This review is addressed to two categories of readers.

- *Hasty readers*, unable to spare the time for reading the entire book, are given the opportunity of acknowledging, so to say, its message.
- *Readers interested in particular questions* can use the review as a means of orientation, indicating the respective sections and more informative than an index.

Release depends, besides slope and coefficient of friction, on coherence with solid rock as basic parameter. By describing the critical configuration of crack length and strength, *fracture mechanics* can be helpful (Sect. 3.1, 3.2).

By lifting or crack opening effects, *hydrostatically pressurised water* can play a key role in the release of single blocks or large masses (e.g. Vaiont, probably also Val Pola; Sect. 2.5, 2.6, 3.3).

Human activity may entail catastrophic events, especially by risky mining (e.g. Piuro, Elm, possibly also Frank) or by wrong interpretation of observed phenomena (e.g. the last stage of Vaiont; Sect. 2.6, 3.3).

Disintegration begins with the first fracture of an essentially coherent mass. Usually due to *static load* in slope variations, it may influence the products of disintegration (resistant blocks after bending, fragile slabs after shearing fracture). Fracture mechanical treatment is often required (Sect. 4.1).

Besides dynamic disintegration, *collisions of particles* with immobile obstacles reduce momentum and kinetic energy of a mass. Collisions between particles reduce only the kinetic energy, thus justifying the term of "internal" (not momentum-bound) kinetic energy. They entail simultaneously spreading and equalisation of particle velocities (Sect. 4.2).

Disintegrated masses, anisotropic by virtue of *gravitation and friction* (active despite the shortness of collisional contacts), are reluctant against relative displacements between particles, thus contributing to preserve the sequential order of components (Sect. 5.1).

Both *coherent and disintegrated masses* have pros and cons with respect to far reach. Pros of coherent motion: no losses of internal kinetic energy; ability to bridge depressions and to shear off asperities; no need for lateral confinements to be kept together. Pros of disintegrated motion: easier passing through narrow clearances and meanders; extra reach owing to longitudinal spreading (especially in distal talus) (Sect. 5.1).

A rather *sudden stop* of a disintegrated mass, as repeatedly observed (e.g. Elm), is an argument in favour of Coulombian friction (Sect. 5.2).

Rolling, frequent in rockfalls, has *no chance in a large mass*, except near the surface (e.g. in the form of large boulders, "floating" on top of the rest). In particular, roller bearings usually are excluded by crushing and – for disintegrated masses – by self-arresting mechanisms (Sect. 5.3).

Bouncing, a frequent mechanism, perfectly *suited for quantitative treatment* (especially if the projectiles are large enough to neglect aerodynamic drag), sometimes seduces to daring calculations yielding unrealistic velocities (e.g. Huascarán) (Sect. 2.7, 5.3).

Repeated rebounding, a particularly economic means of locomotion, works only under very special circumstances. Normally the energetic gain of air travelling is compensated in the next collisional ground contact (Sect. 5.3).

The restricted importance of rolling and repeated rebounding in the displacement of large masses stresses the *importance of sliding* (Sect. 5.3).

Unlubricated sliding rock on rock comprises energy dissipation independent of velocity (e.g. scraping, abrasion of obstacles) and proportional to the square of velocity (e.g. forced detours, acceleration of abraded material). So for expressing the resulting braking force two terms are suggested: a constant and a velocity-quadratic one (the second sometimes is negligible, see under velocimetry hereafter) (Sect. 5.4, 6.2).

Unlubricated sliding underlies various *secondary effects*, in particular wedging (increased friction in sharp-edged cross-sections) and size effects (e.g. abrasion of asperities which follows the "area-to-volume" rule; more efficient track-making by thick masses) (Sect. 5.4, 5.7).

Quasi-laminar motion of a disintegrated mass in several layers is incompatible with Coulomb's rule: the sliding surface with the lowest coefficient of friction (usually near to the bottom) sooner or later forces all other surfaces into a relative standstill ("rule of lowest μ dominance") thus helping to preserve the sequential order of components (Sect. 5.4, 5.5).

Lubrication frequently occurs by pressurised water. This effective, load-sharing mechanism is due to the weight of a rockslide which, moving over a water-saturated valley fill, compresses it and thereby pressurises the water, thus forcing it to take part in supporting the weight (Sect. 2.2, 5.5).

Lateral and proximal attenuation of pressurised water lubrication (due to escape of water) may form *pull-apart voids* (cracks, funnels) through which pressurised water rapidly ascends (as long as they are open) and may fill them with particles carried along, thus yielding a local gradation coarse on bottom (Sect. 2.2, 5.7).

The absence of an effective mechanism of compression and the high effectiveness of escape (laterally as well as through a mass which is non-resistant against extension) exclude *lubrication by air* after take-off from a "jumping hill". The concept of an airfoil-shaped giant rock glider can be reduced to absurdity by calculation of the required velocity (Sect. 2.3, 5.5).

Self-lubrication by thermally transformed material of ground and/or mass (melted: ice, crystalline rock; dissociated: carbonate; dehydrated: gypsum, serpentine; evaporated: water) is effective owing to its occurring exactly at the locations requiring lubrication (e.g. Köfels: field evidence of gneiss complemented by artificial friction-melting with marked lubricating effect). Some processes are difficult to trace a posteriori as certain materials leave no durable products (Sect. 2.2, 2.4, 2.7, 5.5).

Self-lubrication by solid-state particles, possible on steep slopes (e.g. rockfalls), is not an energetically plausible mechanism for rockslides, all the more so for long run-out events (Sect. 5.5).

Fluidization by *water* is possible in a fine-grained material. It then behaves like a viscous fluid and thus entails a reversal of sequential order of components. Exception: a thin layer of fluidised material at the bottom of a mass may act as a lubricant (e.g. Goldau, Gros Ventre) (Sect. 3.1, 5.5, 5.6).

Fluidization by *air or solid state particles* can be shown as being less economical than the respective modes of lubrication. So a discussion in more detail is superfluous (Sect. 2.3, 5.5, 5.6).

Acoustic fluidisation by the noise of the moving mass, though theoretically well-founded, is limited in its validity by the uncertainty of continuous transfer of acoustic waves between shaken clasts and by an insufficient (though possible) concentration of velocity gradients at the bottom of the mass, necessarily entailing a reversal of the sequential order of components (Sect. 5.6).

Curves in a mainly horizontal plane may serve for velocimetry by superelevation. As a disintegrated mass is able to oscillate around a longitudinal axis, due consideration of forced oscillation is required, i.e. phenomena like time lag, damping, resonance (probable for Huascarán). Besides, the surface of the mass may be convex or concave (Sect. 2.2, 2.7, 5.7).

If a curve lies in a horizontal plane, the increase of total (and therewith also frictional) force resulting from vectorial addition of weight and *centrifugal force* usually is moderate (e.g. Pandemonium Creek). In a vertically oriented curve (due to variations of slope) excessive effects such as taking off (bouncing) and doubling or trebling of friction may occur (e.g. Val Pola, Huascarán) (Sect. 2.2, 2.5, 2.7, 5.7).

The reach of spreading clasts in distal and lateral zones not covered by the bulk of a mass is particularly far if *a light clast is kicked by a heavy one* and if, in addition, it is hit in a phase of running back (resulting velocity may exceed that of the heavy clast by more than 2 : 1) (Sect. 4.2, 5.7).

Velocimetry en route (e.g. by superelevation etc.) is a rare opportunity and may be handicapped by physical flaws. Therefore current a posteriori start-to-stop velocimetry uses either a constant coefficient of friction (Heim/Müller-Bernet) or a resistance with a velocity-quadratic and a constant term (Körner). Both methods are based on the (wrong) assumption of constant coefficients over the entire travel. So the superiority of the twin-terms method is at least dubious (Sect. 2.2, 2.5, 2.6, 2.7, 6.2).

The use of a *variable frictional resistance*, based on en route velocimetry, improves accuracy and practically equalises the results of both methods – an argument in favour of the single-coefficient approach (Sect. 6.2).

For both methods substantial errors – frequent despite their triviality – result from the "*original sin*", i.e. the use of overall dimensions instead of the path of the centre of gravity of the mass (Sect. 2.2, 2.5, 2.7, 6.2).

Reach prediction by the size effect, mathematically optimised on the basis of many events and represented by a plot of Fahrböschung vs. mass volume (Scheidegger function), is feasible, though encumbered by wide scatter. Replacing Fahrböschung by apparently better parameters like "excessive travel distance" offers no physically justified advantage (Sect. 6.4).

A size-independent *tribological quality parameter* ("standard Fahrböschung"), obtained by referring the Fahrböschung of an event to a standard volume, allows for direct comparison (and detection of similarities) between large and small events and thus can serve as a tool for "prediction by similarity" (Sect. 6.4).

In parallel with velocimetry, subdivision of a track into characteristic sections (e.g. naked rock, glacier ice, water-saturated valley fill, etc.) and mathematical *deduction of local coefficients* from a large number of known Fahrböschung values, after appropriate preparatory work, may become a promising basis for prediction (Sect. 6.4).

Once more in parallel with velocimetry, also the replacement of Fahrböschung by the inclination of the path covered by the *centre of gravity of the mass* is promising in terms of improving accuracy, again at the cost of additional evaluation of many events (Sect. 6.2, 6.4).

Obviously *combinations* of the three last-mentioned methods are not excluded (Sect. 6.2, 6.4).

Flood waves may be particularly dangerous if a relatively *small mass of water* is directly launched by a far larger rockslide mass. It can run up and run back on the front slope of the moving rock mass, thus attaining almost twice the velocity of the rock mass (e.g. Val Pola). Even worse can be the case of water forcibly expelled between an immobile steep wall of rock and the approaching rockslide mass (e.g. Vaiont) (Sect. 7.1).

References

First-mentioned authors are listed with the details of the respective publication, other authors with arrows indicating the first-mentioned authors. By way of exception, for a publication with 13 authors only the first-mentioned is presented.

Abele G (1974) Bergstürze in den Alpen. Wissenschaftliche Alpenvereinshefte 25, München

— (1984) Schnelle Felsgleitungen, Schuttströme und Blockschwarmbewegungen in den Alpen im Lichte neuerer Untersuchungen. In: Ruppert (ed) Geographische Strukturen und Prozessabläufe im Alpenraum. Münchener Studien zur Sozial- und Wirtschaftsgeographie 26:165–179

— (1991a) Der Fernpassbergsturz. Österreichische Geographische Gesellschaft, Zweigverein Innsbruck, Jahresbericht 1989–1990, 22–32

— (1991b) Durch Bergstürze mobilisierte Muren und durch Muren transportierte Bergsturzmassen. Österreichische Geographische Gesellschaft, Zweigverein Innsbruck, Jahresbericht 1989–1990, 32–39

— (1990–1994) Bergsturz nelle Alpi (Nicoletti PG: Italian translation of Abele 1974). Rapporti interni 289–291, 298, 303–305, 359, 366, 409, 419, 425, 436, 437, Istituto di Ricerca per la Protezione Idrogeologica, Cosenza

— (1997a) Influence of glacier and climate variation on rockslide activity in the Alps. European Science Foundation, Project: European Palaeoclimate and Man, Special Issue 1–6, Stuttgart Jena Lübeck Ulm

— (1997b) Rockslide movement supported by the mobilization of groundwater-saturated valley floor sediments. Zeitschrift für Geomorphologie, N.F., 41, 1:1–20

Alden WC (1928) Landslide and flood at Gros Ventre, Wyoming. Engineer's Transactions 76:347–361

Ampferer O (1939) Die geologischen Hypothesen über die Formung des Talraumes zwischen Umhausen und Längenfeld im Ötztale. Akademie der Wissenschaften Wien, Sitzungsbericht Mathematisch-Naturwissenschaftliche Klasse, Abteilung 1, 148:123–140

Ascher H (1952) Neuer Sachbestand und neue Erkenntnisse über das Bergsturzgebiet von Köfels. Geologie und Bauwesen 19:128–134

Ashby WR (1961) General system theory and the problem of the black box. In: Mittelstaedtt (ed) Regelungsvorgänge in lebenden Wesen. München, 51–62

ASTM (1974) Standard E 399-74

Azimi C, Desvarreux P, Giraud A, Martin-Cocher J (1982) Méthode de calcul de la dynamique des chutes de blocs. Bulletin de Liaison, Laboratoire Ponts et Chaussais 122, Ref. 2698:93–102

Bagnold RA (1954) Experiments on a gravity-free dispersion of large solid spheres in a Newtonian fluid under shear. Proceedings of the Royal Society 225 A:49–63

— (1956) The flow of cohesionless grains in fluids. Proceedings of the Royal Society 229 A:235–297

Bak P, Chen K (1989) The physics of fractals. Physica D 38:5–12

— ⟶ Wiesenfeld

Baker ⟶ Francis

Belitz ⟶ Stötter

Bertrand ME (1757) Mémoires historiques et physiques sur les tremblements de terre. La Haye

Bhandari RK (1990) Practical lessons in investigation, instrumentation and monitoring of landslides. In: Bonnard (ed) Landslides. Rotterdam Brookfield, 1435–1457

— ⟶ Hutchinson

Bishop AW (1973) The stability of tips and spoil heaps. Quarterly Journal of Engineering Geology 6:335–476

Bollo ⟶ Hacar-Benitez

Bonani ⟶ Ivy-Ochs

Bonnard ⟶ Noverraz

Borella ⟶ Lyons

Bowden FP, Tabor D (1964) The friction and lubrication of solids. Part 2, Oxford
Bozzolo D (1987) Ein mathematisches Modell zur Beschreibung der Dynamik von Steinschlag. Doctor's Thesis 8490 ETH, Zürich
—, Pamini R, Hutter K (1988) Rockfall analysis - a mathematical model and its test with field data. In: Bonnard (ed) Landslides. Rotterdam-Brookfield, 555–560
Broili L (1967) New knowledges on the geomorphology of the Vaiont slide slip surfaces. Rock Mechanics and Engineering 5:38–88
— (1974) Ein Felssturz im Grossversuch. Rock Mechanics, Supplement 3:69–78
Brückl E (1998) Geophysikalisch-geomechanische Ursachen-Wirkungsmodelle für Hangbewegungen. Österreichische Akademie der Wissenschaften, IDNDR-Projekt 12/97, Jahresbericht (1997), Institut für Theoretische Geodäsie und Geophysik, TU Wien
— ⟶ Heuberger
Brückner ⟶ Penck
Brune ⟶ McKenzie
Brunner ⟶ Purtscheller
Büdel J (1977) Klima-Geomorphologie. Berlin Stuttgart
Bull WB, King J, Kong F, Moutoux T, Phillips WM (1994) Lichen dating of coseismic landslice hazards in Alpine mountains. Geomorphology 10:253–264
Buss E, Heim A (1881) Der Bergsturz von Elm den 11. September 1881. Zürich
Caloi P, Spadea MC (1960a) Serie di esperienze eseguite in sponda sinistra a monte della diga del Vaiont. Unpublished
—,— (1960b) Indagini geosismiche eseguite presso la stretta del Vaiont durante la campagna estiva 1959. Unpublished
—,— (1961) Indagine geosismica condotta nel mese di Dicembre 1960 a monte della diga del Vaiont in sponda sinistra. Unpublished
—,— (1962) L'attivita sismica nella valle del Vaiont e zone limitrofe dal Maggio 1962 al Settembre 1963. Unpublished
Campbell CS (1989) Self-lubrication for long run-out landslides. The Journal of Geology 97, 6:653–665
Cancelli A, Mancuso M, Notarpietro A (1990) A short description of the 1987 Val Pola rockslide in Valtellina. Alps '90, Alpine Landslide Practical Seminar, 6th ICFL Conference, Supplemento a. 79a:15–22, Milano
Chen K ⟶ Bak
Chen LH ⟶ Rigney
Chowdhuri RN (1980) Landslides as natural hazards - mechanisms and uncertainties. Geotechnical Engineering 11:135–180
Clague ⟶ Evans
Costa JE (1991) Nature, mechanics, and mitigation of the Val Pola landslide, Valtellina, Italy, 1987–1988. Zeitschrift für Geomorphologie, N.F. 35, 1:15–38
Cotton CA (1960) Geomorphology. Wellington
Crosta G (1990) A study of slope movements caused by heavy rainfall in Valtellina (Italy). Alps '90, Alpine Landslide Practical Seminar, 6th ICFL Conference, Supplemento a, 79b:247–258, Milano
— (1991) Studio di movimenti in massa. Modellisazione teorica e sperimentale con osservazioni e rilievi di campagna. Doctor's Thesis, Università degli Studi di Milano
Crozier MJ (1991) Determination of paleoseismicity from landslides. In: Bell (ed) Landslides. Rotterdam, 1173–1180
Cruden DM (1976) Major rock slides in the Rockies. Canadian Geotechnical Journal 13:8–20
—, Antoine P (1984) The slide from Mt. Granier, Isère et Savoie, France, on November 24, 1248. Proceedings of the 4th International Symposium on Landslides, Toronto, 475–481
—, Krahn J (1978) Frank rockslide, Alberta, Canada. In: Voight (ed) Rockslides and Avalanches. Amsterdam Oxford New York, 97–112
Czichos H (1968) Properties and metrology of surfaces. Institution of Mechanical Engineers; Proceedings 182:21–108
— (1971) Festkörperreibung - Teilgebiet der Tribologie. Umschau 4:116- 120
Dal Piaz G (1948) Sulla struttura geologica della valle del Vaiont agli effetti degli smottamenti dei fianchi che possono derivare dal progettato invaso e dalle oscillazioni del livello del lago. Unpublished
— (1953) Geological map and photos annexed to report of 1948. Unpublished
— (1960) Esame delle condizioni geologiche del bacino del Vaiont e della zona d'imposta della diga di sbarramento quali risultano allo stato attuale dei lavori. Unpublished
De Nardi A (1965) Il bacino del Vaiont e la frana del Monte Toc. "l'Universo", Revista dell'Istituto Geografico Militare 45:1
Dent JD (1985) Flow properties of granular materials with large overburden loads. Script of Department of Civil Engineering/Engineering Mechanics, Montana State University, Bozeman, MT 59717

De Simoni B, Migani M, Moia F (1990) Analisi preliminare delle registrazioni della caduta della frana di Val Pola da alcune stazioni sismiche italiane e straniere. Conferenze di Meccanica e Ingegneria delle Rocce, Torino, 1988, Memoria 7:1–5

Desvarreux ⟶ Azimi

Deutsch A, 12 co-authors (1994) The impact-flood connection: does it exist? Terra Nova 6:644–650

Douglass ⟶ Voight

Doukhan ⟶ Leroux

Drever ⟶ Scott

Eden WJ (1975) Mechanism of landslides in Leda clay with special reference to the Ottawa area. Proceedings of 4th Guelph Symposium on Geomorphology, DBR Paper 713:159–169

Egger J (1880) Geschichte Tirols von den ältesten Zeiten bis in die Neuzeit, Vol. 3. Innsbruck

Eichinger A ⟶ Roš

Eisbacher GH (1977) Rockslides in the Mackenzie Mountains, District of Mackenzie. Geological Survey of Canada, Paper 77-1A:235–241

— (1979) Cliff collapse and rock avalanches (Sturzstroms) in the Mackenzie Mountains, Northwestern Canada. Canadian Geotechnical Journal, 16, 2:309–334

—, Clague JJ (1984) The ancient (1806) Goldau rockslide (Bergsturz). Alps '90:156–159

Enegren ⟶ Evans

ENEL (Ente Nazionale per l'Energia Elettrica) (1964) Relazione sulle cause che hanno determinato la frana nel serbatoio del Vaiont. Roma

Ennemoser ⟶ Purtscheller

Erismann TH (1979) Mechanisms of large landslides. Rock Mechanics 12:15–46

— (1980) Der Gefahrbereich grosser Bergstürze. EMPA 1880–1980, Dübendorf St. Gallen, 270–277

— (1988) Modes and consequences of first rock fracture in landslides. Zeitschrift für Geomorphologie, N.F. 32, 3:257–272

— (1989) The strongest cable testing machine in the world. International Journal of Materials and Product Technology 4, 3:273–282

—, Heuberger H, Preuss E (1977) Der Bimsstein von Köfels (Tirol), ein Bergsturz-"Friktionit". Tschermaks Mineralogisch-Petrographische Mitteilungen 24:67–119

— ⟶ Kienholz

Escher von der, Linth A (1845) Beiträge zur Kenntnis der Tyroler und Bayerischen Alpen. Neues Jahrbuch für Mineralogie, Geognosie, Geologie und Petrofacten-Kunde, 536–561

Evans SG, Clague JJ, Woodsworth GJ, Hungr O (1989) The Pandemonium Creek Rock Avalanche, British Columbia. Canadian Geotechnical Journal 26:427–446

—, Hungr O, Enegren EG (1994) The Avalanche Lake rock avalanche, Mackenzie Mountains, Northwest Territories, Canada: description, dating, and dynamics. Canadian Geotechnical Journal 31, 5:749–768

Fiebiger ⟶ Kienholz

Fliri F (1992) Trientl als Naturforscher. In: Der Mistapostel, ein Leben für den Bauernstand - Adolf Trientl, 1817–1897. Innsbruck

Flotron ⟶ Wegmann

Francis PW, Baker MCW (1977) Mobility of pyroclastic flows. Nature 270:164–165

—, Gardewig M, Ramirez CF, Rotherey DA (1985a) Catastrophic debris avalanche deposit of Socompa Volcano, Northern Chile. Geology 13:600–603

—, —, —, — (1985b) Catastrophic debris avalanche deposit of Socompa Volcano, Northern Chile. Universitad de Chile, Santiago, Departemento de Geologia, Communicacion 35:89–71

Frank ⟶ Ivy-Ochs

Funk M (1999) Fels- und Eisstürze auf der Südseite des Mont-Blanc Massivs. In: Fachtagung Naturgefahren: Gletscher und Permafrost, CENAT/VAW (ETH) 1998. ETH Zürich

— ⟶ Wegmann

Gadi ⟶ Lajtai

Gardewig ⟶ Francis

GEOTEST, BWW (Bundesamt für Wasserwirtschaft) (1992) Der Bergsturz von Randa. Videotape, GEOTEST AG, CH-3052 Zollikofen, Birkenstrasse 15

Ghiglino Antunez L (1970) Alud de Yungay y Ranrahirca del 31 de Mayo de 1970. Revista Peruana de Andinismo y Glaciologia 9:4–88

Giraud ⟶ Azimi

Girsperger S (1974) Geologische Untersuchung der Breccien bei Glarus. Diploma Thesis, ETH, Zürich, 71–79

Glicken ⟶ Voight

Goetz P (1981) Systematics of acoustic fluidization mechanism and its application to catastrophic landslides. Master of Science Thesis, SUNY, Stony Brook

Goetz P, Melosh HJ (1980) Experimental observation of acoustic fluidization in sand. Abstracted in EOS 61:373
Goguel J (1969) Le rôle de l'eau et de la chaleur dans les phénomènes tectoniques. Revue de Géographie Physique et Géologie Dynamique 2, 2:153–163
Govi M (1989) The 1987 landslide on Mount Zandila in the Valtellina, northern Italy. Landslide News, Japan Landslide Society 3:1–3
Gramberg J (1989) A non-conventional view on rock mechanics and fracture mechanics. Rotterdam Brookfield
Griffith AA (1920) The phenomena of rupture and flow in solids. Phil. Transactions of the Royal Society 221, Series A:163–198
Griggs RF (1922) The Mageik Landslide. In: The Valley of Ten Thousand Smokes. National Geographic Society, 135–147
Guest JE (1971) Geology of the farside crater Tsiolkovsky. In: Fielder (ed) Geology and Physics of the Moon. Amsterdam, 93–103
Habib P (1967) Sur un mode de glissement des massifs rocheux. Conseil de la Recherche, Academie des Sciences 264, Serie A:151–153
— (1975) Production of gasous pore pressure during rock slides. Rock Mechanics 7:193–197
— (1976) Note sur le rébondissement des blocs rocheux. Rockfall Dynamics and Protective Works Effectiveness, ISMES Proceedings 90:123–125
Hacar-Benitez MA, Bollo MF, Hacar-Rodriguez MP (1977) Bodies falling down on different slopes – dynamic study. 9. ISSMFE Proceedings 2, 3/18:91–94
Hacar-Rodriguez ⟶ Hacar-Benitez
Haefeli R (1967) Zum progressiven Bruch in Schnee, Boden, Fels und Eis. Felsmechanik und Ingenieurgeologie 5, 1:17–37
Harrison JV, Falcon NL (1937) The Saidmarreh landslip, Southern Iran. Journal of Geography 89: 42–47
—,— (1938) An ancient landslip at Saidmarreh in southwestern Iran. Journal of Geology 46:296–309
Heierli W, Merk A, Temperli A (1985 (1981): Schutz gegen Steinschlag. Bundesamt für Strassenbau, Forschungsarbeit 6/80, Bern
Heilmann P, Rigney DA (1981) An energy-based model of friction and its application to coated systems. Wear 72:195–217
Heim A, (1882a) Der Bergsturz von Elm. Zeitschrift der Deutschen Geologischen Gesellschaft 34: 74–115
— (1882b) Über den Bergsturz von Elm. Zeitschrift der Deutschen Geologischen Gesellschaft 34: 435–439
— (1932) Bergsturz und Menschenleben. Zürich
— ⟶ Buss
Heissel W (1938) Gesteinsaufschmelzungen bei Ötz im Ötztal (Tirol). Jahrbuch der Geologischen Bundesanstalt 88:207–215
Heuberger H (1966) Gletschergeschichtliche Untersuchungen in den Zentralalpen zwischen Sellrain- und Ötztal. Wissenschaftliche Alpenvereinshefte 20
— (1975) Das Ötztal. Bergstürze und alte Gletscherstände, kulturgeographische Gliederung. Innsbrucker Geographische Studien 3:213–249
— (1994) The giant landslide of Köfels. Mountain Research and Development 14, 4:290–294
— (1996) Das Ereignis von Köfels im Ötztal (Tirol) und die Sintflut-Impakt-Hypothese. Umwelt-Mensch-Gebirge. Festschrift B. Messerli, Jahrbuch der Geographischen Gesellschaft Bern 59:271–280
—, Brückl E (1993) Reflexionsseismische Messungen am Bergsturz von Köfels. Geologie des Oberinntaler Raumes, Schwerpunkt, Blatt 144, Landeck, Arbeitstagung Geologische Bundesanstalt, 156–158
—, Masch L, Preuss E, Schröcker A (1984) Quaternary landslides and rock fusion in Central Nepal and in the Tyrolean Alps. Mountain Research and Development 4:345–362
— ⟶ Erismann
— ⟶ Ivy-Ochs
Hewitt K (1988) Catastrophic landslide deposits in the Karakoram Himalaya. Science 242:64–67
Hillemeier B (1976) Bruchzähigkeit von Naturstein aus dem Ötztal. Unpublished report, by courtesy of Prof. Dr. HK Hilsdorf, Institut für Baustofftechnologie, Universität Karlsruhe (TH)
Horn ⟶ Storzer
Howard K (1973) Lunar avalanches. Lunar Science 4:386–387
Howe E (1909) Landslides in the San Juan Mountains, Colorado. U.S. Geological Survey, Professional Paper 67, Washington
Hsü KJ (1975) Catastrophic debris streams (sturzstroms) generated by rockfalls. Geological Society of America, Bulletin 86:129–140

— (1978) Albert Heim: Observations on landslides and relevance of modern interpretations. In: Voight (ed) Rockslides and avalanches. Amsterdam Oxford New York, 70–93
Hubbert MK, Ruby WW (1959) Role of fluid pressure in mechanics of overthrust faulting, I. Bulletin of the Geological Society of America 70:115–166
— —→ Ruby
Huber A (1982) Felsbewegungen und Uferabbrüche an Schweizer Seen, ihre Ursachen und Auswirkungen. Eclogae Geologicae Helveticae 75, 3:563–578
Hungr —→ Evans
Hutchinson JN (1988) Morphological and geotechnical parameters of landslides in relation to geology and hydrogeology. In: Bonnard (ed) Landslides. Rotterdam Brookfield, 3–35
— (1993) Types of rapid gravitational subaerial mass movements and some possible mechanisms. CEMAGREF Workshop, Grenoble
—, Bhandari RK (1971) Undrained loading, a fundamental mechanism of mudflows and other mass movements. Géotechnique 21, 4:353–358
Hutter K (1991) Two- and three-dimensional evolution of granular avalanche flow - theory and experiments revised. Acta Mechanica, 167–181
—, Koch T (1991) Motion of a granular avalanche in an exponentially curved chute: experiments and theoretical predictions. Philosophical Transactions of The Royal Society A, 334:93–138
—, Szidarovski F, Yankowitz S (1986) Plane steady shear flow of a cohesionless granular material down an inclined plane: a model for flow avalanches. Acta Mechanica 63:87–112 and 65:239–261
— —→ Nohguchi
— —→ Savage
Hutton J (1795) Theory of the Earth. Edinburgh
Irwin GR (1957) Analysis of stresses and strains near the end of a crack traversing a plate. Journal of Applied Mechanics 24, 3:361–364
ISMES (1990) Technology and management of a real-time monitoring system. Field Workshop on Landslide, Valtellina, Comments to Alps '90, Alps 24:1–17
— (1990) Technology and management of a real-time monitoring system. The lesson from Val Pola landslide. Field Workshop on Landslide, Valtellina, Alps '90
Ivy-Ochs S, Heuberger H, Kubik PW, Kerschner H, Bonani G, Frank M, Schlüchter, Ch (1998) The age of the Köfels event. Relative, ^{14}C and cosmogenic isotope dating of an early holocene landslide in the Central Alps (Tyrol, Austria). Zeitschrift für Gletscherkunde und Glaziologie 34, 1:57–68
Jäckli H, Kempf T (1975) Der Felssturz von Amden vom 21. Januar 1974. Geologisches Institut der ETH Zürich, Mitteilung 207
Janda —→ Voight
Jätzold R (1971) Luftbild: Die verschüttete Stadt Yungay, Peru. Erde 102:108–117
Jerz H, Poschinger A von (1995) Neuere Ergebnisse zum Bergsturz Eibsee-Grainau. Geologica Bavarica 99:383–398
Keller J (1981) Carbonatic volcanism in the Kaiserstuhl alkaline complex. Journal of Volcanology and Geothermal Research 9:423–431
Kelly M (1979) A prehistoric catastrophic rock avalanche in Holsteinsborg, West Greenland. Bulletin of the Geological Society of Denmark 28:73–79
Kenney TC (1967) Stability of the Vaiont slope. Felsmechanik und Ingenieurgeologie 5, 1
Kent PE (1966) The transport mechanism in catastrophic rock falls. Journal of Geology 74:79–83
Kerschner —→ Ivy-Ochs
Keusen —→ Wegmann
Kienholz H, Erismann TH, Fiebiger G, Mani P (1993) Naturgefahren: Prozesse, Kartographische Darstellung und Massnahmen. In: Barsch, Karrasch (ed) Geographie und Umwelt, Verhandlungen des Deutschen Geographentages 1991 in Basel 48:293–312
Kiersch GA (1964) Vaiont Reservoir Disaster. Civil Engineering, March 1964
King —→ Bull
Kinzl —→ Welsch
Klebelsberg R von (1935) Geologie von Tirol. Berlin
— (1949) Das Ötztal, Natur und Bild. Jahrbuch des Österreichischen Alpenvereins 74:5–21
Kleinmann —→ Storzer
Koch —→ Hutter
Kojan E, Hutchinson JN (1978) Mayunmarca rockslide and debris flow, Peru. In: Voight (ed) Rockslides and avalanches. Amsterdam Oxford New York, 315–361
Kong —→ Bull
Körner HJ (1976) Reichweite und Geschwindigkeit von Bergstürzen und Fließschneelawinen. Rock Mechanics 8:225–256

— (1977) Flow mechanisms and resistances in the debris streams of rock slides. Bulletin of the International Association of Engineering Geology 16:101–104
— (1983) Zur Mechanik der Bergsturzströme vom Huascarán, Peru. In: Patzelt (ed) Die Berg- und Gletscherstürze vom Huascarán. Innsbruck, 71–110
Kovari K (1990) Methods of monitoring landslides. In: Bonnard (ed) Landslides. Rotterdam Brookfield, 1421–1433
Kranz W (1938) Beitrag zum Köfels-Problem: Die "Bergsturz-Hebungs-Sprengtheorie". N. Jahrbuch für Mineralogie, Geologie und Paläontologie 80, Beilage-Band, Abteilung B:113–138
Krauter E (1988) Applicability and usefulness of field measurements on unstable slopes. In: Bonnard (ed) Landslides, vol 1, Rotterdam Brookfield, 367–373
— (1994) Hangrutschungen und deren Gefährdungspotential für Siedlungen. Geotechnische Rundschau 46, 7/8:422–428
Kristan-Tollmann E, Tollmann A (1994) The youngest big impact on Earth deduced from geological and historical evidence. Terra Nova 6:209–217
Krumdieck MA (1984) On the mechanics of large landslides. Proceedings of the 4th International Symposium on Landslides, Toronto, 539–544
Krüse K (1940) Beiträge zur Kenntnis der Radioaktivität der Mineralquellen Tirols. Mittelung des Reichsamtes für Bodenforschung 1:70–80
Kubik ⟶ Ivy-Ochs
Kurat G, Richter W (1968) Ein Alkalifeldspat-Glas im Impaktit von Köfels im Ötztal (Tirol). Naturwissenschaft 55:490
—, — (1972) Impaktite von Köfels, Tirol. Tschermaks Mineralogisch-Petrographische Mitteilungen 17:23–45
Lahodynsky ⟶ Lyons
Lajtai EZ, Gadi AM (1989) Friction on a granite to granite interface. Rock Mechanics and Rock Engineering 22:25–49
Law KT, Lumb P (1978) A limit equilibrium analysis of progressive failure in the stability of slopes. American Geotechnical Journal 15, 1:113–122
Ledésert ⟶ Schneider
Lehmann O (1926) Die Verheerungen in der Sendlinggruppe. Denkschriften der Akademie der Wissenschaften in Wien, Wien Leipzig, 259–299
Leroux H, Doukhan JC (1993) Dynamic deformation of quartz in the landslide of Köfels, Austria. European Journal of Mineralogy 5:893–902
Liebowitz H (Ed.) (1968–1972) Fracture. 7 volumes, New York London
Lliboutry L (1975) La catastrophe de Yungay (Pérou). In: Neiges et glaces: Actes du Colloque de Moscou, aout 1971, IASH 104, Gentbrugge, 353–363
Lucchitta BK (1978) A large landslide on Mars. Geological Society of America, Bulletin 89:1601–1609
Lyons JB, Officer CB, Borella PE, Lahodynsky R (1993) Planar lamellar substructures in quartz. Earth and Planetary Science Letters 119:431–440
Mancuso ⟶ Cancelli
Mani ⟶ Kienholz
Martin G (1963, 1965) Die Verschüttung von Plurs (Piuro) im Bergell und die Deutung ihrer geologischen Ursachen durch Friedrich Rolle. Jahresbericht der Naturforschenden Gesellschaft Graubünden, 91
Martin-Cocher ⟶ Azimi
Masch L, Preuss E (1977) Das Vorkommen des Hyalomylonits von Langtang, Himalaya (Nepal). N. Jahrbuch für Mineralogie, Abhandlungen 129 (3):299–311
—, Wenk HT, Preuss E (1985) Electron microscopy study of hyalomylonites - evidence for frictional melting in landslides. Tecnophysics 115:131–160
— ⟶ Heuberger
Maukisch ⟶ Stötter
McConnell RG, Brock RW (1904) Report on the great landslide at Frank, Alberta, Canada. Canadian Departement of the Interior, Annual Report 1902–1903, Part 8
McKenzie D, Brune JN (1972) Melting on fault planes during large earthquakes. Geophysical Journal, R Astr Soc 29:65–78
McSaveney MI (1978) Sherman glacier rock avalanche, Alaska, U.S.A. In: Voight (ed) Rockslides and avalanches, Amsterdam Oxford New York, 197–258
Mellor M (1968) Avalanches. US Cold Regions Research and Engineering Laboratory, Monograph III-A 3d
Melosh HJ (1979) Acoustic fluidization: a new geologic process? Journal of Geophysical Research 84, B13:7513–7520
— (1983) Acoustic fluidization. American Scientist 71:158–165

— ⟶ Goetz
Mencl V (1966) Mechanics of landslides with non-circular slip surfaces with special reference to the Vaiont Slide. Géotechnique 16:329–337
Merk ⟶ Heierli
Miccadei ⟶ Nicoletti
Migani ⟶ De Simoni
Miller DJ (1960) Giant waves in Lituya Bay Alaska. Geological Survey, Government Printing Office, Professional Paper 354-C:51–86
Milton DJ (1965) Fused rock from Köfels, Tyrol. Tschermaks Mineralogisch-Petrographische Mitteilungen 9:86–94
Moia ⟶ De Simoni
Montandon F (1933) Chronologie des grands éboulements alpins, du début de l'ère chrétienne à nos jours. Matériaux pour l'étude des calamités 32:271–340
Moutoux ⟶ Bull
Muir J (1912) The Yosemite. The Natural History Library, New York
Müller L (1964) The rock slide in the Vajont Valley. Felsmechanik und Ingenieurgeologie 2, 3–4:148–212
— (1968) New Considerations on the Vaiont Slide. Felsmechanik und Ingenieurgeologie 6:1–91
Müller W, Rummel F (1987) Fracture toughness of granites. The Geothermal Fracture Project, Institut für Geophysik, Ruhr-Universität Bochum, Manuscript, 1–15
Müller-Bernet ⟶ Heim (1932, not quoted as co-author)
Naylor ⟶ Rigney
Nicoletti PG (1989) On the mobility of sturzstroms (rock avalanches). Societá Geologica Italiana, Bolletino 108:419–429
—, Parise M, Miccadei E (1993) The Scanno rock avalanche (Abruzzi, South-Central Italy). Bollettino della Società Geologica Italiana 112:523–535
—, Sorriso-Valvo M (1991) Geomorphic controls of the shape and mobility of rock avalanches. Geological Society of America, Bulletin 103:1365–1373
Nievergelt ⟶ Schindler
Noda E (1970) Water waves generated by landslides. Journal of Waterways, Harbours, and Coast Engineering Division; American Society of Civil Engineers, WW4, 835–855
Nohguchi Y, Hutter K, Savage SB (1989) Similarity solutions for a finite mass granular avalanche with variable friction. Continuum Mechanics and Thermodynamics 1:239–265
Notarpietro A (1990) Geologic outline of Valtellina (Italy). Alps '90, Alpine Landslide Practical Seminar, 6th ICFL Conference, Milano, Supplemento a. 79a:11–14
— ⟶ Cancelli
Noverraz F, Bonnard Ch (1991) L'écroulement rocheux de Randa, près de Zermatt. In: Bell (ed) Landslides. Rotterdam Brookfield, 165–170
Officer ⟶ Lyons
Parise ⟶ Nicoletti
Paris PC (1962) The growth of cracks due to variations in load. Doctor's Thesis, University of Lehigh
Patzelt G (Ed.) (1983) Die Berg- und Gletscherstürze vom Huascarán, Cordillera Blanca, Peru. Arbeitsgemeinschaft für vergleichende Hochgebirgsforschung, 6, Innsbruck
—, Poscher G (1993) Der Tschirgant-Bergsturz. In: Geologische Bundesanstalt, Arbeitstagung 1993: Geologie des Oberinntaler Raumes, Wien, 208–213
Pavoni, N (1968) Ueber die Entstehung der Kiesmassen im Bergsturzgebiet von Bonaduz-Reichenau (Graubünden). Ecologae Geologicae Helvetiae, 61, 2:494–500
Penck A (1925) Der postglaziale Vulkan von Köfels im Ötztale. Preussische Akademie der Wissenschaften, Sitzungsbericht 12:218–225
—, Brückner E (1909) Die Alpen im Eiszeitalter, vol. 1. Leipzig
Phillips ⟶ Bull
Pirchl ⟶ Purtscheller
Pichler A (1863) Die vulcanischen Reste von Köfels. Zur Geognosie Tirols, II. Jahrbuch der Kaiserlich-Königlichen Geologischen Reichsanstalt Wien 13:591–594
Plafker G, Ericksen GE, Fernandes Concha J (1971) Geological aspects of the May 31, 1970, Peru earthquake. Bulletin of the Seismological Society of America 61, 3:543–578
—, — (1978) Nevados Huascarán avalanches, Peru. In: Voight) Rockslides and avalanches, vol. 1, Amsterdam Oxford New York, 277–314
Poschinger A von, Haas U (1997) Der Flimser Bergsturz, doch ein warmzeitliches Ereignis? Bulletin für angewandte Geologie 2, 1:35–46
—, Thom P (1995) Neue Untersuchungsergebnisse am Bergsturz Hintersee, Ramsau (Berchtesgadener Land). Geologica Bavarica 99:399–411

Preuss E (1974) Der Bimsstein von Köfels im Ötztal, Tirol: die Reibungsschmelze eines Bergsturzes. Jahrbuch des Vereins zum Schutze der Alpenpflanzen und -tiere 39:85–95
— → Erismann
— → Heuberger
— → Masch
Purtscheller F (1971) Ötztaler und Stubaier Alpen. Sammlung geolog. Führer, 53, Berlin
—, Pirchl T, Sieder G, Stingl V, Tessadri T, Brunner P, Ennemoser O, Schneider P (1995) Radon emanation from giant landslides of Köfels (Tyrol, Austria) and Langtang Himal (Nepal). Environmental Geology 26:32–38
Ramirez → Francis
Reithofer → Hammer
Richter → Kurat
Rigney DA, Chen LH, Naylor MGS (1984) Wear processes in sliding systems. Wear 100:195–219
— → Heilmann
Roš M, Eichinger A (1949) Die Bruchgefahr fester Körper bei ruhender - statischer - Beanspruchung. Eidgenössische Materialprüfungs- und Versuchsanstalt, Bericht 172, Zürich
Rotherey → Francis
Rouse WC (1984) Flowslides. In: Brunsden, Prior (ed) Slope instability. New York
Ruby WW, Hubbert MK (1959) Role of fluid pressure in mechanics of overthrust faulting, II. Bulletin of the Geological Society of America 70:167–205
— → Hubbert
Russel RJ (1927) Landslide lakes in the Northwestern Great Basin. University of California Publications in Geography 2, 7:231–254
Salm B (1966) Contribution to avalanche dynamics. International Symposion "Snow and Ice", Davos 1965, JASH Publication 69:199–214
Samalikova M (1977) Linear weathering crusts as one of the factors giving rise to landslides in crystalline rock. Bulletin of the International Association of Empirical Geology 16:231–233
Savage SB (1979) Gravity flow of cohesionless granular materials in chutes and channels. Journal of Fluid Mechanics 92:53–96
—, Hutter K (1989) The motion of a finite mass of granular material down a rough incline. Journal of Fluid Mechanics 199:177–215
— → Nohguchi
Scheidegger AE (1970) Theoretical Geomorphology. 2. ed., (Mechanics of slope formation). Berlin Heidelberg New York, 73–154
— (1973) On the prediction of the reach and velocity of catastrophic landslides. Rock Mechanics 5:231–236
— (1984) A review of recent work on mass movements on slopes and on rock falls. Earth-science Review 21:225–249
— (1998) Tectonic predesign of mass movements, with examples from the Chinese Himalaya. Geomorphology 26:37–46
Scheller E (1970) Geophysikalische Untersuchungen zum Problem des Taminser Bergsturzes. Doctor's Thesis 4560, ETH, Zürich
Schindler C, Nievergelt P (1990) Einführung in Geologie und Petrographie. Geologisches Institut, ETH, Zürich
Schlüchter → Ivy-Ochs
Schneider J-L, Wassmer P, Ledésert B (1999) La fabrique interne des dépôts du sturzstrom de Flims (Alpes Suisses): caractéristiques et implications sur les méchanismes de transport. Earth and Planetary Sciences 328:607–613
Schneider → Purtscheller
Schnitter G, Weber E (1964) Die Katastrophe von Vaiont in Oberitalien. Wasser- und Energiewirtschaft 56, 2:2–3
Schröcker → Heuberger
Scott JS, Drever HI (1953) Frictional fusion along a Himalayan thrust. Proceedings of the Royal Society 65b, Part 2:121–142
Selby J (1987) Rock slopes. In: Anderson, Richards (ed) Slope stability. Chichester New York Brisbane Toronto Singapore, 475–504
Selli R, Trevisan L (1964a) Studio geologico della frana del Vaiont. Allegato a: Relazione della Commissione d'Inchiesta sulla sciagura del Vaiont
—, — (1964b) Caratteri e interpretazione della frana del Vaiont. Giornale di Geologia 32:1
Semenza E (1966–1967) Sintesi degli studi geologici sulla frana del Vaiont dal 1959 al 1964. Memorie del Museo Tridentino di Scienze Naturali, A 29–30. 16, 1
Sharpe CFS (1938) Landslides and related phenomena. Columbia University Press, New York, 76–78
Shreve RL (1966) Sherman Landslide, Alaska. Science 154:1639–1643

— (1968a) Leakage and fluidization in air-layer lubricated avalanches. Geological Society of America, Bulletin 79:653–658
— (1968b) The Blackhawk Landslide. Geological Society of America, Special Paper 108
Sieder ⟶ Purtscheller
Simstich ⟶ Stötter
Sorriso-Valvo ⟶ Nicoletti
Spadea ⟶ Caloi
Spray JG (1987) Artificial generation of pseudotachylyte using friction welding gear: simulation of melting on a fault plane. Journal of Structural Geology 9, 1:49–60
Stadelmann J (1983) Zur Dokumentation der Bergsturzereignisse vom Huascarán. In: Patzelt (ed) Die Berg- und Gletscherstürze vom Huascarán. Innsbruck, 51–70
Statham I (1979) A simple dynamic model of rockfall. ISMES International Colloquium on Physical and Geomechanical Models, Bergamo, 237–258
Stingl ⟶ Purtscheller
Stiny J (1939) Gedanken über das Maurach von Umhausen-Köfels-Längenfeld. Zentralblatt für Mineralogie und Geologie, Abteilung B, 209–216
Storzer D, Horn P, Kleinmann B (1971) The age and the origin of Köfels structure. Austria. Earth and Planetary Science Letters 12:238–244
Stötter J, Maukisch M, Simstich J, Belitz K (1996) Auswirkungen des zeitlich-räumlichen Wandels der Permafrostverteilung im Suldental (Ortlergebiet) auf das Gefährdunngspotential durch Erosionsprozesse. Internationales Symposion INTERPRAEVENT, Tagungspublikation 1:447–457
Straub S (1997) Predictability of long runout landslide motion: implications from granular flow mechanics. Geologische Rundschau 86:415–425
Stutzer O (1937) Die Talweitung von Köfels im Ötztal (Tirol) als Meteorkrater. Zeitschrift der Deutschen Geologischen Gesellschaft 88:523–525
Suess FE (1937) Der Meteor-Krater von Köfels bei Umhausen im Ötztale, Tirol. N. Jahrbuch für Mineralogie, Geologie und Paläontologie, Abh, 72, Beilage B and Abteilung A, 98–155
Surenian R (1988) Scanning electron microscope study of shock features in pumice and gneiss from Köfels (Tyrol, Austria). Geologische und Paleontologische Mitteilungen Innsbruck 15:135–143
— (1989) Shock metamorphism in the Köfels structure (Tyrol, Austria). 52nd Annual Meeting of the Meteorological Society, Vienna
— (1989) Shock metamorphism in the Köfels structure. 52nd Meeting of the Meteoritical Society, Wien, Abstracts and Program, 234–235
Szidarovski ⟶ Hutter
Tabor ⟶ Bowden
Tang ⟶ Wiesenfeld
Temperli ⟶ Heierli
Ter-Stepanian G (1977) Types of compound and complex landslides. Bulletin of the International Association of Empirical Geology 16
Terzaghi K (1960) From theory to practice in soil mechanics. London
Tessadri ⟶ Purtscheller
Tianchi L (1983) A mathematical model for predicting the extent of a major rockfall. Zeitschrift für Geomorphologie 27, 4:473–482
Tollmann ⟶ Kristan-Tollmann
Trevisan ⟶ Selli
Trientl A (1895) Die Bimssteine von Köfels. Tiroler Landeszeitung 6, 5, Imst
Tuttle ⟶ Wyllie
Varnes DJ (1978) Slope movement types and processes. In: Schuster, Krizek (eds) Landslides, Analysis and Control. Special Report 176, National Academy of Sciences, Washington DC, 11–33
VAW (Versuchsanstalt für Wasserbau und Glazialogie) (1991) Neues Gerinne der Vispa als Folge des Bergsturzes von Randa. Annual Report of VAW (ETH), Zürich
Vischer DL (1986) Rockfall induced waves in reservoirs. Water Power and Dam Construction, September 1986
Visser WA (1980) Geological Nomenclature. Royal Geological and Mining Society of the Netherland. The Hague Boston London
Voellmy A (1955) Über die Zerstörungskraft von Lawinen. Schweizerische Bauzeitung 73, 12/15/17/19, 159–165/212–217/246–249/280–285
Voight B (1973) The mechanism of retrogressive block-gliding, with emphasis on the evolution of the Turnagain Hights landslide, Anchorage, Alaska. In: De Jong, Scholten (eds) Gravity and Tectonics. New York, 97–121
—, Janda RJ, Glicken H, Douglass PM (1983) Nature and mechanics of the Mount St. Helens rockslide-avalanche of 18 May 1980. Géotechnique 33:243–273

Völk H (1989) Die Bergsturzkatastrophe im Veltlin 1987. Die Geowissenschaften 7, 1:1–9
Wassmer ⟶ Schneider
Watson RA, Wright HE (1967) The Saidmarreh landslide, Iran. Geological Society of America, Special Paper 123:115–139
Weber ⟶ Schnitter
Wegmann M (1998) Frostdynamik in hochalpinen Felswänden am Beispiel der Region Jungfraujoch-Aletsch. Versuchsanstalt für Wasserbau (ETH), 161, Zürich
—, Funk M, Flotron A, Keusen HR (in press): Movement studies to forecast the time of breaking off of ice and rock masses. Proceedings of INDR Conference on Early Warning Systems for the Reduction of National Disasters, 1998, Potsdam
Weiss EH (1964) Vaiont - Geologische Betrachtungen zur Felsgleitung in den Stausee. Steirische Beiträge zur Hydrogeologie
Welsch W (1983) Begleitworte zum Höhenlinienplan 1 : 25 000 der Bergsturzmure vom Huascarán am 31. Mai 1970. In: Patzelt (ed) Die Berg- und Gletscherstürze vom Huascarán. Innsbruck, 31–50
—, Kinzl H (1970) Der Gletschersturz vom Huascarán (Peru) am 31. Mai 1970, die größte Gletscherkatastrophe der Geschichte. Zeitschrift für Gletscherkunde und Glazialgeologie 6:1–2, 181–192
Wenk ⟶ Masch
Wiesenfeld K, Tang C, Bak P (1989) A physicist's sandbox. Journal of Statistical Physics 54, 5/6:1441–1458
Wildberger A (1988) The landslide of September 8th, 1986, near Giswil (Obwalden, Switzerland). In: Bonnard (ed) Landslides, vol. 2. Rotterdam Brookfield, 1381–1384
Wolff KF (1977) Die Dolomiten und ihre Sagen, 14th edition. Innsbruck
Woodsworth ⟶ Evans
Wright ⟶ Watson
Wyllie PJ, Tuttle OF (1960) The system CaO-CO_2-H_2O and the origin of carbonatites. Journal of Petrology 1:1–46
Yankowitz ⟶ Hutter
Yin Kunlong (1990) Frictional heat in landslide process. 6th International IAEG Congress, 1769–1773

Index

N

O

P

Q

R

Y

Z

Printing (Computer to Film): Saladruck, Berlin
Binding: Stürtz AG, Würzburg

MIX

Papier | Fördert gute Waldnutzung

FSC® C083411

Kolibri 360 GmbH
Ferdinand-Jühlke-Straße 7
99095 Erfurt, Deutschland
produktsicherheit@kolibri360.de